DISCOVERIES AND INVENTIONS

OF THE

TWENTIETH CENTURY

FIG. 271.—R.34, THE MACHINE WHICH CROSSED THE ATLANTIC. (See page 362.)

Front.

DISCOVERIES
AND INVENTIONS
OF THE
TWENTIETH CENTURY

BY

EDWARD CRESSY

Third Edition Revised and Enlarged
Profusely Illustrated

LONDON
GEORGE ROUTLEDGE AND SONS, LIMITED
NEW YORK: E. P. DUTTON AND CO.
1930

PRINTED IN GREAT BRITAIN BY
STEPHEN AUSTIN AND SONS, LTD., HERTFORD.

PREFACE TO THE FIRST EDITION

IT was not without misgiving that the author accepted the
invitation to write a book which should be in the nature of a
sequel to Robert Routledge's *Discoveries and Inventions of the
Nineteenth Century*. Progress in recent years has been so
extraordinarily rapid that the generous basis upon which the
original work was planned could not be followed within the
limits of a single volume. Nor would the task have been simplified
by confining it to a description of the more striking discoveries
and inventions which have been made during the last fourteen
years, for each achievement is one in a long series and involves
no inconsiderable amount of explanation in order to render it
intelligible to the non-technical reader.

Having regard to all the circumstances, it was decided to
deal with the characteristic features of development in certain
selected fields of enterprise during the last twenty-five years.
Thus the first five chapters discuss the revival of water power,
economy in the use of fuel, modern steam engines, gas, oil, and
petrol engines, and the generation and distribution of electricity.
These are followed by chapters on electric lighting and heating,
new processes in the manufacture and treatment of steel, some
typical modern devices in the engineering workshop and the
factory, and the extraordinary number of manufacturing
processes which have their birth in the electric furnaces. From the
highest temperatures which man has, so far, been able to produce
the book passes to a consideration of the artificial production
of cold and its applications in the manufacture of ice, cold
storage on land and sea, and the liquefaction of gases. This
chapter is succeeded by one dealing with the interesting facts
which have recently been discovered relating to the fertility of
the soil and the yield and quality of wheat.

v

One of the most characteristic features of the twentieth century is the improvement in transport and communication, and Chapters XII to XVII contain some account of railways, electric traction, motor-cars, modern ships, aeroplanes and airships, and wireless telegraphy. The constitution and some of the weapons of a twentieth century navy are described in Chapter XVIII ; Chapter XIX deals with the photography of colour and of bodies in motion ; and the book closes with a brief account of the recent marvellous discoveries relating to radium, electricity, and matter.

While the plan adopted is open to criticism, it has enabled a wide field to be covered, a fairly coherent picture to be drawn, and a limited amount of explanation to suffice. Numerous cross-references render immaterial the order in which the chapters are read. The terms " discovery " and " invention " have been interpreted liberally, so as to include results of human enterprise which, though not embodying any new principle, yet rank as great achievements and are rendered possible by other results which fall legitimately under these headings. There appeared, moreover, to be a distinct advantage in presenting any discovery or invention in close association with its practical relations. The aim, purpose, and value are thus emphasized, and the whole scheme contributes to sanity of outlook.

Considerations of space and the necessity of linking up the twentieth century with the past have led to the exclusion of much, even within the limited field covered, that should rightly have been included. But for this it seemed easier to offer an apology than to find a remedy. The book is written for those, young and old, who wish to have a non-technical account of the great scientific and material triumphs which man has achieved and is achieving in their own day ; and it seemed desirable to give the first place to those theories, facts, and accomplishments which are now exercising the greatest influence upon human life. For science exists not so much to tickle the intelligences of the few as to brighten the lot of the many.

The author desires to acknowledge his indebtedness to such

indispensable journals as *Engineering, The Engineer, Cassier's Engineering Monthly, Science Progress,* and to standard works on the various subjects considered in the volume. Thanks are especially due to the numerous firms and public and private individuals who have contributed to the work either illustrations or special information. Without the assistance which has been so willingly rendered the task would have been greater and the details fewer and less accurate. Certain sections have been read and corrected in proof by the Director of the National Physical Laboratory, Mr. V. G. Converse, the Chief Engineer of the Ontario Power Co., Messrs. Bellis and Morcom, and Messrs. Barr and Stroud. To these gentlemen—and to his friend Mr. Alfred Harris, who has read the whole of the proofs—the author owes a larger measure of gratitude. But though credit for any merit the book may possess is necessarily distributed, the author must accept responsibility for the defects and submit himself to the kind indulgence of his readers.

E. C.

PREFACE TO THE SECOND EDITION

THE first edition was published in September, 1914, and the intense activity in discovery and invention during the four years of the Great War, as well as its specialized direction, have rendered the task of producing a second edition a formidable one. While some chapters have been but little altered, the greater part of the book has been wholly rewritten. Chapters I to XI correspond in matter and scope with the first eleven chapters in the first edition, though there are considerable additions. Chapter XII, which gives an elementary account of some of the achievements of modern chemistry, more especially in relation to physics and biology, is new. Chapters XIII to XX correspond with Chapters XII to XX of the old edition with the omission of Chapter XVIII—Ships of War and their Weapons. So much has been done in this field and so little is available for publication that it was thought desirable to omit the chapter rather than to give an account which would inevitably be imperfect. The gyro-compass, however, has been retained, and included under Modern Ships in Chapter XVI.

The author is indebted to many friends, private firms, and public institutions for information or illustrations, and he trusts that due acknowledgment has been made. But it is not easy in all cases, and at the moment of writing, to trace many pieces of information to their source, and he owes more general acknowledgments to Sir A. D. Hall's *The Soil*, Burton's *Physical Properties of Colloidal Solutions*, the Reports on *The Chemistry of Colloids* issued first by the British Association and later by the Department of Industrial Research, Rideal and Taylor's *Catalysis in Theory and Practice*, Harrow's *Vitamines*, Baird's *British Airships—Past, Present, and Future*, Fleming's *The Thermionic Valve and its Development in Radiotelegraphy and Telephony*, Crowther's *Ions, Electrons, and Ionising Radiations*, *Nature*, and the Engineering Supplement of *The Times*.

E. C.

SEPTEMBER, 1922.

PREFACE TO THE THIRD EDITION

IN preparing a third edition the text has been thoroughly revised and the number of illustrations has been increased. Statistics have been brought up to date wherever it seemed important that this should be done. The new matter is mostly scattered throughout the chapters. Thus reference is made to the great development of the marine oil engine, the growth of the artificial silk industry, the slotted wing for aeroplanes, the screened-grid valve, beam wireless, phototelegraphy, and television, among other advances that have arisen, or have become important, since the last edition. The author desires to acknowledge his indebtedness to the firms which have supplied new illustrations and special information.

E. C.

SEPTEMBER, 1929.

CONTENTS

LIST OF ILLUSTRATIONS

LIST OF ILLUSTRATIONS

DISCOVERIES AND INVENTIONS

OF THE

TWENTIETH CENTURY

CHAPTER I

THE REVIVAL OF WATER POWER

PROBABLY one of the most important steps ever taken by primitive man in his unconscious efforts to escape from savagery was the discovery of the wheel. The fact that rolling produced less friction than sliding was but dimly recognized : the mechanical principle involved was perhaps but vaguely distinguished. There were no patent laws to protect the inventor, no legal formularies upon which he need enter, no manufacturers to whom licences might be issued and from whom royalties might be obtained. He was not absorbed by visions of untold luxury and ease. But he must soon have grasped the fact that here was a contrivance that would facilitate locomotion and increase his power over his surroundings. For this last, after all, represents the aim and destiny of mankind since the world began—an aim which is still paramount, though modern life is so complex that few know their bearings outside the small circle in which they live. This fortunate discoverer, together with he who first produced fire, were the forerunners of the engineers and manufacturers, the scientific discoverers and inventors of to-day. The wheel made it easy to move huge weights and to cover great distances, and when it was applied to spinning it transferred part of the burden of providing clothing from the animal to the vegetable kingdom. Rude skins gave place to finely woven fabrics, and the tiller of the soil vied with the hunter and the shepherd in covering man's nakedness.

At first the wheel was driven by manual toil or by the use of beasts, but when, after many centuries, wind and water were used, man saw opening up a wider vista which promised speed of production and more leisure to him who could harness the natural elements to his service. Was there joy when the first wheel turned in the wind, or a mad clapping of hands when one of these rough contrivances first creaked beneath the force of a mountain stream? We shall never know. In those days man was too much occupied with maintaining his existence. The art of speech was probably incapable of exact description; the arts of drawing and writing too crude to permit of accurate record. And perhaps it is as well that some of these early events should be left to the imagination, so they may acquire a sanctity that fancy weaves about them and which exact knowledge might destroy.

It is hardly possible to realize that until the middle of the eighteenth century wind and water were the only means of obtaining power from the prodigal forces of Nature. Clothing, tools, weapons had been made, houses and ships had been built, and international trade had arisen, by hand labour and a few relatively unimportant waterfalls. The ruins along the narrow valleys east and west of the Pennine Chain indicate the birthplaces of the British textile industries, where once fitful streams drove the looms that wove the fabrics for which Lancashire and Yorkshire have become famous.

England, however, is not rich in large waterfalls: puny streams could only aid in a small way the development of the factory system and were unable to compete with the steam-engine; so the industries vanished from the hill-sides and re-appeared amidst the sharp hiss of steam instead of the murmur of falling water. And if the suppression of water power had been universal and permanent this chapter need not have been written. But it was neither. In other lands there are streams and waterfalls so large that those who have not seen them can have little conception of their real size and only a vague impression of the power they represent. These acquired a greater value when the progress of knowledge had shown how electricity could be produced and distributed, and during the last twenty years their value has been still further enhanced by the discovery of electrical manufacturing processes. What this has meant to relatively poor countries like Norway and Sweden can readily be imagined. There the peasant by arduous toil wins a frugal existence from the soil. Sweden has its iron ores, but is dependent largely upon timber for fuel; and Norway has its fisheries and forests. But in these northern latitudes the summer is short, and

FIG. I.—THE RUNNER OF A PELTON WHEEL.

FIG. 2.—A LARGE PELTON WHEEL.

To face page **2.**

FIG. 3.—A DOUBLE VORTEX TURBINE WITH
END COVER REMOVED.

To face page 3.

there is little time for harvest, while the relative absence of manufactures means an absence of money to purchase luxuries from other lands.

TURBINES AND WATER-WHEELS

Broadly speaking, power is obtained from water by two types of machines, and the one chosen depends upon whether a high or a low fall is available. In the former case a Pelton wheel is used. From Fig. 1 it will be seen that this consists of a disc mounted on a shaft, having a number of cups fixed round the edge. These are known as buckets, and they have a ridge in the centre which splits the jet as it impinges upon them. The surface is so shaped that the water glides round without splashing and runs out at the lower edge. When the water issues from the jet it has a velocity which depends upon the height of the surface above the wheel. If the wheel were prevented from rotating the water would have its velocity reversed owing to the shape of the cups, and by virtue of this velocity it would still be capable of doing practically the same amount of work. If again the wheel were to rotate so that the velocity of the buckets was equal to that of the jet, no work would be done on the wheel. Now the greatest efficiency will be obtained when the water falls from the buckets with all its original velocity taken out of it, and this will be the case when the buckets move with half the velocity of the jet. For the water will be flung back at a velocity which just balances the difference between the velocities of the jet and the buckets, and will fall exhausted into the well below.

There must therefore be a definite relation between the height of the fall and the speed of the rim of the wheel. If low speeds are required then a large wheel must be used, but if high speeds are desirable a smaller wheel may be employed. With a large volume of water two or occasionally three jets may play upon one wheel, or two wheels may be fixed side by side on the same shaft. They are made so small as to give no more than $\frac{1}{5}$ horse-power, and so large as to give 16,000 horse-power, and they can be enclosed in a casing or remain open. As an example, consider the wheel shown in Fig. 2, which is 20 feet diameter and was erected at a South Wales tinplate works by Messrs. Gilbert Gilkes & Co., of Kendal. The available fall is 100 feet, and 200 horse-power is obtained at 36 revolutions per minute. The unusually large size of the wheel, which weighs 11 tons, is due to the necessity for a slow speed to drive a rolling mill, and as the power required varies enormously as the metal passes into or out of the rolls, a 50-ton fly-wheel is fixed at the side of the Pelton wheel to equalize the motion. It is interesting to note that the actual velocity of the jet is about 70 feet per second, and of the buckets about 37 feet per second. This is very nearly in accordance with the conditions laid down in the last paragraph. There are two wheels, and the water is conveyed to each through a 39-inch riveted steel pipe.

Where the fall is low or very large quantities of water have to be dealt with, the Pelton wheel is replaced by a turbine. This is a wheel with curved blades, enclosed in a casing. The water usually enters at the circumference of the latter, is deflected upon the blades by guides, and discharged at the centre. Fig. 3 is an illustration of the double-vortex turbine which is the parent of all inward-flow turbines of to-day. The blades on the wheel lie across the path of water on its way to the centre and freedom, and are elbowed to one side, thus causing the wheel to rotate. There are so many modifications to meet differences in the head and quantity of water to be dealt with that it is not possible to illustrate them in detail. Turbines are made with horizontal or vertical shafts, and will work with a head of only 3 feet. For heads of less than 16 feet the vertical type is used, but in all other cases the horizontal type is preferable.

SOME WATER-POWER INSTALLATIONS

One of the most interesting examples of the importance of water power is the new industry for the manufacture of Norwegian saltpetre by the method described in Chapter IX. The source of power consists of three lakes—Maarvand, Mösvand, and Tinnsjö —situated in Southern Norway (Fig. 4).

The principal power houses are at Rjukan, between Lake Mösvand and Lake Tinnsjö, where no less than 295,000 horse-power is now developed. The water for the first of the two stations is taken from above the Rjukan Falls to a point 970 feet below in ten steel tubes 5 feet diameter (Fig. 5). These are riveted in the upper sections, but in the lower where the pressure reaches 420 lb. per square inch the plates, 1 inch thick, are welded together. The water enters ten sets of Pelton wheels each developing from 14,000 to 19,000 horse-power (Fig. 6), and then passes through another tunnel to a second power house

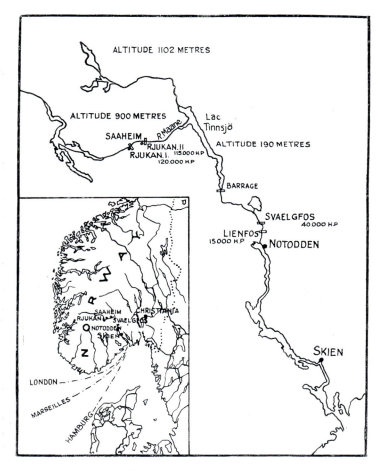

ALTITUDE 1102 METRES

ALTITUDE 900 METRES

Lac
Tinnsjö

SAAHEIM

R.Maane

RJUKAN.II
RJUKAN.I. 115.000 H.P
120.000 H.P

ALTITUDE 190 METRES

BARRAGE

SVAELGFOS
40.000 H.P

LIENFOS
15.000 H.P

NOTODDEN

SAAHEIM
RJUKAN
SVAELGFOS
CHRISTIANIA
NOTODDEN
SKIEN

LONDON

MARSEILLES

HAMBURG

SKIEN

FIG. 4.—MAP SHOWING SOURCES OF POWER FOR THE
NORWEGIAN NITRATE FACTORIES.

To face page 4.

FIG. 5.—PIPES WHICH SUPPLY WATER TO THE RJUKAN POWER HOUSE.

To face page 5.

three miles away and 909 feet lower opened in 1914. These supply electricity to the nitrate factory at Saaheim.

Another factory, at Notodden below Lake Tinnsjö, is supplied from two power houses at Lienfos and Svaelgos (Fig. 7). At the former station the fall is about 55 feet and there are four turbines of 5,000 horse-power each ; at the latter the fall is over 160 feet and the four turbines are of 10,000 horse-power each. The total horse-power employed by the company in 1923 was 360,000. The amount of nitrogen abstracted from the air increased from 53·5 tons in 1908 to 2,600 tons in 1920.

Perhaps a clearer picture of what the rise of such an industry means to a country like Norway will be realized when it is recalled that the population is only 2,500,000. The first factory was opened in the summer of 1903, and employed only four people. Eight years later the number of employees was nearly 1,500. Notodden was a village of 500 people ; it has now more than 5,000 inhabitants. Saaheim was a district tenanted by a few poor farmers and supporting not more than fifty souls all told ; in 1916 it was a thriving town with a population of 10,000. Railways have been constructed, steamboat services have been established on Lake Tinnsjö and the river which flows from this lake into the Baltic. In less than ten years a naked wilderness had been clothed in the mantle of civilization.

Not the least noteworthy feature of the new industry is that the Company have taken steps to house the workpeople in comfort and at a reasonable cost. Moreover, by arrangement with the local banks they may ultimately own their houses by payments which cover the cost with 5 per cent interest. This care and solicitude for the health and comfort of those whose labour is as necessary as the power, and upon whose sympathetic and conscientious co-operation so much depends, is a characteristic feature of every branch of modern industry in which there has been great rapidity of growth ; and it is a little curious that the chemical industries—soap, soda, cocoa, etc., were the pioneers in this respect.

This is of course the largest of many similar water-power plants in the Scandinavian peninsula. Thus at Odda on the Sondrefjord is a large factory established by an English company for the manufacture of calcium carbide and nitrolim, a substance to which reference is made in Chapter X. Here, in a spot previously only known to a few tourists, Lake Tyssedal has been dammed higher up in the mountains and the water is conveyed in tunnels to six Pelton wheels of 4,600 horse-power each. These drive dynamos, and current is transmitted nearly four miles to Odda, where it is supplied to the electric furnaces, and does all the work of the factory. Cranes, conveyors, crushing-machines are all electrically driven, and none of the material is touched by hand except to charge it into the furnaces. In 1913 only 23,000 horse-power was utilized, but the water supply is capable of giving 75,000 to 80,000.

In Sweden again advantage is rapidly being taken of the hundreds of upland lakes, and the waterfalls and torrents which convey the overflow to the sea. Twelve falls on the Dal River alone are capable of yielding 175,000 horse-power. To the north of this area there are ten more rivers equal or superior to the Dal, and in Central and South Sweden there are many more. The electricity is distributed over wide areas, and lights towns, drives machinery, and provides the motive power of an increasing portion of the State railways.

The same development is going on in Switzerland and those adjoining countries into which the Alps penetrate. But it is in America perhaps that the widest use of " white coal " is made. All along the Pacific coast the streams which run from the watersheds of the Sierra Nevada have been harnessed for many years. Originally the water here was used for " placer " mining. Gold, for example, occurs in loose sand and gravel, and the miners constructed canals high up on the hill-sides, which were fed by streams from the winter snows. From these canals the water was led through pipes and the jet was directed against the loose gravel of the lower slopes. The gold and sand were then separated in troughs of running water in which the heavier metal settled, and the earthy material was washed away.

Incidentally this locality was the birthplace of the Pelton wheel. The buckets or cups used on the rim of the earlier wheels were single—they had no central ridge. A carpenter named Pelton engaged in repairs noticed on one occasion that a wheel became displaced so that the jet struck the edge of the buckets. He observed that the water falling on the inner edge curved round the surface with less splashing than in the ordinary wheel, and that the wheel ran faster. So he constructed the wheel with divided buckets which now bears his name.

But no water-power plants make such an appeal to the

FIG. 6.—THE RJUKAN POWER HOUSE.

To face page 6.

FIG. 7.—THE POWER HOUSE AT SVAELGOS

To face page 7.

imagination as those which have been established in the neighbourhood of Niagara. No fewer than five companies are now diverting a tiny fraction of the upper river through their turbines and discharging it below the falls, without appreciably diminishing their grandeur, and then distributing their power electrically over many hundreds of square miles. As every schoolboy is aware, the Niagara River forms the spout through which the surplus waters of Lake Erie overflow into Lake Ontario (Fig. 8). In the 36 miles of its length, it falls through 326 feet, of which 216 feet is in the falls and the rapids just above them. Where the latter commence, about half a mile from the edge of the cliff, the river is divided into two portions by Goat Island, giving a fall 1,000 feet wide on the American side, and the famous Horseshoe Fall 2,600 feet wide on the Canadian side. The American fall is 167 feet high, while owing to the rapids which occur chiefly on the Canadian side of Goat Island the Horseshoe Fall is about 8 feet less. The quantity of water pouring over these two lips is almost incomprehensible. It has been estimated at 222,400 cubic feet per second, or nearly a cubic mile a week. Expressed in units of weight and power, this represents 22,000,000 tons an hour, and is equivalent to 5,000,000 horse-power.

Some idea of the importance of Niagara Falls as a centre for the production of power may be gathered from the distribution of the population of the two countries between which they lie. Probably few realize from their school study of Geography that if a circle 500 miles in radius be struck from Niagara as a centre, this circle will include three-quarters of the population of Canada, and half the population of the United States. For it encloses Toronto, Ottawa, Montreal, and Quebec ; New York, Philadelphia, Washington, Pittsburg, Detroit, Cincinnati, Chicago, Milwaukee, and Buffalo. The power supplied at present extends to a radius of over 200 miles, but the whole area includes a network of railways, including five trunk lines, the Erie Canal, and all of the great lakes except the western half of Lake Superior.

The largest, the most recent, and the most perfect of the five installations which tap the vast resources of the Niagara River is that which was constructed by the Ontario Power Company, and the writer is indebted to the courtesy of the chief engineer for the particulars and illustrations which follow. The original charter was granted by the Dominion Parliament as long ago as 1887, but apparently no progress was made until the present

owners took possession thirteen years later. Constructional
work was commenced in 1902, and power was first supplied
in 1905.

Water is taken from the river at a point on the Canadian
shore, about a mile above the crest of the Horseshoe Fall, and
just above the rim of the first cascade of the upper rapids. The
intake works consist of a dam nearly 600 feet long stretching
out in a down-stream direction nearly parallel to the main
current ; and a submerged wall or dam connecting the outer end
on the intake with the shore. The forebay thus formed is shown
in Fig. 9. Water enters through twenty-five openings in the
intake dam, situated 9 feet from the surface, and extending to
the bottom of the river, which is here 15 feet deep. The floating
debris and ice is mostly deflected by the upper portion of the
intake dam, and water from the bottom of the river only is
taken.

In the comparative calm of the outer forebay any ice or
debris which has crept beneath the barrier from the turbulent
river beyond, rises to the surface and is either washed away over
the submerged wall or trapped by a concrete curtain at the
screen-house, which hangs 5 feet below the surface. The area of
the outer forebay is 8 acres, and its depth is from 15 to 20 feet.
The inner forebay has an area of 2 acres and a depth of 20 to
30 feet. In the tranquil waters of this basin the last remnants of
floating material rise to the surface and are prevented from
passing to the pipe lines and turbines below.

The power station is situated at the foot of the cliff on the
Canadian shore just below the falls and is a solid concrete
structure with walls from 9 to 12 feet in thickness. The water
is first led through three conduits laid under the Queen Victoria
Park until it reaches a point on the cliff above the power house
6,000 feet away. One of these conduits is 18 feet diameter,
consisting of a steel tube covered with concrete, and terminating
in an overflow chamber through which surplus water can escape
by a spiral tunnel to the lower river. Another, shown in process
of construction in Fig. 10, has the same sectional area but is
oval in shape, having a horizontal diameter of $19\frac{1}{4}$ feet, and
a vertical diameter of $16\frac{1}{2}$ feet. It is built entirely of ferro-
concrete—that is of concrete having steel bars embedded in
it, and consists of a shell 18 inches thick, strengthened by a
continuous saddle. This conduit terminates in a circular concrete

FIG. 8.—BIRD'S-EYE VIEW OF NIAGARA FALLS.

To face page 8.

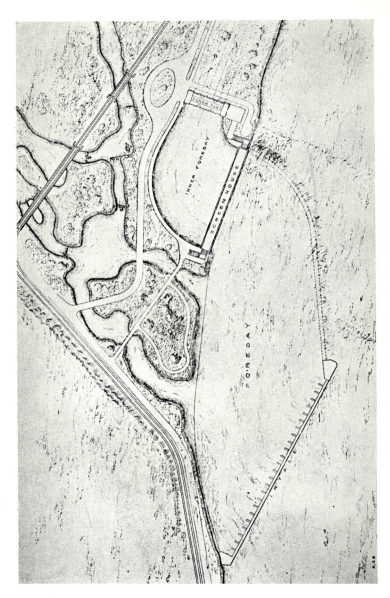

FIG. 9.—THE INTAKE WORKS OF THE ONTARIO POWER CO.

To face page 9.

surge tank 75 feet in diameter, which serves to store excess of water when the load on the turbines is reduced. If some plan of this kind were not adopted enormous forces would be developed by the sudden stoppage of thousands of tons of moving water. The tank serves the additional purpose of supplying water to the turbines when that in the conduit is just beginning to move.

Beneath the lower ends of the conduits near to the overflow chamber and surge tank, are valve chambers carved out of the solid rock and having arched concrete roofs to support the conduits. These chambers are about 300 feet long, 10 feet high, and 16 feet wide. Here the water passes through valves into the penstocks or steel tubes 9 feet diameter, which convey it to the turbines. Each valve is operated by a 30 horse-power electric motor which opens or closes it in four or five minutes.

The power possessed by this mass of water filling the two conduits over a mile long and moving with a velocity of 12 to 15 feet per second can hardly be realized, and water pipes 9 feet in diameter are outside the range of ordinary experience. To absorb this power the turbines and dynamos must be enormous, especially as the fall is not more than, say, 190 feet. Far smaller machines are possible where a great head of water is obtainable, as at Rjukan, because a higher velocity is attained by the water in its descent. Some idea of the size of the machinery will be gathered from Fig. 11, which shows the interior of the power house. The size of the man in the foreground brings out sharply the huge dimensions of the machines he controls.

The plant was purchased in 1917 by The Hydroelectric Power Commission of Ontario. This is a co-operative municipal owner-ship enterprise which supplies electricity for light and power throughout the whole province. It controls a number of generating stations grouped in such a way as to provide, economically, definite areas. By far the largest is the Niagara group which includes the works just described, the Toronto Power Station purchased in 1920, and the Queenstown-Chippawa plant which was constructed by the Commission and opened in 1921. No less than 850,000 horse-power is obtained from the Niagara River by these three generating stations, and over 1,000,000 horse-power is distributed by the Commission to more than 550 municipalities in the province of Ontario.

Power from Niagara is provided for the electric furnaces

employed in the reduction of iron, copper, and other ores, and the manufacture of cement, calcium carbide, nitrate of lime, carborundum, and graphite, in Port Colborne, Welland, Niagara Falls, Thorold, and Chippawa, Ontario; and Lockport, New York. The tramway systems in Syracuse, Rochester, Canadaigua, Geneva, Lackawanna, and Hamburg, and the inter-urban railways, Syracuse, Lake Shore and Northern, Syracuse and South Bay, Syracuse and Auburn, Rochester and Syracuse, Rochester and Geneva, Rochester and Mt. Morris (Erie Railroad), Buffalo, Lockport, and Rochester, Buffalo and Hamburg, and Buffalo and Lake Erie, are operated wholly or in part by the power from this centre.

But this is only half the tale. The electric current from the same source drives the machinery of the Canadian Steel Foundries at Welland, and of the Lackawanna Steel Company, which employ 7,000 men. It turns the rolling mills of the Seneca Iron and Steel Company, pumps the water at Depew and Lackawanna, supplies the repair shops of the New York Central and Hudson River Railway, and the Delaware, Lackawanna, and Western Railroad Company, crushes stone and grinds lime at Akron, Pekin, and Oakfield, and runs the shops of the American Locomotive Company at Dunkirk. For 300 miles east and west, and over 100 miles north and south, the transmission lines radiate, carrying the latent power vested in a tiny fraction of the waters which thunder through the rocky gorge in their passage to Lake Ontario, the St. Lawrence, and the sea.

In these examples we see an approach to the ideal arrangement of centralized production of power, to which reference will be made from time to time throughout this volume. Incidentally, it will be clear, that the term cheap water power is liable to be misunderstood, for there is usually a vast expenditure to be undertaken in dams and pipe lines before the energy of filling water can be profitably utilized. But so far it is the only source of power which is reasonably constant, and the use of which does not lead to exhaustion of natural capital. Moreover, with improvements in the production and transmission of electricity, and the discovery of new methods of manufacture in which electricity is the prime agent, a new era has arisen in which industrial prosperity is no longer dependent upon or measured by the cheapness of coal. During the next hundred years the areas in which manufacturing industries are congregated most

FIG. 10.—CONDUIT NO. 2 IN PROCESS OF CONSTRUCTION.

To face page 10.

FIG. 11.—THE POWER HOUSE OF THE ONTARIO POWER CO.

thickly will not only be situated upon the coalfields, but also in those districts where water pursues its most vigorous progress towards the sea.   And the beautiful places on the earth formerly known only to the tourist, the simple shepherd, or the hunter, will have their fastnesses invaded, and their silence, broken now only by the roar of the waters, will reverberate softly to the hum of the turbine wheels.

CHAPTER II

THE extent to which all manufacture and transport, and all businesses are paralysed during a coal strike is an indication of the complete change which has come over the conditions of life within the last 150 years. A temporary stoppage of the supply of fuel throws all the machinery of existence out of action, and reveals the magnitude of the debt which civilized nations owe to the men who win precious fuel from the earth's storehouse.

THE PERILS OF THE MINE

Hardly any industrial operation excites an interest so fluctuating as that of mining. Carried on, for the most part in districts remote from the larger towns, by men who spend a third of their lives underground, it is a case of " out of sight, out of mind ". A man or two may be buried beneath a falling roof, or mangled by a runaway train without comment ; but every now and then the world is startled by an appalling disaster in which scores of men lose their lives. And so, from early days, when industry began to cry out for more and yet more coal, inventors have been busy devising all sorts of methods and appliances for the prevention of accidents. The Davy safety-lamp, familiar to all, is the parent of scores of others, the fruits of widening knowledge and lengthening experience ; and in a hundred ways the perils which beset the miner have been met and countered. But in spite of all, disastrous explosions still occur, though with far less frequency than if the only precautions were those of even 50 years ago.

These explosions are due to a gas called marsh gas, or fire-damp, having the formula CH_4. It is inflammable, violently explosive when mixed with air and ignited, and is given off in large quantities from many varieties of coal. Naked lights may be used in the Forest of Dean, where no gas is evolved, but in most other coalfields the safety-lamp is a necessity. But all mines must be ventilated by forcing air through them with a fan, and the quantity of air must be sufficient to keep the percentage of gas below a dangerous level. The mine is " examined " at regular intervals by a " fireman " who can estimate approximately the percentage of gas present by the size of the faintly luminous " cap " which hovers above the flame of his lamp. This depends upon the size of the flame, and the necessity for some special training for this work will be apparent from Fig. 12, which shows the cap over large and small flames. That on the right is due to $4\frac{1}{2}\%$ and that on the left to $3\frac{1}{2}\%$ of gas. It will be observed that the smaller percentage gives exactly the same size of cap over the larger flame as the larger percentage gives over the smaller one.

Explosions have occurred, however, in cases in which it is extremely doubtful whether gas has been present in dangerous quantity, and attention has been drawn to another possible cause. Many varieties of coal produce fine dust, which settles in the roadways—on roof, and sides, and floor. For a number of years there has been a controversy as to the relative importance of gas and dust in producing explosions, and the question is still one which gives rise to a lively difference of opinion. But there is no doubt that a mixture of coal-dust and air is explosive, and that even if an explosion is started by gas the disturbance creates clouds of dust which cause secondary explosions and spread the disaster over a wider field than was originally affected.

Consequently the rules of the Home Office require the ventilating current to be reversed periodically in order to remove dust from the lee side of timbering and crevices ; and the roadways have to be watered to keep the dust from rising. More inspectors have been appointed, and firemen, whose duty it is to visit the workings and report the presence of gas or defects in the ventilation, are required to possess a certificate of competency. The plan of spreading fine stone-dust in the roadway is also being tried. This becomes mixed with the coal-dust and renders it less explosive.

Unfortunately the disastrous effects of an explosion do not end with the explosion itself. The chief products of combustion of fire-damp or coal-dust are carbon monoxide, CO, and carbon dioxide, CO_2. The second of these causes suffocation ; the first is a poison. It is the dreaded " after-damp " of the miner. Those who survive the explosion are therefore in imminent danger and it becomes imperative to restore the ventilation with the

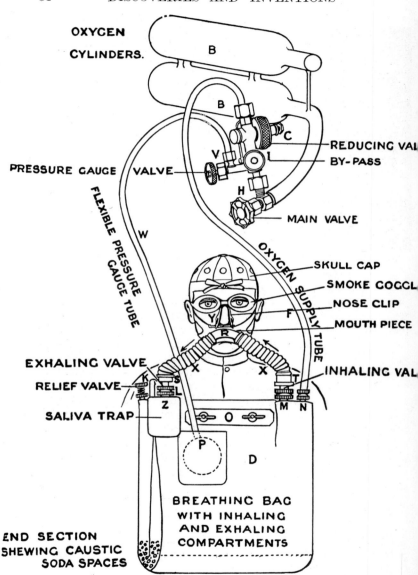

Fig. 15. DIAGRAM OF THE "PROTO" RESCUE APPARATUS.

FIG. 12.—FLAME CAPS ON A MINER'S LAMP.

FIG. 13.—THE "PROTO" RESCUE APPARATUS: FRONT VIEW.

FIG. 14.—THE "PROTO" RESCUE APPARATUS: BACK VIEW.

To face page 15.

least possible delay. For even if the fan, which drives air through the workings, has escaped injury, falls of roof may have blocked up some of the roadways, or the explosion may have torn down doorways and provided a short cut for the air. But if it is dangerous for men in the pit at the time, it is equally dangerous for others to go down and effect repairs or render first aid. The work of the rescue party is therefore a labour of desperate heroism, and not infrequently attended by additional loss of life.

During the last twenty years it has been found possible to reduce the danger to which members of rescue parties are exposed by providing them with respirators which fit over the mouth and nose. The wearer is supplied with oxygen from two small steel bottles strapped across his back, and the way in which the appliance is constructed and works will be easily understood from Figs. 13, 14, and 15. The bag in front contains sticks of caustic soda which absorbs the carbon dioxide produced during respiration. The bottles contain ten cubic feet of oxygen gas, which is sufficient for two hours' strenuous labour.

In all the important colliery districts special rescue stations have been established. These are buildings in which small groups of men can be trained in the kind of work they may be called upon to do, and in such an atmosphere that would be produced by an explosion or a fire in the mine. There are now thousands of men accustomed to wearing the apparatus, able by its aid to penetrate smoke and foul air, remove any who have been overcome, and effect such repairs as may be necessary to restore ventilation.

THE GASIFICATION OF COAL

Coal, burnt in ordinary grates, or directly under a boiler, is not a very economical fuel, and it is far less wasteful to convert it into gas. The production of gas by heating coal in fireclay retorts is, of course, a century old, but the gas obtained in this way is chiefly valuable as an illuminant and the coke left in the retorts has only a limited use as a fuel. It retains too little volatile matter to burn well in household grates, and it is not hard enough for metallurgical purposes. A great advance was made when Frederick Siemens invented, in 1857, a gas producer for use with his process of making steel. For in 1878, J. Emerson Dowson designed a producer for supplying gas for factories and domestic purposes, and at the York meeting of the British

Association, in 1881, he showed a small gas-plant driving a three horse-power Otto gas-engine. Sir Frederick Bramwell, the eminent engineer, even went so far as to prophesy that in fifty years the gas engine would have replaced the steam engine as a source of power.

The gas producer consists of a cylindrical furnace which is charged with coke, or with anthracite coal which contains a high percentage of carbon. Air is admitted at the bottom, where the oxygen immediately combines with the carbon, of which the coke is mainly composed, to form carbon dioxide, or CO_2. In its upward passage the carbon dioxide reacts with the red hot coke to form carbon monoxide, CO, which will burn in air with a very hot flame. Producer gas has, however, a low calorific value or heating power, owing to the fact that four-fifths of the air which is admitted to the furnace consists of nitrogen. A product of greater heating power is obtained by using steam instead of air, when a mixture of carbon monoxide and hydrogen is obtained. Unfortunately, the steam lowers the temperature of the furnace and the only way to keep up the supply is to force steam and air in alternately. Water-gas, as the mixture of carbon monoxide and hydrogen is called, is frequently made in this way for mixing with rich coal-gas ; or is itself enriched with oil-gas and used instead of coal-gas. When the furnace is blown up by air, the gas, being chiefly carbon dioxide and nitrogen, is allowed to escape, and the intermittent working of this producer has been against its wider adoption.

By using air and steam together the temperature of the furnace can be maintained and the semi-water-gas contains hydrogen and a lower percentage of nitrogen than producer gas.

But these producers require either coke or anthracite, and while the former is cheap only to the manufacturer of town gas, who obtains it as a by-product, the latter is a very expensive fuel. Consequently, when Dr. Ludwig Mond invented, in 1889, a process which would work with cheap coal-slack, then obtainable at 5/- or 6/- a ton, another step in economy was taken. But the advantage of the process did not rest entirely upon the cheapness of the fuel, for the use of coal instead of coke, and the use of a large quantity of steam—$2\frac{1}{2}$ lb. for each lb. of coal consumed—enabled ammonia to be obtained as a by-product. Only one-fifth of the steam goes to form carbon monoxide and hydrogen, but the excess served to keep the temperature low and increased the production of ammonia to 80 lb. a ton, whereas the ordinary gas-works process yielded only 30 lb. a ton. Further, the excess of steam was used to warm the incoming air, so that the heat in that which escaped was not entirely wasted.

A large plant was erected in 1895 by the South Staffordshire Mond Gas Company, and supplies gas through a network of pipes to works and factories over an area of 123 square miles. Provision was made for 32 producers, and the main distributing pipes are 3 ft. in diameter. This is an admirable example of the centralized production of energy, and the value of the process is not restricted to large scale production. Centralization is invariably the more economical. Twenty years ago it was considered too costly to instal plant for recovering by-products of a smaller size than 3,000 horse-power ; but there are many of only a few hundred horse-power in use to-day.

Most of the gas manufactured in this way is used in gas engines, and for heating metallurgical furnaces, and a smaller quantity for heating steam boilers. Its suitability for gas engines has led to a modification in the ordinary method of working. Instead of *forcing* air and steam through the red-hot fuel, the gas engine is made to *draw* its own supply. A good type of suction gas producer, made by Crossley Bros., is shown in Figs. 16 and 17.

From the figures, which are almost self-explanatory, it will be seen that the temperature may be raised at first by a small fan on the extreme right of Fig. 16. Once the fuel is hot, the air and steam are drawn through by the suction of the engine. Coke is fed in at the top through a hopper, which is so constructed as to prevent the escape of gas. Water drops into a system of tubes contained in a cylinder at the side of the furnace, where it is converted into steam by the hot gases. Air passes through these tubes, takes up moisture, and enters the red-hot material at the bottom. The escaping gases pass through a tower filled with coke, over which water trickles, to cool them, and to remove dust, etc., and are then dried by filtering through sawdust on the way to the engine.

Such a producer will consume anything that can be burnt in one of the ordinary type. Like them, it was first used to consume coke or anthracite at 15/- to 30/- a ton. It will, however, produce gas from bituminous slack at less than a third the price, while for places where coal is dear or unobtainable, it can be constructed to work on sawdust, wood refuse, rice husks, olive-oil residues,

c

tannery refuse, cotton seed, mealie cobs, or any other waste material that is available.

There is a source of danger in the use of producer gas arising

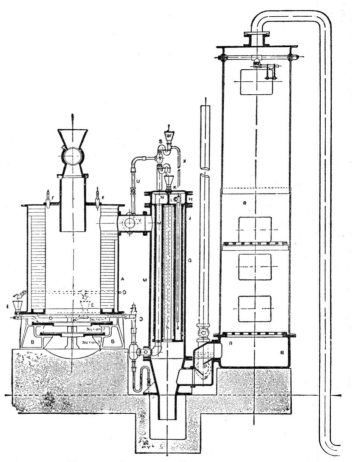

Fig. 17. SECTION OF GAS PRODUCER.

from the poisonous character of carbon monoxide. When breathed in minute quantities, it causes an effect which lasts for a long time, and which is, therefore, cumulative. If a person is

FIG. 16.—SUCTION GAS PRODUCER.

To face page 18.

FIG. 18.—A FEATHERED SENTINEL.

To face page 19.

exposed to it daily each dose is added to the previous dose, and the result may be serious. So small an amount in the air as 0.15% is distinctly dangerous, and anything above 0.03% will produce unpleasant symptoms. Unfortunately, the gas has no smell, and thus cannot be detected in time to prevent mischief. In engine houses and sheds, where producers are installed, there is always liability to an escape, and when the man in charge enters in the morning he may inhale a considerable quantity. It is usual to have one or more sentinels, in the shape of small animals or birds, to give warning of danger. A mouse or canary is affected by the gas in about one-tenth of the time required for a man, and quickly shows signs of stupor when it is present in minute quantities. A mouse may lie down as though asleep, but a bird sleeps on its perch, falls off only when stupefied, and is therefore a safer guide. The illustration (Fig. 18) shows the canary on duty in the producer-house of the University of Liverpool.

Let us now glance at the broad effects of the gasification of coal. In the first place, the value of a fuel lies in the quantity of heat which a given weight of it will produce. The heat is measured by the extent to which it will raise the temperature of a given weight of water, and the unit of heat is that quantity which will raise one pound of water through one degree Fahrenheit. It is called a British Thermal Unit. A pound of coal of fairly good quality will produce on burning 14,000 units of heat. In no possible way can the whole of the heat-producing power of the coal be utilized. There is always a loss. Some is radiated to surrounding objects, some is used in converting certain constituents of the coal into liquids and gases, and some is lost through the escape of particles of solid carbon. If the waste by radiation is to be reduced, the process of burning must be rapid, for the longer it lasts the greater is the amount lost in this way. But there is a limit to the rapidity with which a solid fuel can be burnt, because the air can come into contact only with the surface of the pieces. In this respect gas has obvious advantages in that it can be mixed intimately with air and the combustion proceeds both more rapidly and with greater uniformity than with a solid fuel. Moreover, it is more fully under control.

There is, however, a way of using coal which secures uniform and rapid combustion under complete control. The fuel is pulverized, dried, supplied to the furnace by a blast of air. During the last ten years this method has been developed considerably for raising steam in boilers. It enables small coal, which is hardly suitable for any other purpose, to be used, and it possesses most of the merits of a gaseous fuel.

Let us now review the various methods of using coal. Firstly, it may be burnt in a raw state, either in lumps or in a pulverized condition. By this method all the by-products—ammonia, tar, benzol, etc. are lost. Secondly, it may—indeed, some of it must, be converted into metallurgical coke, producing gas, and permitting the by-products to be collected. Thirdly, it may be converted into town gas, with production of household coke or semi-coke, and recovery of the by-products. Fourthly, it may be converted wholly into gas with recovery of by-products. The first method is the least and the last method the most economical. But metallurgical coke is necessary, many people prefer to use solid fuel, and gas-works coke is relatively cheap and can be used where hard metallurgical coke would be impossible.

But the general tendency must be in the direction of economy. The large super-power stations which are being built consume enormous quantities of raw coal. They waste the by-products. They do great damage to growing crops in their neighbourhood, and cannot therefore be healthy for human beings. They convert only about 14% of the energy in the coal into useful work. A gas engine converts 30%, but there are mechanical difficulties of constructing these engines of large size. So far as large scale production of power is concerned a country which possesses coalfields must continue for the present to use raw coal. A large power station is much more economical than a small one. But the use of raw coal for domestic purposes has no justification. If gas were used, and if the use of electricity could be extended, the black pall that hangs over thickly populated manufacturing districts would disappear, the grime of the city would cease to exist, the open grate with its smoky chimney and its ashes would give place to the electric radiator or the gas fire, labour would be saved, and life would be cleaner, brighter, healthier than it is to-day.

THE STORY OF OIL

Mineral oil is to-day so common and has such a variety of uses that we are apt to forget how recently it has been discovered in large quantities or how rapid has been the growth of its production. For centuries " rock-oil " has been collected as it oozed from the ground ; but the quantity was so small that practically the whole output was consumed in the locality, and people remote from oil-bearing strata fed their lamps with fuel from animal or vegetable sources. The discovery that petroleum could be obtained in more generous measure from deep underground reservoirs was made by Colonel Drake who, in 1859, sunk a well at Oil Creek in Pennsylvania. The first cargo reached London in 1861, and the annual output of 1,000,000 tons rose to 9,000,000 tons in 1891—all from Pennsylvania and Ohio. The charges made by carters and railway companies became so exorbitant as the industry flourished that long lines of pipes were laid from the oilfields to the ports, and special tank steamers were built into which the oil could be pumped from the terminal reservoirs. These were followed by tanks mounted on railway trucks and road vehicles, so that small consumers could buy oil which had been conveyed in bulk almost to their doors.

The extraordinary success of the Pennsylvania fields and the development of the oil and petrol engines encouraged prospectors to search for other oil-bearing areas, and soon Kentucky, Tennessee, Colorado, Indiana, and Illinois began to contribute to the world's supply. Then West Virginia, Texas, California, and Oklahoma developed the industry, so the whole North American continent seems to have been saturated with oil. The search has been carried across the border into Mexico and one firm alone has the right to sink wells over 75,000 square miles of territory.

The second largest oil-producing country in the world is Russia, which yields nearly 10 million tons per annum. Though the original wells at Baku are becoming exhausted, there are large tracts of land which have yet to be tapped. When Russia recovers from the chaos which reigns at present, it is highly probable that all the output of the Caucasian oil-fields will be needed for home consumption. Roumania and Galicia produce more than $3\frac{1}{2}$ million tons per annum, and find a ready market in Germany. Great Britain has, so far, no oil worth mentioning, and is dependent upon the product of Persia and Burma, and what the Americans and other countries can spare.

Petroleum occurs in certain porous layers of the earth's crust, just in the same way that water collects in porous sandstones. It frequently contains in solution gaseous substances, so that when the well reaches the required depth the oil is forced out in a fountain several hundred feet high. Some " gushers " pour out thousands of gallons a day for weeks or months after they are first tapped, but the pressure gradually decreases until the oil has to be pumped to the surface. As thus obtained it is an evil-smelling liquid, varying from colourless through shades of brown to black. It differs in composition in different localities, and there is a corresponding variation in the methods of purification and the products obtained.

The crude oil is a mixture of many hydrocarbons, or bodies consisting of hydrogen and carbon. Some of these are light, highly-inflammable liquids, which become gaseous at the ordinary temperature ; others are heavier, but still inflammable liquids ; others are yet heavier liquids, thick and treacly in appearance, less inflammable, but of great value for lubrication ; while still others are greasy or waxy solids at ordinary temperatures. Each of these is suited to its particular purpose, and the method of separation is based upon the principle that every pure substance boils at a definite temperature under a given pressure. If, therefore, a mixture like crude petroleum is heated, the constituents of lower boiling-point distil over first, and if the receiver in which the liquids are collected is changed from time to time, fractions boiling between certain limits of temperature are obtained.

Two methods are employed. In one, the vessel containing the crude oil is heated gradually and as the lighter liquids pass off the temperature in the still rises. The vapours are cooled by passing through several hundred feet of pipe, over which cold water flows, and ultimately run into a receiver which can be changed as occasion requires. In practice, the actual temperature is not observed. The distilled oil flows into a box with glass sides, and the man in charge can tell from the appearance and rate of flow when oil is to be directed into a fresh receiver. This is known as the intermittent process.

In the other, or continuous, process, the oil is pumped in succession through a series of stills of successively higher temperatures. Passage through the first causes the oils of lower boiling-point to evaporate ; passage through the second separates the group of substances having a higher boiling-point, and so on. With the first process the best yield of illuminating oil is obtained, and with the second the best yield of lubricating oil.

The products, in the order in which they are obtained, are as follows :—

1. Gases—solidifying near the freezing point of water.
2. Clear, colourless light oil—naphtha.
3. Yellow illuminating oil—kerosene or paraffin.
4. Lubricating oils.
5. Paraffin wax.
6. Coke, pitch or asphalt.

The gases which come off first are allowed to escape into the air, or are used to heat the stills. The naphtha is redistilled and gives

(a) Gasolene or petrol.
(b) Commercial naphtha.
(c) Benzine.

The first of these is the substance so largely used in the engines of motor-cars and aeroplanes. The last is used for dry cleaning, and should not be confused with benzene, a coal-tar product which is sometimes used for motor-cars owing to the present high price of petrol. A similar process of redistillation is applied to the illuminating oils by which the different qualities are separated.

If in the original process a high yield of illuminating oil is required, a plan known as "cracking" is adopted when two-thirds of this oil has come over. It consists in raising the temperature of the furnace quickly, and causing some of the lubricating oils to decompose, thus increasing the yield of oil suitable for giving light. Should a higher yield of lubricating oil be required, superheated steam is driven through the liquid in the still in order to encourage the oils of higher boiling-point to evaporate without decomposition. It should be observed that there is a marked difference between American and Russian methods partly because the American oils vary so much and partly because, while the American desires kerosene or lubricating oils, the Russian refiner seeks a high yield of the residue, or *astatke*, for fuel.

This process of " cracking " is likely to become very important now that the lighter fractions are so much in demand for motor-cars. It would appear that there are many less valuable heavy oils that yield a high percentage of light oil on being subjected suddenly to a high temperature. In some cases the tendency to form acetylene under these conditions may be prevented by carrying out the operation in the presence of hydrogen gas.

The lubricating oils and the paraffin wax both are further refined before they come on the market. The high speeds, high pressures, and high temperatures employed in modern engines have imposed severe conditions upon the oils which are required to reduce friction, and the separation of these into grades suitable for different purposes has become a fine art.

Before considering the special use of oil as a fuel it will be interesting to glance at the great variety of services which petroleum products render to mankind. Of the 200 substances that have their origin in raw petroleum, the illuminating oils have surely the oldest and widest interest. In all the far corners of the earth, where the advantages of town life do not exist, they add to the light of day and well-nigh double the hours which man can give to his labours. They supplement the beams of the Arctic moon, and dispel the gloom of the tropical night. They illuminate the sick room and diminish the terrors of darkness. In a thousand and one ways they contribute to man's comfort, and aid him in his fight against time and circumstances.

The lighter products are valuable solvents for rubber. Cloth may be rendered waterproof by a thin layer of rubber, which, when dissolved in naphtha can be applied with a brush. As the naphtha evaporates a continuous skin of rubber remains, which is light and impervious to rain. So, in a similar way, resins can be dissolved, forming varnishes, which on drying give a bright, hard surface that acts as a preservative of the material upon which it is laid. The readiness with which it dissolves fats and other substances not soluble in water, causes benzine to be used in extracting grease from leather, in dry cleaning, and in extracting oil from the seeds of plants. It is also used in the manufacture of jute, the fibre that is woven into the coarse canvas or " scrim ", which is so largely employed for packing bales of cotton and other fabrics. Finally paraffin is mixed with water or lime wash for spraying fruit trees to destroy insect pests.

From the heavier samples come vaseline, which is closely allied to the lubricating oils, petroleum jelly, and similar substances. Paraffin wax, obtained from the heavier varieties by freezing, and purified by six or seven successive processes, is used as an insulator for electrical work, for candles, in the manufacture of matches, for lining barrels to render them water-tight, and—for chewing-gum ! Look where you will, and there is some product of petroleum to meet a necessity or provide a comfort.

But most of these substances are by-products, and the enormous activity in the oil industry arises from its value as a fuel. From what has been said about gaseous fuel, it will be apparent that the best way to burn a liquid fuel is to convert it into vapour, or at all events into a fine state of division. In using oil, therefore, in a furnace or under a boiler it is necessary to convert it into a fine spray, and this is usually effected by forcing it through a special nozzle which breaks it up into fine particles. These form an intimate mixture with the air supply, and when the latter is properly adjusted rapid and complete burning results. The cost of the lighter oils prevents their use for this purpose except on a small scale. The heavier oils, which are cheaper, do not flow freely, and they must be heated and then forced through a nozzle by a jet of steam or compressed air. Such a nozzle is called an atomizer, because it breaks up the jet into extremely fine particles.

The fact that heavy grades of petroleum or even coal-tar can be and are used in this way has had an enormous effect on the oil industry. The Californian oils, for example, are heavy, contain but a small proportion of the lighter constituents, and do not pay to refine. The value of the oil from this State therefore depends very largely upon its use as a fuel. In marked contrast oil from Mexico and the East Indies yields a very valuable proportion of petrol.

The special value of a liquid fuel in steam-raising depends upon the fact that the flame immediately reaches its maximum temperature—ignoring for a moment the cooling effect of the furnace. In a coal fire, on the other hand, some time must elapse before it is hot enough to raise steam. Many fire-engines are now supplied with oil-fired boilers, which enable them to get up steam with great rapidity.

Besides burning it beneath boilers, however, oil is used in

enormous quantities in the internal-combustion engines described in Chapter IV, and for the details of its employment in this way that chapter must be consulted. It may, however, be stated here that while formerly the chief demands were for the middle fractions—the illuminating and lubricating oils—the petrol and heavy oil engines have created an enormous market for the lighter and heavier products respectively. Moreover, it should be noted that while America produces three-fourths of the world's supply, it is not the only country which sends oil to Great Britain. Large quantities come from Mexico, Persia, Roumania, the Dutch East Indies, and other countries.

For burning under boilers and in the Diesel engine crude grades of heavy oils, and even tar, can be used ; and these are obtained by distilling oil shales and coal. So we are by no means dependent upon oil wells for oil fuel. It must be remembered that the production of oil from all sources is far less than the quantity of coal available. Mr. Dugald Clerk has calculated that not more than 20 per cent of the world's power could be produced in this way.

ALCOHOL AS A FUEL

The rise in the price of petrol has led to a search for substitutes especially suitable for use in small motors, and it may have puzzled some readers to know why so much stress should have been laid on alcohol. The fact is that alcohol costs very little to manufacture. Practically all plants contain starch or cellulose —in fact, the latter is their chief constituent—and both starch and cellulose produce sugar either in the natural processes which accompany plant growth, or by artificial fermentation. Further, sugar yields alcohol when the living ferment yeast is grown in it. It is clear, therefore, that while some forms of vegetable life would produce more alcohol than others, this liquid, which will burn and can be used in internal-combustion engines, could be obtained in enormous quantity if required.

But the question raised by the use of alcohol is of far wider significance than appears at first sight. Timber is a slow-growing form of fuel, and its use is attended with disadvantages, to which reference has already been made ; and alcohol can be prepared cheaply from any kind of quick-growing vegetation that absorbs carbon dioxide from the air to build up the cellulose of its framework or the starch of its cells. This may not appeal very strongly to those who live in thickly populated countries where land is dear and needed for raising food, but it may appeal to the colonial farmer, who sees an opportunity of clothing profitably the vast acres around him.

Coal and petroleum, on the other hand, are not, so far as we know, in process of formation at the present time in any part of the earth's crust, and the use of these kinds of fuel is a continual drain upon capital. The materials which the plants take from the soil can be returned to it, but there is no way of replacing coal in a mine or of renewing the oil in an exhausted well. If in time the ancient store of natural fuel should give out, then so far as we can tell now there would remain as sources of power only wind, water, and such combustible material as could be grown after the demand for food has been satisfied. The hungry man does not break his fast on firewood and small coal, but the thirsty man often drinks an unnecessary amount of alcoholic liquid, which is a really valuable source of power. And it is possible to imagine that the housewife of the future may feed the kitchen fire with whisky and warm the drawing-room with effervescing champagne.

CHAPTER III

WHEN James Watt, in 1769, improved the crude and clumsy contrivance that worked by steam, he created the driving force by which the industrial revolution of the eighteenth century was achieved. In the century which has elapsed since his time the material conditions of life have altered to a greater extent than in the previous 1,700 years. A new civilization has arisen, so different from any which have previously existed in the history of the world that man has hardly yet grasped the significance of the change, and can only see " as in a glass, darkly ", the possibilities of the coming years.

For more than a century the steam engine had a clear field. The production of power is under more complete control than from a waterfall whose volume varies with the seasons. The great manufacturing towns sprang up on or within easy reach of the coalfields. Knowledge of electricity, the possibilities of which had been seen by Faraday in 1832, passed through a long period of infancy, and by the time that efficient generators of large size were a commercial success the steam engine was firmly established. Not until after 1876 did the internal combustion engine appear on the scene, and for twenty years it did little more than supplement in a humble way the efforts of the giant that had altered the habits and customs of the civilized world.

To no country was the time and circumstance of Watt's improvement so important as to our own. From that period until the twentieth century, Great Britain had been comparatively free from war. The great continental nations, on the other hand, had been frequently embroiled, and it was during the Napoleonic wars that we laid the foundation of an industrial supremacy that opened up to us the markets of the world. With generous natural resources, a unique geographical position, and vast colonial possessions, this country was able to take advantage of scientific discovery and mechanical invention, and not only to initiate a new era in the progress of man, but to hold her place even after other countries had entered the field. A just pride

28

in the army and its weapons, in the navy, in the merchant
service, in internal transport, in manufacture, should be tempered
by the reflection that the seed from which all these have sprung
is the mechanical invention of a Glasgow instrument maker
nearly 150 years ago.

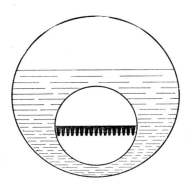 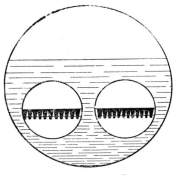

Fig. 19. TRANSVERSE SECTION   Fig. 20. TRANSVERSE SECTION
OF CORNISH BOILER.   OF LANCASHIRE BOILER.

In the engines which man uses to wrest from coal the stored-
up energy of the prehistoric sun the line of progress of the last
century has been to secure more power from each pound of fuel

Fig. 21. THREE TYPES OF VERTICAL BOILERS.

used. The steam-engine is a heat engine. The coal in burning produces heat—each pound of coal giving about 14,000 units. This heat is taken to the engine in the form of hot steam, and when the steam passes out of the engine it is cooler. The useful portion of this cooling is due to the expansion of the steam in forcing the piston backwards and forwards, and the rest is more or less unavoidable loss. The higher the temperature of steam to begin with, and the lower its temperature at the end, the greater will be the amount of work done, provided that the losses do not increase in the same proportion. If therefore the greatest amount of heat is to be obtained from the coal, it is necessary to consider two sets of losses—those which occur in the boiler, and those which occur in the engine. Let us consider the boiler first.

THE MODERN BOILER

The diagrams in Figs. 19 to 23 represent the chief types of boiler in use some forty years ago. In all cases the hot gases pass through a number of tubes or flues to the chimney. If these tubes are large in diameter, as in the Cornish or Lancashire boiler, the hot gases in the middle of the flue do not come into contact with the walls, and the heat they contain escapes with them up the chimney. To prevent this, wide flues have water tubes across them, which not only intercept the hot gases, but encourage more rapid circulation of the water. Return flues are built in the brickwork on either side of the boiler, and the Lancashire type is a very efficient form of steam generator for continuous operation. For intermittent working it is wasteful because of the mass of brickwork which has to be heated every time the boiler is required.

The efficiency of a boiler is measured by the quantity of water it will convert into steam per pound of fuel consumed. It must have a large heating surface, a fiercely burning fire, and be capable of withstanding high pressures. Increase in heating surface has been secured by arranging that a portion of the water is exposed to the fire in narrow, inclined tubes amongst which the hot gases flow on their way to the chimney. On account of their relatively small diameter—3 or 4 inches—they may be made of thin material and yet be strong enough to resist the high pressure to which they are subjected. The water in these tubes takes up heat rapidly, decreases in density, and rises through

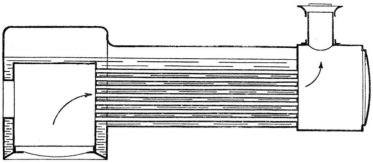

Fig. 22. Locomotive Boiler.

the upper ends into a cylinder or drum which contains the main body. Cool water then flows from the drum into the other ends of the narrow tubes and the circulation is kept up. In this way, not only is the water in the tubes heated quickly, but it moves on quickly to make room for cooler water from the drum.

Very frequently boilers on land can be equipped with chimneys of such a height that the natural draught is sufficient to maintain rapid combustion, but forced draught is coming into greater use. The air for this purpose is usually supplied by a fan, which forces it directly into the furnace ; but on ships the fan is placed outside the stokehold, which is closed up so that the men work under the pressure which drives the furnaces. The practice on

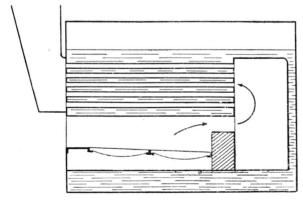

Fig. 23. Scotch Marine Boiler.

locomotives, invented by George Stephenson, was to allow the exhaust steam to pass up the chimney, but this is far too wasteful to be used for stationary or marine engines under modern conditions. More especially, fresh water at sea is so scarce that every ounce passing through the engine is condensed, freed from oil, and returned to the boiler.

A saving is effected in large boiler installations by the use of " economizers" which consist of nests of tubes, through which the feed-water passes, arranged between the boiler and the smoke-stack. A quantity of heat which would otherwise be lost is caught and returned to the boiler, which has less heat to supply than if the water was fed in cold.

The amount of steam at a given temperature that can be produced per pound of coal depends a good deal on careful stoking. If the fire is allowed to burn low, and is then choked with a heavy charge of coal, much smoke will be produced, the pressure of the steam will vary, and the boiler will be inefficient. Such irregularity is avoided in large installations by the use of mechanical stokers. The coal is fed into a hopper in front of the boiler and is carried into the furnace on a wide chain belt or pushed in by a ram which moves backwards and forwards. By this means a steady supply of fuel is provided without opening the doors and allowing a sudden inrush of cold air.

A further device, though this affects the efficiency of the engine rather than that of the boiler, may be mentioned here. In most boilers it is practically impossible to draw off dry steam, i.e. steam free from small drops of water ; and this water serves no useful purpose in the production of power. The presence of water in the steam is known as " priming " and has to be reduced as far as possible, either by a special device which removes the water, or by superheating steam by passing it through tubes contained in the flues on its way to the engine, as in Fig. 24. It is possible to give it a temperature considerably higher—by 100° or 200° F.—than the temperature in the boiler. The tiny drops of water are converted into steam and its volume increases. The thread of steam in the hot tube is drawn out and lengthens towards the cylinder, which it fills with less weight than would be required at a lower temperature. Not only are the defects of priming eliminated, but the increase of temperature produces the same effect as an increase of pressure, and the engine uses less steam per horse-power. Superheating is no new device, but contrivances

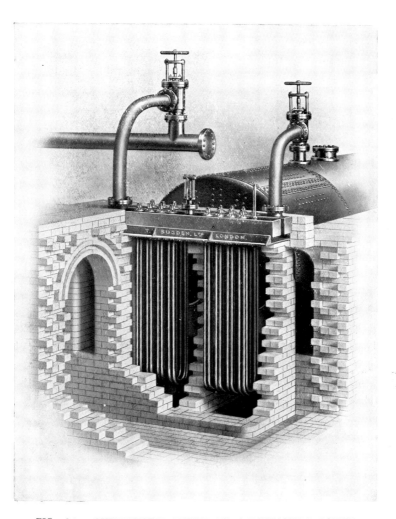

FIG. 24.—SUPERHEATER FITTED TO A LANCASHIRE BOILER.

To face page 32.

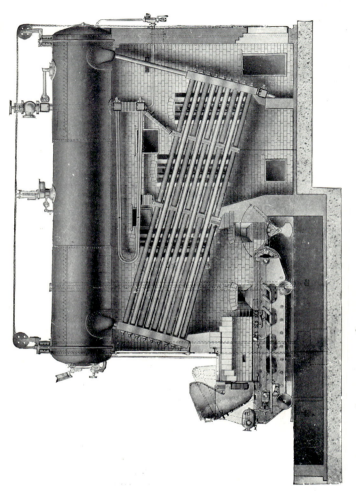

FIG. 26.—BABCOCK & WILCOX WATER TUBE BOILER WITH SUPERHEATER AND MECHANICAL STOKER.

To face page 33

for effecting it have improved a good deal in recent years, and metallic packing with non-carbonizing cylinder oils have rendered a higher degree of superheat possible. It is now applied to every type of engine—stationary, marine, and locomotive—and it may be said generally that a saving of 1 per cent of fuel is effected by every 10 degrees of superheat.

There are several very interesting methods of automatically regulating the supply of feed-water to a boiler. Under ordinary

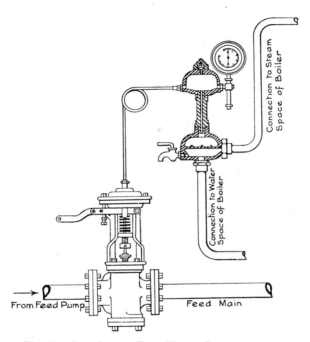

Fig. 25. THE CROSBY FEED-WATER REGULATOR.

circumstances it is the business of the man in charge to keep an eye on the water-gauge, and to adjust the supply from the feed-pump whenever necessary. There is one level which gives the best results in practice, and a constant level in any case leads to less priming, more uniform pressure, and generally to more regular working. If this can be taken out of the hands of a man

D

and put under the control of a machine, so much the better. The particular form selected for illustration is that made by the Crosby Steam Gauge and Valve Company, which is shown diagrammatically in Fig. 25. The tube between the valve which admits the water from the pump to the boiler, and the bulb, which has a partition across the middle, are filled with distilled water. Any change of temperature in the lower half of the bulb, under the partition, will cause this water to expand or contract, and thus to open or close the valve. The bulb is fixed so that the partition is at the desired level of the water in the boiler, and the tubes connect the lower half with the steam space and the water space respectively. If the water-level in the boiler rises ever so little, then water from the lower part of the boiler comes into contact with the partition, cools it, and closes the valve. But if the water-level in the boiler sinks, steam enters the bulb, warms up the distilled water through the partition and opens the valve. It is difficult to imagine a more beautiful contrivance than this. When steam is being drawn from the boiler, the valve is rarely completely closed or open, but executes a slight movement according to the rate of evaporation. With unerring accuracy it feels the pulse of the boiler, and responds to the faintest variation of level. The machine does what no human being should have to do : by sheer concentration upon one mechanical detail it executes its duty with perfect reliability. It has no variety of initiative to be destroyed ; and the man has.

For rapidity of steam raising and general efficiency there is nothing to beat the modern water-tube boiler, of which two examples will be described. A section through the land form of the Babcock and Wilcox boiler, in Fig. 26, shows very clearly the arrangement of inclined tubes fixed at right angles to the stream of hot gases, and connected at each end with the drum at the top. It also shows the baffle-plate by which the hot gases, having passed between the upper halves of the tubes are directed in turn between the lower halves. The U-shaped tubes, fixed horizontally just below the drum, form the super-heater. In front is shown the hopper into which the coal is fed, and below is the mechanical stoker mounted on a truck so that it can easily be withdrawn from the furnace. The coal falls from the hopper on to a chain belt, which passes round toothed rollers at each end of the carriage, and feeds the coal gradually on to the grate. The grate is fitted with rocking levers which,

FIG. 27.—THE YARROW BOILER.

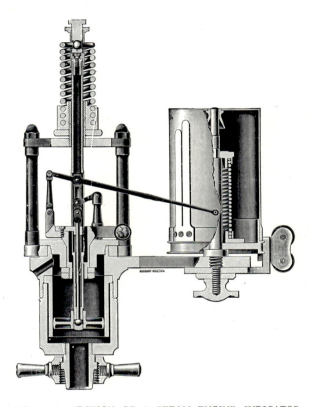

FIG. 30.—SECTION OF A STEAM ENGINE INDICATOR.

To face page 35.

moving backwards and forwards, prevent the formation of clinker and keep the fire-bars clear of ashes.

The Yarrow boiler illustrated in Figs. 27 and 28 is the outcome of many experiments made by Mr. (now Sir) A. F. Yarrow, the well-known engineer and shipbuilder, who has done so much for the scientific development of shipbuilding and marine engineering.

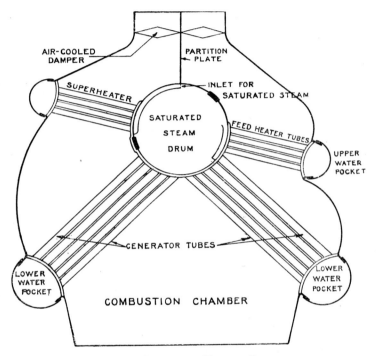

Fig. 28. SECTION OF YARROW BOILER.

It consists of two lower drums and an upper drum, with which the lower drums are connected by tubes arranged on each side of the furnace. A superheater is fixed between the tubes and the casing on one side, and a feed-water heater in a similar position on the other. The feed-water enters the upper drum at the side and is deflected by a plate down the outer row of tubes to the lower drum, so that it does not mix immediately with the main body of hot water which is being converted into steam. The heat can be cut off from the feed-water heater or

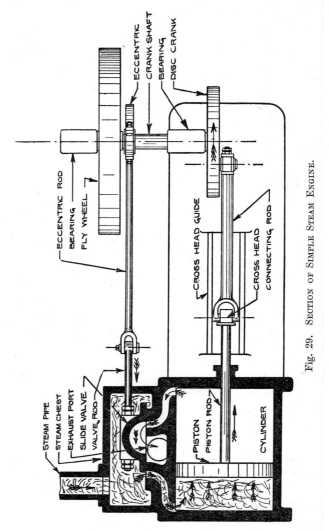

Fig. 29. Section of Simple Steam Engine.

superheater by dampers on each side of the uptake leading to the chimney. In the original boiler the lower drums were roughly

D-shaped, being more or less flattened on the side which received the tubes. During the war, however, these were found to be unreliable, and were replaced by drums of circular section, with a thicker plate to hold the tubes.

The Babcock and Wilcox boiler is made in two forms, one for land and the other for marine purposes ; the Yarrow is only one form. One or the other is used in nearly every navy in the world. They enable steam to be raised quickly and produced rapidly, and like all other boilers can be adapted to burn oil-fuel. They represent the last word, for the time being, in boiler construction. Thirty years ago the production of one horse-power required 6 to 7 square feet of boiler heating surface ; to-day, it requires 3 to 4 square feet. Formerly 3 or 4 pounds of coal were required per horse-power per hour ; now the same power can be obtained for from 1 to $1\frac{1}{4}$ pounds. A very similar degree of economy can be secured by the Stirling, White-Foster, Thorneycroft and other forms. The two last-named are used mainly on launches, yachts, and torpedo boats.

THE RECIPROCATING ENGINE

Before considering the modern improvements in the steam engine it will be desirable to recall briefly how the engine works. Referring to Fig. 29 the steam enters the steam-chest, and when the crank is in the position shown, it passes through the back port into the cylinder, and presses the piston forward. *Before* the piston has reached the end of its stroke, the valve moves so as to admit steam on the other side of the moving piston to steady it ; then the back port is put into communication, through the hollow in the underside of the slide valve, with the exhaust port. The steam entering at the front of the piston now forces it back until, at the end of the stroke, it is allowed to escape through the exhaust. Since the time of Watt it has been the custom, in all engines in which a high efficiency is required, to condense the steam issuing from the exhaust, either by passing it through a nest of tubes conveying cold water (surface-condenser), or by leading it into a chamber containing a spray of cold water (jet-condenser). In either case an air pump is used to reduce the back pressure on the piston.

The object of admitting steam in front of the moving piston is to prevent shock, by forming a " cushion " which reduces the speed of the piston gently. The object of cutting off steam early in the stroke is to utilise as much as possible of the heat energy in the steam. Expansion produces cooling, and the heat which disappears corresponds, when allowance has been made for that used in raising the temperature of the cylinder, to the work done on the piston. If the steam is cut off at one-third stroke, it expands to three times the volume admitted ; if at a quarter stroke to four times ; if at one-fifth to five times, and so on. The disadvantage of too great a range of expansion in an ordinary cylinder is that the condensation occurs upon the cylinder walls and thus is re-evaporated again. Generally the cylinder has a steam jacket to prevent condensation. Further, the steam has to pass out through the same ports by which it entered. Unless, therefore, the steam emerges with a very high velocity, congestion will occur in the ports if the ratio of expansion

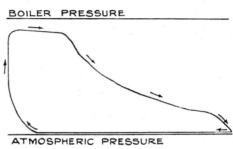

BOILER PRESSURE

ATMOSPHERIC PRESSURE

Fig. 31. INDICATOR DIAGRAM.

has been high, and an excessive back-pressure would be produced on the piston. An expansion of more than five or six has been found to be undesirable.

In order to see why expansive working is economical it is necessary to understand what is going on in the cylinder. Suppose a piston is fitted in a tube connected with one end of the cylinder, and held down by a spring which will allow it to move up or down as the pressure in the cylinder varies. If a pencil is fixed to this piston it will trace on a paper held against it a line representing in length the difference of pressure which occurs during the stroke. Suppose next that the paper is mounted on a small drum which is connected to the crosshead (or, piston rod) of the engine by a string in such a way that it rotates as the piston moves. A fixed pencil pressed against this drum would trace a line on the paper representing the length of the stroke to scale, and the rate at which this line was drawn at any point would correspond to the speed of the piston at that stage of its journey. But if, instead of the fixed pencil, the pencil registering the changes of pressure were used the line traced on the paper would indicate both changes of pressure and the corresponding movement of the piston. Such an arrangement is called an indicator (Fig. 30) and the figure traced on the paper is called an indicator diagram (Fig. 31).

The shape of the diagram furnishes information as to the variation of pressure throughout the stroke, and the rapidity with which steam enters or leaves the cylinder. Its area represents to scale the work done by the stream. The problem of the engineer, therefore, is so to adjust the initial pressure, cut-off, and movements of the valves as to obtain a maximum area for a given weight of steam. In Fig. 31, the cut-off is at quarter stroke. Remembering that the boiler pressure and the stroke are fixed, it will be seen that if there was no cut-off—if steam was admitted at full pressure to the end of the stroke—the area of the diagram and, therefore, the work done would not be twice as much though four times the weight of steam would have been used.

Many of the improvements in the first hundred years were improvements in valves, and in the methods by which they were operated. Friction was reduced, steam was admitted more quickly and allowed to escape more quickly, and the point of cut-off could be varied to meet different conditions of working. The old D-shaped slide valve (Fig. 29) is difficult to keep steam-tight without unduly increasing the friction, and has been replaced in marine engines and high speed engines for electric power stations by the piston valve. In this case the valve chamber is like another cylinder, to which steam is admitted first, and the movement of two pistons on one rod open and close the ports between the two. Again the desirability of opening and closing the ports quickly has been met by the use of drop valves and valves of the Corliss type. These are often seen on large engines used for driving mills or factories.

A very interesting type now being made is the Uniflow engine. This has a very thick piston, the thickness being nearly half the length of the stroke. The exhaust ports are situated in the middle of the cylinder and are put into communication with each and alternately by the movement of the piston. This avoids the reversal of flow which ordinarily occurs when the steam, having forced the piston to the end of its stroke, escapes through the same opening by which it entered.

The problem of obtaining a large amount of work depends upon higher initial pressure and lower pressure of exhaust, or in other words upon the range of expansion. There are, however, several disadvantages in expanding steam to more than five times its original volume in one cylinder ; so the compound triple expansion and quadruple expansion engines were devised in which the steam passes successively through two, three, or four cylinders of increasing size to accommodate its increased volume.

The use of a condenser to reduce the back-pressure was Watt's greatest gift to the steam engine. The increase of efficiency by expanding the steam and condensing it in a vacuum is so great that it justifies the use of air-pumps to remove the exhaust steam from the engine, and water-pumps to circulate the cooling water.

While the various improvements which have been described have been adopted in both locomotive and marine engines, these have retained to a large extent their original form. With stationary engines, however, there is a marked tendency to replace the horizontal by the vertical type and to employ high speeds. The latter necessitates reliable material and unimpeachable workmanship. But it also introduces certain mechanical difficulties which require special means to overcome them. The first of these is lubrication. When two surfaces are rubbing together they soon become hot, unless they are separated by a film of oil. With high speed engines very large forces are called into play, and a thin oil would be squeezed out. Again, at high speeds, the film is liable to be broken and cavities formed. Both these dangers are avoided by forcing the oil between the surfaces by a small pump driven from the engine shaft.

The next problem is that of vibration. As the piston moves backwards and forwards, it alternately pushes and pulls the crank. This produces alternating pushes and pulls in the frame or foundation which connects the bearings with the cylinders, and when these alternations are taking place 600 or 700 times a minute a good deal of vibration may be produced.

But a more serious vibration may arise from another cause.

The weights of the rotating parts are not equally distributed round the shaft. The crank-pin and connecting-rod are moving round the shaft—now in front, now beyond, now above, now below. If a stone is whirled round at the end of a string the latter is stretched tightly, and if the stone is heavy or is whirled round very rapidly the string will break. The force exerted outwards by a rotating body is given by the formula.

$$\frac{WV^2}{gr}$$

where W is the weight, r is the radius of swing, V is the velocity in feet per second, and g is the gravitation constant ($= 3.22$). Suppose the weight to be 100 lb., the radius 20 in., and the number of revolutions per minute 300, the force on the bearings due to the rotating parts would be nearly two tons. This may squeeze out the lubricants, cause over-heating, and even burst the bearings. In order to avoid this, the sides or slabs of the crank are continued backwards and expanded in the shape of a fan in such a way as to balance as nearly as possible the rotating parts on the other side of the shaft. In this way an approximate solution can be found. With two cranks at right angles the problem is more difficult. In the locomotive the reader will have observed that the space between two spokes of the driving wheel is filled in. These solid masses of metal prevent in some measure the excessive vibrations that are liable to occur at high speeds.

While many horizontal engines are still made, the type *par excellence* of modern reciprocating engine is a high-speed, totally enclosed vertical engine with forced lubrication. The advantage of a vertical over a horizontal engine of the same power in the matter of space is well shown in Fig. 32. Again, the vertical position results in more even wear of the cylinder liner and stuffing boxes. High speed gives the steadiest running, total enclosure keeps out dust and grit, and oil fed into the bearings and over other rubbing surfaces under pressure from a small pump renders the engine practically fool-proof.

A good example of this modern tendency is that made by Messrs. Bellis and Morcom, of Birmingham. This is called a quick-revolution rather than a high-speed engine, because the stroke is relatively short, and the linear speed of the piston is not greater than that in slow-speed, long-stroke engines. The

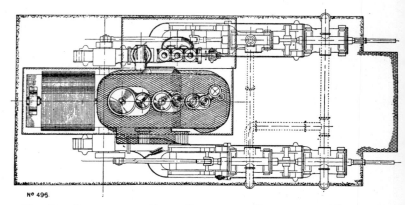

Nº 495.

Fig. 32.   COMPARISON OF SPACE OCCUPIED BY VERTICAL AND HORIZONTAL
ENGINE OF THE SAME POWER.

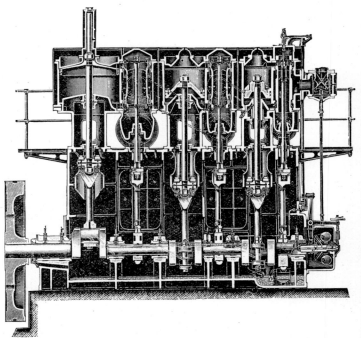

Fig. 33.   SECTION OF BELLIS AND MORCOM "QUICK-REVOLUTION" ENGINE,

general arrangement will be clear from Fig. 33, which represents a " C " type compound engine, with the high pressure cylinder on the right hand and the low pressure cylinder on the left-hand. A single piston valve serves both cylinders. A small pump forces oil from the well under the crank-shaft between all rubbing surfaces at a pressure of 10 lb. to 20 lb. per square inch. On the right of the shaft is the governor which operates a valve at the steam inlet on the right of the high-pressure cylinder, and thus regulates the speed of the engine to within 3 per cent. at full load and 10 per cent. for momentary removal of load.

The first engine to be provided with forced lubrication is still running at the Birmingham works for ten or twelve hours a day. Much has been written of the marvellous reliability of a modern watch, but when it is stated that one of these engines installed in a chemical works ran for 99·77 per cent. of the total number of hours in a year, making 85,000,000 revolutions from July 1st to November 30th without a stop, and required no repairs or adjustments, some idea will be gained of the accuracy of workmanship, durability of material, and reliability of a modern steam-engine.

THE STEAM TURBINE

The type of engine which has been described has both advantages and disadvantages. It is as efficient as a steam engine can be over a wide range of load, and it is capable of being readily adjusted to special conditions. It is the concentrated essence of a century of invention directed to the attainment of efficiency without modifying the principle of action. But in large engines there are heavy masses of metal in the piston, piston-rod, crosshead, and connecting-rod, which move at high speeds and have their direction reversed many times a minute. Part of the energy of the steam is used in setting these in motion, and part in bringing them to rest preparatory to setting them in motion again in the opposite direction. In fact, a reciprocating engine is wasteful in starting and stopping a portion of its own moving mass. Moreover, the effect of the connecting-rod on the crank varies throughout the stroke, reaching a maximum only when the two are at right angles. Consequently engineers have endeavoured, from the beginning, to obtain a direct rotary force upon the shaft, without the intervention of piston, connecting rod, or crank. The result of their efforts is represented by the steam turbine.

The simplest form is that invented by Dr. Gustaf de Laval,

and its action is explained by Fig. 34. The disc has a number of curved vanes fitted radially near its outer edge, and overlapping like the laths of a venetian blind. The steam is directed upon these by four, six or more nozzles, one of which is shown transparent in the figure, in such a way that it impinges upon the blades and causes the wheel to spin round. The whole arrangement is enclosed in a case through which the shaft passes, so that the steam can be drawn off after it has gone through the wheel and either discharged into the air or condensed.

There are several scientific principles of great interest involved. The first of these determines the shape of the nozzles, one of which is shown in section in Fig. 35. It will be observed that the size of the opening increases as the mouth is approached. If steam is allowed to escape from a narrow opening into a region of much lower pressure, it is "throttled", and has only a moderately high velocity. If, however, the opening expands towards the mouth the steam expands, and acquires a very high velocity; hence though the weight of steam may be very small it is able to exert considerable force upon anything which stands in its path. Each blade therefore receives an impulse from the jet of steam which issues from the nozzles with a velocity of 3,000 or 4,000 feet per second.

If the wheel be prevented from rotating the steam will issue on the other side of the wheel with the same velocity that it left the nozzle, but this velocity will be in another direction—the direction in which the paths between the vanes point on the exhaust side of the wheel. Suppose the wheel to be rotating so that the vanes are moving as fast as the steam is issuing from the nozzle, the steam then will exert no force upon them at all. It should be clear therefore that there is some velocity between nothing and the velocity at which the steam is issuing at which the greatest amount of useful work will be done, and this is nearly half the velocity of the issuing steam.[1]

The velocity of steam expanding through a nozzle of the type shown is very high, and may easily reach 3,000 or 4,000 feet a second. This means that the vanes ought to move at 1,500 to 2,000 feet per second, or 90,000 to 120,000 feet per minute ! In the case of a small machine with a wheel only 6 inches in diameter this would involve, theoretically, a speed of nearly 80,000 revolutions per minute. In actual practice the speed ranges from 30,000 revolutions per minute in the small turbines to

---

[1] Compare the Pelton wheel, p. 3, which is an "impulse" water turbine.

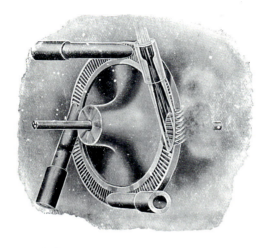

FIG. 34.—DISC AND NOZZLES OF
IMPULSE TURBINE.

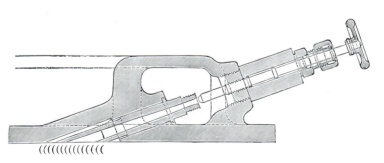

FIG. 35.—SECTION OF NOZZLE OF IMPULSE TURBINE.

To face page 44.

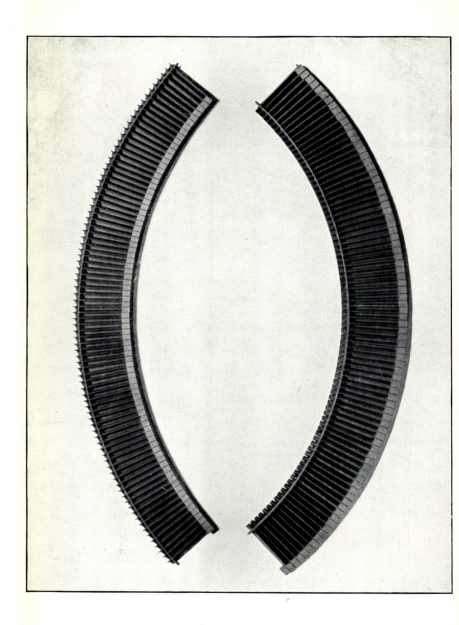

To face page 45

9,000 revolutions per minute in the larger ones. Such an enormous velocity cannot be applied directly to any machine, and the power has to be transmitted through toothed gearing.

From what has been said about vibration and balancing on pp. 40–41, it will be clear that the turbine brings into play a series of problems from which the reciprocating engine is relatively free. The centrifugal forces in the wheel cause large stresses which tend to burst it, and the best possible material must be used. Moreover, no amount of care will result in an accuracy of workmanship that gives perfect balance, and the tiniest fraction becomes serious at these high speeds. Some compensation has, therefore, to be sought which will render such small inaccuracies as are unavoidable free from danger ; and this has been found in an interesting property of rotating shafts. If a thin spindle is rotated at a gradually increasing speed it begins to bend and whirl instead of rotating in a straight line. This is most marked at one particular speed, which depends upon the length and stiffness of the shaft. At higher speeds than this the shaft stops whirling and settles down to steady motion, just as a top " goes to sleep " at high speed. It will be observed that in Fig. 34 the wheel is mounted on a slender shaft, and this is of such dimensions—only $\frac{1}{4}$-inch diameter for 5 horse-power, and only $1\frac{1}{4}$-inch diameter for 300 horse-power—that the " critical speed " at which the greatest whirling takes place is below that at which the turbine is designed to run. The case surrounding the wheel allows for a little play so that the turbine can be run up to its steady condition without the blades being torn off.

While de Laval's turbine has been described first on the ground of its simplicity, it was later in point of time than the one which is now to be considered. The Hon. (now Sir) Charles Parsons filed his first patent for a reaction turbine in 1884, and in 1885 a machine was constructed which, though rotating at 18,000 revolutions a minute, gave great satisfaction. In its modern form it consists of a drum upon the outer surface of which are fixed circular rows or rings of blades, and the casing in which the drum is enclosed also carries rings of blades, projecting inwards, which fit with very small clearance between successive rings on the drum. The shape of the blades and their appearance on the drum are shown in Figs. 36 and 38, while Fig. 37 shows the fitter fixing them in place. Steam enters the first ring of fixed blades and is directed by them upon the first ring of moving blades at a proper angle. The drum is not parallel, and successive rings of blades increase in diameter from the high pressure to the low pressure end, where the steam leaves. The steam, therefore, passes through a larger and larger space, and the expansion takes place as it goes between the blades. The practical consequence of this is that the expansion is split up into a number of stages and the reaction turbine rotates at a lower speed than the original impulse turbine.

From these two fundamental types several forms have been evolved. The Rateau turbine, for example, is an impulse turbine with a number of discs on the same shaft, but running in separate chambers through which the steam passes in turn. By this plan the velocity is split up into a series of stages. Some turbines partake of the character of both types, they have a disc and a drum. Superheated steam acts on the disc and is then expanded through fixed and moving blades of the Parson's type.

The words impulse and reaction are borrowed by analogy from the theory of water turbines, and the simple explanation which has been given is not very satisfactory. We shall, therefore, inquire a little more closely into the theory, and a very elementary knowledge of mechanics will enable the reasoning to be followed. The reader should, however, keep in mind Newton's three laws of motion, which may be stated as follows :—

1. Force is that which changes or tends to change a body's state of rest or uniform motion in a straight line.
2. Change of motion is proportional to the impressed force.
3. To every action there is an equal and opposite reaction.

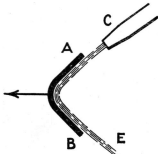

Fig. 39. Diagram to explain Principles of Impulse Turbine.

Suppose a jet of steam C impinges upon a blade A B as in Fig. 39. If the blade is smooth, friction may be neglected, the water will not change its direction. If the blade is fixed the new direction will be that indicated in the Figure, and since force is required to effect this, there will be a tendency for the blade to move in the direction of the arrow. If the blade moves, the steam will follow it up and this motion in the direction of the arrow will reduce the speed which the steam would have *towards the right* if the blade were fixed.

FIG. 37.—FIXING THE BLADES OF A REACTION TURBINE.

To face page 46.

FIG. 38.—THE COMPLETED ROTOR OF A REACTION TURBINE.

To face page 47.

Now the quantity of motion or " momentum " in a body is expressed by the product M × V, where M is the mass and V is the velocity of the body, and it represents a force. If M is in pounds weight and V in feet per second, the force is given in poundals and must be divided by 32·2 (the constant of gravitation) to bring it to pounds. So that if M lb. of steam flowing over the blade has its velocity *to the right* reduced from V_1 to V_2, the force exerted on the blade must be the difference of the momentum before and after the change, or

$$F = \frac{MV_1}{32·2} - \frac{MV_2}{32·2} = \frac{M(V_1 - V_2)}{32·2}$$

This is the principle of the impulse turbine. It must be noted (1) that there is no change in pressure during the whole time that the steam impinges on the blade, (2) if the blade is fixed there is a change in direction only, and (3) if the blade is moving there is a change in *both* the magnitude and direction of the velocity.

Consider next the reaction turbine. Suppose the vessel A, Fig. 40, suspended so that it can swing freely, is filled with steam

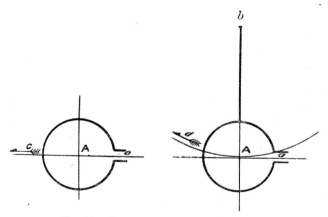

Fig. 40. DIAGRAM TO EXPLAIN REACTION.

at 200 lb. on the square inch. Since gases exert pressure equally in all directions the pressure will be 200 lb. on the square inch all over the interior of the vessel. But if it escapes at the nozzle there will be a force tending to make the vessel swing backwards in accordance with Newton's Third Law. This force depends upon the shape of the nozzle. If the sides are parallel, the outward pressure is never more than 0·58 of the inside pressure, or in this case 116 lb. per square inch.   The pressure tending to drive the vessel backward would then be 200 − 116 lb. = 84 lb. per square inch. An expanding nozzle reduces the outward pressure, so that if it is properly designed the backward thrust may become 185 lb. per square inch—the difference between 200 lb. and the pressure of the atmosphere.

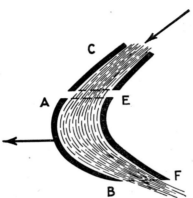

Fig. 41.   Diagram to explain Principle of Reaction Turbine.

If now in Fig. 41 CE represent the fixed and AB, EF the moving blades of a reaction turbine, the change in the direction of motion of the steam will tend to force the moving blades backwards, and as these blades move the steam will tend to follow them up so far as the action is the same as the impulse turbine. But the shape of the blades is such that the pressure at B is less than at A, and the speed of the steam over the surface of the blades increases.   This difference of pressure causes the blades to move backwards, just as in the case of the suspended vessel in Fig. 40.   In a reaction turbine, (1) the steam falls in pressure and therefore expands as it passes through the moving blades ; (2) the work done in consequence of the change of momentum which the steam undergoes (a) by change of magnitude and direction of its velocity, in impinging upon and following up the blade, and (b) by the change of velocity as a result of expansion within the vanes. As a change of pressure occurs the reaction turbine is often called a " pressure " turbine. Finally it may be noted that expansion takes place in both fixed and moving blades of a pressure turbine, and in the fixed blades only of a velocity turbine.

Though invented nearly forty years ago, the turbine was slow in development.   The high speed of the de Laval type was a disadvantage for many purposes, and it is only during the last ten years that silent and efficient gearing, capable of transmitting large powers, has been available. The first Parsons' turbine was of only 6 horse-power and until 1888 the largest was only 150. Then for six years the validity of the patents was in question and progress was hampered. But by 1908 machines of 10,000 horse-power were at work, while the steam consumption had been reduced from 20 lb. per horse-power per hour to 9 lb.

The largest turbine constructed in this country up to 1913 was built by C. A. Parsons and Co., for the Commonwealth Edison Co. of Chicago. It drives an electric generator producing 25,000 kilowatts (= 35,000 horse-power) on a steam consumption of only 8·1 lb. per horse-power per hour. There are two turbines on one shaft, one for high pressure steam and one for low pressure. The total length is 76 ft., width 18 ft., and height 10 ft. The turbines contain 112 rows of blading and individual blades vary in length from $2\frac{3}{4}$ to 19 inches. After completing its work the steam flows into the condenser through an opening 21 ft. long and 12 ft. wide. But machines are now made up to 208,000 kilowatts or about 250,000 horse-power.

One of the best examples of a modern turbine based upon the impulse principle is the Westinghouse-Rateau.[1] This consists of a series of discs mounted on one shaft and separated by similar discs or diaphragms connected to the casing. Both discs and diaphragms are provided with a row of blades, those in the latter forming the nozzles. The construction is illustrated in Figs. 42, 43, and 44. It will be observed that the diaphragm at the high pressure end has only just a sufficient number of openings to pass the required amount of steam. The general arrangement will be understood from Fig. 45, which is the lower half of a longitudinal section. These machines have been built in sizes from 40 to 12,000 horse-power and, compared with reaction turbines or reciprocating engines, occupy extraordinarily little space.

A very important improvement has recently been made in turbines of this type. In the machines hitherto designed the diameter of the drum or disc and the length of the blading has had to be increased considerably towards the low-pressure end in order to accommodate the increased volume of the expanded steam. For this purpose also the diameter of the casing is increased and there is a big bulge at the lower-pressure end. The new design is illustrated in Fig. 46, which represents part

[1] Now called Metropolitan-Vickers Rateau.

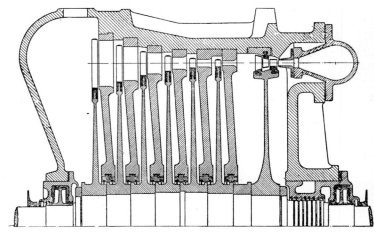

Fig. 45.   Section of Lower Half of Metro-Vickers Rateau Turbine.

of a section of the Metropolitan-Vickers Multiple Exhaust Turbine, of 17,000 horse-power, made for the Liverpool Corporation. It will be seen that beyond the eleventh disc two alterations are made. The fixed guides direct a portion of the steam through the outer portion of the blades into an annular exhaust space, and the blades are different in shape in the upper and lower

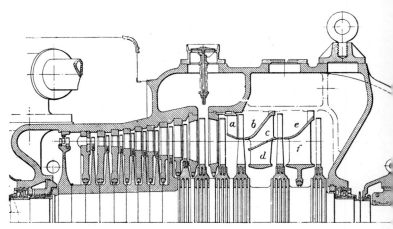

Fig. 46.   Multiple Exhaust of Metro-Vickers Rateau Turbine.

FIG. 42.——HIGH-PRESSURE DIAPHRAGMS OF
METRO-VICKERS RATEAU TURBINE.

FIG. 43.——LOW-PRESSURE WHEEL OF
METRO-VICKERS RATEAU TURBINE.

To face page 50.

FIG. 44.—CONSTRUCTION OF LOW-PRESSURE WHEEL OF METRO-VICKERS RATEAU TURBINE.

To face page 51.

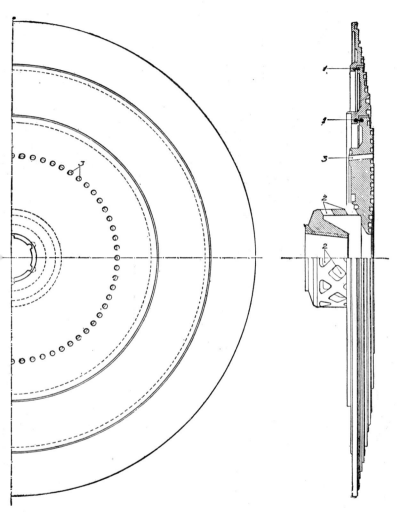

Fig. 48. CONSTRUCTION OF DISC OF BRUSH LJUNGSTROM REACTION TURBINE.

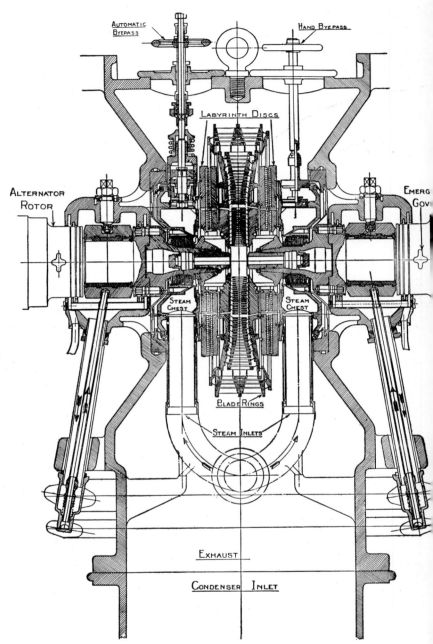

Fig. 49.   SECTION OF BRUSH LJUNGSTROM REACTION TURBINE.

FIG. 47.—ROTORS OF BRUSH LJUNGSTRÖN REACTION TURBINE.

To face page 52.

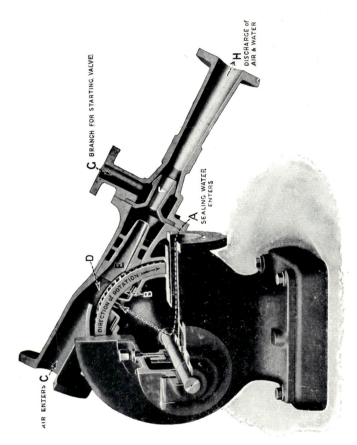

FIG. 51.—METRO-VICKERS LEBLANC ROTARY AIR PUMP.

To face page 53.

portions which are separated by a ring. At each stage, therefore, beyond the eleventh a portion of the steam is expanded down to the condenser pressure, and got rid of, while the remainder passes through blading of such a shape that it does no work. This promises to be one of the most important improvements that have been made in turbine design for some years.

The two types which have been described are known as parallel flow turbines, because the steam flows in a direction parallel to the mainshaft. About 1912 Ljungström, a Swedish engineer, invented a radial-flow turbine which has been considerable developed by the Brush Electrical Engineering Co., of Loughborough. In this machine, there are two rotors which run in opposite directions. Each rotor consists of a disc on the end of a shaft with rings of blades, like squirrel cages, fixed on the face and projecting between similar rings of blades on the face of the other disc. Steam is admitted at the centre and in passing between the rings of blades it causes them to spin round in opposite directions. The principle is that of a reaction turbine. Fig. 47 shows two of the rotors for a 3,000 kilowatt machine, and it will be seen how, when they are pushed close up to one another, the rings of blades overlap. Fig. 48 shows how the disc is constructed. It is built up in sections connected by expansion rings (1, 1) to prevent distortion arising from variation of temperature, and one face is grooved to receive the caulking rings by means of which the blade rings are attached to the disc. Steam is admitted to the centre of the blade system, through the openings, marked (2) in the hub, while openings marked (3) admit extra steam to the outer rings when the turbine is taking an overload. Fig. 49 shows a section through the complete machine. It is arranged to drive two electric generators in opposite directions.

This turbine is highly efficient, using less than 9 lb. of steam per horse-power per hour. Since only low-pressure steam comes into contact with the turbine casing no lagging is required. The very light construction of the blade rings ensures uniformity of temperature and absence of strain. At present it has been manufactured mainly in the smaller sizes, but there is no reason to believe that there is any limit in this respect, though the fact that the power is delivered through two shafts revolving in opposite directions, while not disadvantageous for electrical driving, may reduce its utility for other purposes.

Though the turbine can be used for any purposes, it is particularly valuable for driving electric generators and for the propulsion of ships.   In the former case its high speed and uniformity of running are its main recommendations. With high speeds, high voltages can be secured from a generator of relatively small dimensions, while the turbine itself takes up less space

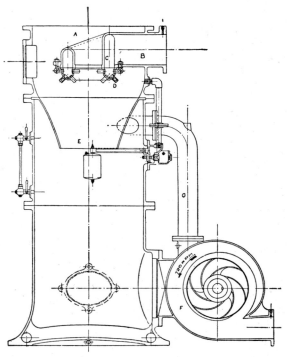

Fig. 50.   Multiple Jet Condenser.

than a reciprocating engine of the same power.   For marine propulsion its advantages are reduction in weight, space, first cost and upkeep, high efficiency, and the fact that the condensed steam is not contaminated with oil, so that it can be returned to the boilers with less trouble in cleaning. The absence of parts gives a free exit to the steam and the back-pressure is reduced with less work from the pumps. It has, however, a serious draw-back in that it runs only in one direction, and a reversing turbine has to be fitted on the same shaft to enable the ship to go astern.

A turbine will work very efficiently with low pressure steam, and considerable improvements have been made in air-pumps to reduce the pressure in the condensers. One of the most interesting of these is the Metropolitan-Vickers Leblanc Multiple Jet Condensing Plant, the general arrangement of which is shown in Fig. 50. The steam to be condensed enters the top of the condenser at A. The injection water for condensing it enters at B, passes into the distributor C, and is discharged through nozzles at D. Lower down in the condenser is a cone E, which promotes the mixing of water and steam. The condensed steam is removed by the centrifugal pump F.

The Leblanc Rotary Air Pump is shown in Fig. 51. It consists of a Pelton wheel, with the curvature of the blades reversed, so that if water enters at the side it is thrown outwards through the spaces between the blades. This water, known as the " sealing water ", enters the pump casing at A and passes through the fixed guide nozzle B to the reverse Pelton wheel, which ejects it into the discharge cone E in the form of thin sheets which move forward with high velocity. Each of these sheets form a small piston, and air entering from the air suction branch C is entrapped between them at D, compressed in the cone E, and driven into the diffuser H, where it acquires sufficient velocity to overcome the atmospheric pressure.

CHAPTER IV

GAS, PETROL, AND OIL ENGINES

IF you have ever heard a gas engine work you will know that it has a cough, and the more regularly it coughs the better it is working.

Throughout the first half of last century a number of men spent their lives trying to make an engine which would work by burning a mixture of gas and air behind a piston, but none of them really succeeded until 1860. In that year Lenoir, a Frenchman, designed an engine which would work, and of these a number was sold. His triumph was short-lived, however, for in 1876 Dr. Otto patented an engine that is the parent of the gas-engine of to-day.

Very few of these early inventors knew quite what they wanted, and they had only vague ideas as to the method of obtaining it. But though they were unsuccessful, they paved the way for others, and designed many of the features which were utilized by those who followed them. As results of their work, combined with improvements in the manufacture of steel, are the modern submarine boat, the motor-car, and the aeroplane engine, it will be worth while to inquire why the gas-engine has developed and how it works.

The source of power in any heat-engine is the fuel. The greater the amount of heat produced by the fuel that is used in the engine, and the less that is allowed to escape from it, the more efficient does the engine become. In the ordinary steam-engine the heat produced by the burning coal is very largely wasted. Some of it goes up the chimney, some of it is radiated from the large surface of the boiler and steam pipes. It is clear that if the fire could be made to burn *inside* the cylinder, less heat would be able to get away until it had done the work required of it. But there is another advantage. No solid or liquid fuel burns so readily and so completely as a gas, which can be intimately mixed with exactly the amount of air required for

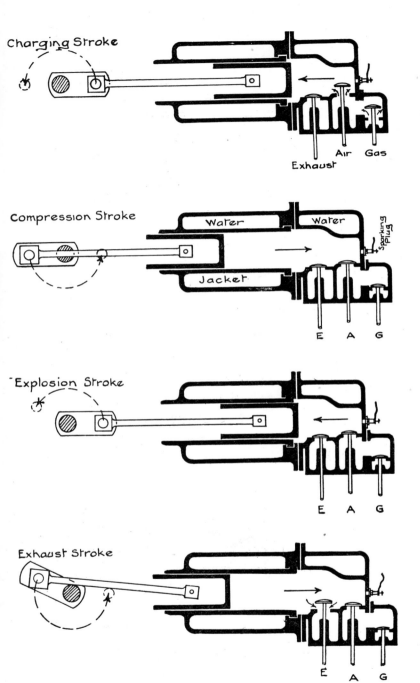

Fig. 52. DIAGRAM TO SHOW THE ACTION OF A GAS-ENGINE.

[*page* 57

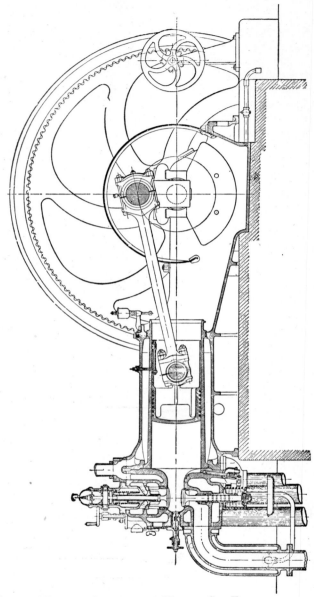

Fig 53. SECTION OF A MODERN GAS-ENGINE.

page 58]

its combustion. So what inventors have aimed at is to produce an engine in which the heat shall be liberated inside the cylinder and in which the combustion is as regular and perfect as can be.

A skeleton diagram of a gas-engine is given in Fig. 52. Suppose the piston is in the position shown in top figure. As it moves outward the valve G opens and admits gas, while the valve A opens and admits air. In this way the cylinder is filled with the mixed gases, and if the valves have been properly designed this mixture will be that which gives the best results on combustion. The next stroke of the piston compresses the mixture. As it reaches the end and is about to return, the charge is ignited by means to be described later, and the explosion forces the piston outwards. When it returns the exhaust valve opens and the products of combustion are swept out of the cylinder.

This series of operations is repeated every two revolutions of the crank, and is called the Otto Cycle. The engine is only single-acting—the piston is pushed towards the crank, and, as the fly-wheel turns, the crank pushes the piston back again in a sort of " you push me and I'll push you " spirit. But the crank gives two pushes and one pull to the piston's one push, so that for one-quarter of the time the piston drives the crank, and for three-quarters of the time the crank drives the piston. If there were no fly-wheel the crank-shaft would move very rapidly for one half-turn and then stop. But the fly-wheel, once it has started rotating, takes some time to come to rest, so that it carries the crank-shaft round twice, by which time there is a fresh charge of gas and air in the cylinder, and the piston receives another impulse. An engine of this kind is sometimes called a four-stroke engine, because only one stroke in four is a driving stroke, and a four-stroke engine must have a heavy fly-wheel to equalize the motion.

It will be observed that the piston is unlike that generally used in a steam-engine. There is no need for a cylinder cover in front, and a bucket-piston is employed. When the piston makes the driving stroke it produces a good deal of pressure on the cylinder walls, and this form distributes the pressure over a wider area.

The valves are of the " mushroom " type, and are kept on their seatings by springs. They are opened just at the right moment by cams fixed on a shaft which rotates at half the speed of the crank-shaft, and therefore opens each valve once every two revolutions. These statements will be clear from a study of Figs. 53 and 54, and which illustrate one of Messrs. Crossley Brothers' well-known engines of moderate size.

There are two methods of igniting the explosive mixture—a tube or chamber kept hot by a lamp, and an electric spark. The former is gradually giving way for large engines before the electrical method, which has been improved and rendered more reliable in recent years. In the hot-tube method, a narrow tube is kept hot by an external flame, and the explosive mixture is momentarily admitted to it by a valve. Electric ignition will be dealt with on pp. 66–7. All engines—steam, gas, or oil—are constructed to run at a certain speed. If the machinery they are intended to drive is more or less idle (i.e. if the " load " is taken off or reduced), they run away, or " race ", and some form of governor is necessary to keep the speed as constant as possible. The steam-engine governor will be familiar. It cuts off steam when the speed exceeds a certain limit by means of a " throttle " valve. The gas-engine governor is similar in construction but acts by cutting off the gas supply entirely and causing a " miss-fire "—in which case it is called a " hit and miss " governor—or by merely reducing the supply of gas and allowing a weaker mixture to explode. The latter type is displacing the former.

There is one respect in which the internal-combustion engine differs from the steam-engine. The cylinders of the latter need to be kept hot to reduce steam condensation, and to this end the cylinders are often " jacketed " with steam. The internal-combustion engine cylinder, on the other hand, tends to become too hot, and the temperature has to be kept down by a water-jacket. Usually the water circulates round and round through the jackets and a cooler or radiator, the same water being used over and over again.

The gas-engine was originally regarded as suitable for small powers using town gas, which is rather an expensive fuel. Mr. J. Emerson Dowson in 1878 devised a complete plant for producing gas for factory and domestic purposes, and exhibited it at the York meeting of the British Association in 1881, when it drove for the first time a 3 horse-power Otto gas-engine. At that time no engine larger than 20 horse-power was working. The idea of using blast-furnace gases or gases from coke ovens arose in the early 'nineties. The fact that the waste gases of the blast furnaces in the United Kingdom alone are capable if used

FIG. 54.—A MODERN GAS ENGINE.

To face page 60.

FIG. 62.—PART SECTION OF WOLSELEY 20 HORSE-POWER
MOTOR-CAR ENGINE.

1. CRANK CASE.
2. OILBASE.
3. OIL TROUGHS FOR CONNECTING RODS.
4. CRANKSHAFT.
5. CRANKSHAFT BEARING BLOCKS.
6. FLYWHEEL.
7. CYLINDERS.
8. CYLINDER PLUGS.
9. WATER OUTLET PIPE.
10. CONNECTING RODS.
11. CONNECTING ROD BEARINGS.

12. PISTONS.
13. GUDGEON PINS.
14. CAMSHAFT.
15. OIL PUMP DRIVING WHEEL.
16. ENGINE CHAIN COVER.
17. FAN BLADES.
18. FAN CENTRE AND PULLEY.
19. WATER PUMP.
20. MAGNETO MACHINE.
21. STARTING HANDLE.

To face page 61.

in gas-engines of producing 750,000 horse-power is in itself sufficient to attract attention. The result has been a regular and continuous increase in size, so that gas-engines of 1,000 horse-power are quite common in England, on the Continent, and in the United States.

Many of the large engines are double-acting, two-stroke-engines, and this involves an interesting modification. In the ordinary four-stroke engine the burnt gases are not entirely expelled during exhaust, because there must always be room at the back of the piston into which the fresh charge can be compressed. Their presence is undesirable in any case, and in a two-stroke engine they must be cleared out at the end of each explosion stroke. In other words, the cycle must be explosion stroke—compression stroke and the exhaust and admission must take place between these two. To secure this the exhaust ports (no valves are necessary, though they may be used) are in the side of the cylinder, in such a position that they are uncovered by the piston just before the end of the explosion stroke. At that moment a charge of compressed air enters the cylinder through a valve, sweeps out the burnt gases, and provides the air necessary for mixing with the gas for the explosion stroke. This blowing-out of the burnt gases is called "scavenging". The plan was proposed by Dugald Clerk as long ago as 1881, but nobody took it up and it was left for Koerting to apply it to large gas-engines on the continent more than twenty years later.

There is, unfortunately, a limit to the size of gas-engines, owing to the difficulty of keeping the cylinder and piston cool. Large cylinders are not easy to cast without strain, and the great differences of temperature to which they are subject when the engine is working—a gaseous explosion inside and cold water outside—renders them liable to crack. Some makers cast the cylinder in two or four pieces, and bolt them together. But in a large engine it is necessary to cool the piston also, and this involves pipes with joints that must permit of free movement without leakage. Consequently the limit is about 1,000 horse-power per cylinder, and many makers prefer to produce engines with several cylinders each giving no more than 250 horse-power in order to avoid the mechanical difficulties.

The real advantages of the gas-engine lie in the fact that there are no "stand-by losses", it can be used with gas prepared from any kind of fuel, and it is thermally twice as efficient as the steam-engine. It will perform any of the work done by a steam-engine within its range, but it is not so steady in running as a turbine and therefore not quite so suitable for driving electrical generators.

The reader will recollect how the steam turbine dispenses with all the moving parts of a reciprocating steam-engine except those which rotate, and he will now be prepared to hear of a

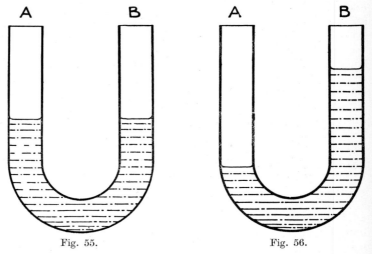

Fig. 55.   Fig. 56.

DIAGRAMS TO EXPLAIN ACTION OF HUMPHREY PUMP.

marvellously simple modification of the gas-engine. On the explosion pump invented by Mr. H. A. Humphrey there is no piston or connecting-rod, no crank or flywheel, and only the simplest of mechanisms for controlling the valves.

Suppose a quantity of water is contained in a wide U-tube shown in Fig. 55. If air be forced into the limb A, the water in that limb will be depressed, and the water in the other limb must rise, as in Fig. 56. On removing the pressure the water will flow back until the height in A is very nearly equal to that at which it stood in B, the difference in height being due to friction. This to-and-fro movement, or oscillation, will go on for some time, the height attained at each swing gradually decreasing. But the time taken for each oscillation will be the same or very nearly so. The smaller the displacement of the water in the first instance the more uniform will the time of the swing be. It depends in any case upon the quantity of water, and can easily be calculated.

Once the water has begun to swing a very slight impulse at the right moment will suffice to keep up the movement. If therefore the end A (Fig. 56) is closed and an explosion of gas and air can be arranged at the moment when the water reaches its highest point in that limb, the water can be kept oscillating for as long as the explosions are maintained. This is the principle

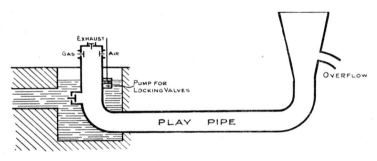

Fig. 57. DIAGRAM OF HUMPHREY PUMP.

upon which the Humphrey pump works. In Fig. 57 the pipe in which the water oscillates, called the play-pipe, is made of cast-iron. It is about 6 feet diameter, and the horizontal portion is about 60 feet long. The right limb is open and funnel-shaped, and it has a discharge pipe through which water can flow into the reservoir. The left-hand limb is closed by the cylinder, and is built into a well or pit supplied with the water to be lifted. The pump is 7 feet diameter and 10 feet long. Round the upper end are placed two sets of valves for the admission of gas and air, while lower down is a valve opening inwards which admits water. At the top is the exhaust valve. When an explosion takes place the water in the play-pipe is driven forward, rising in the water-tower, and overflowing into the reservoir. Once such a body of water has been set in motion it continues to move after the exploded gases have fallen below atmospheric pressure, and water enters the pump from the pit, replacing in the play-pipe that which has been lost from the discharge pipe. The water in the play-pipe then comes back into the pump and forces the waste gases through the exhaust valves. Having effected this, it flows a second time towards the water-tower, creating a vacuum in the cylinder, and drawing in a fresh charge of gas and air. The return of the water compresses the mixture, which is ignited at the proper moment and forces the water towards the tower again.

All the valves are held lightly to their seatings by springs. They open and close automatically in obedience to changes of pressure inside the cylinder, and when not required to be in action they are locked by the operation of a small water motor. It will be observed that the strokes are not equal in length. That due to the explosion is a long one, and that which sweeps out the waste gases is longer still. But the charging and compression strokes are short ones. In the ordinary gas-engine, the strokes are all equal. In the Humphrey pump each one is of a length appropriate to, and determined by, the duty it is required to perform.

Such a pump as has been described will deliver from 12 to 14 tons of water per minute. Those erected at Chingford for raising water from the River Lea are five in number, four of them capable of delivering 40,000,000 gallons and one 20,000,000 gallons of water through a height of 25 to 30 feet every twenty-four hours.

Explosion pumps can be made double-barrelled, and be adapted to give an impulse every two strokes. They can be used as air compressors, in which the moving water acts as a piston, and experiments have been made to apply them to the propulsion of ships. They are simple in construction and therefore low in first cost, require no lubrication, are economical in working, with probably an assured future for large pumping stations; and if the practical difficulties which attend their application to other purposes can be overcome they are bound to exercise a very considerable influence in the production of power. But in any case they involve a new principle and stand out as one of the most remarkable recent engineering inventions in the world.

PETROL-ENGINES

The earlier inventors who struggled with the problem of the gas-engine were not unaware that the substance used in an internal-combustion engine might be supplied in a liquid form, and several of their patents claimed the right to use paraffin or some similar substance in their engines. But it was left for Daimler, who had been for ten years manager of Dr. Otto's gas-engine works, to invent the first practical light oil-engine, and his original motor was produced in 1886. Since then there have been many forms differing mainly in detail, but all, until quite

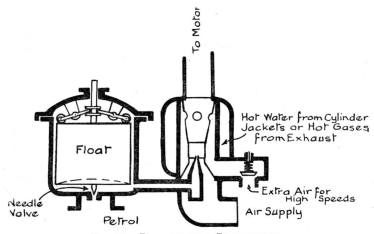

Fig. 58. DIAGRAM OF A CARBURETTOR.

recently, working on the four-stroke cycle described on pp. 57–8. While petrol is the fuel which has been found most satisfactory, others such as benzol are sometimes used, and many attempts have been made to burn alcohol, which, were it not for the heavy duty, would be a cheap and serviceable fuel. The liquid is sprayed into the cylinder or combustion chamber and ignites at the moment when the compression has reached its highest point. The cylinder has to be cooled with water or air. If it is freely exposed, as in the case of a motor-cycle engine, the body has thin fins externally which offer a large cooling surface and water need not be used. With single-cylinder engines a fly-wheel is required to overcome the jerkiness of action, but with several

F

cylinders and cranks set at angles one with another the motion is equalized, and the weight of a flywheel is saved. Apart from variations in general arrangement to suit the conditions under which it will have to work, the chief lines of development have been in carburettors and ignition devices.

The carburettor, Fig. 58, is a device for mixing the petrol vapour with air in the right quantity for complete combustion. There are many forms, but the most usual are provided with a

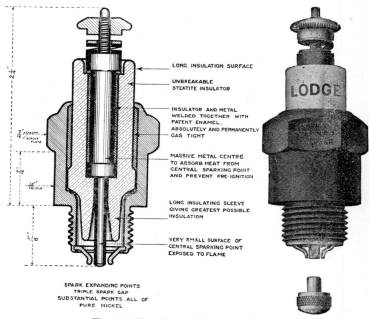

LONG INSULATION SURFACE

UNBREAKABLE
STEATITE INSULATOR

INSULATOR AND METAL
WELDED TOGETHER WITH
PATENT ENAMEL.
ABSOLUTELY AND PERMANENTLY
GAS TIGHT

MASSIVE METAL CENTRE
TO ABSORB HEAT FROM
CENTRAL SPARKING POINT
AND PREVENT PRE-IGNITION

LONG INSULATING SLEEVE
GIVING GREATEST POSSIBLE
INSULATION

VERY SMALL SURFACE OF
CENTRAL SPARKING POINT
EXPOSED TO FLAME

SPARK EXPANDING POINTS
TRIPLE SPARK GAP
SUBSTANTIAL POINTS ALL OF
PURE NICKEL

Fig. 59. The Lodge Sparking Plug.

chamber containing a float, the rise and fall of which regulates the amount of petrol flowing from the tank. The petrol then enters a second chamber, into which it is drawn by the suction of the engine, through a fine jet which converts it into a spray and facilitates an intimate mixture with air. The air enters freely through an open pipe, which permits sufficient to pass for complete combustion under ordinary conditions of working. An additional opening, normally closed by a valve, enables the engine to draw a further supply at high speed.

GAS, PETROL, AND OIL ENGINES 67

In the earlier forms of petrol-engine, the explosive mixture was ignited by a hot tube of nickel or platinum which was open to the interior of the cylinder and closed at the outer end. In this case the tube tended to remain partly full of the waste gases of combustion and ignition was effected when the return of the piston compressed the new explosive mixture into the tube

The survival of the hot tube in both gas and petrol-engines for so many years was due to the ineffectiveness of the apparatus for producing an electric spark. Two forms have been used, the low tension and the high tension. The former was invariably produced by a small dynamo called a magneto, driven from the crank-shaft; the latter from a special constructed magneto or an induction coil. The electricity is led into the cylinder through the sparking-plug, at the inner end of which two metal points connected with the wires were separated by the gap in which the spark was formed. One of the best modern types of plug is shown in Fig. 59. It will be observed that

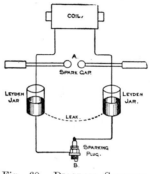

Fig. 60. DIAGRAM SHOWING PRINCIPLE OF LODGE SPARK.

the spark can take place between the central rod and any one of the three points surrounding it.

The chief difficulty hitherto has been the choking up of the plug with oil and dirt, so that the electricity took the easier path and avoided jumping the gap. This defect has been overcome in a very ingenious way by Sir Oliver Lodge, who employs a special kind of spark. The current produced by an ordinary magneto machine or an induction coil merely jumps across the gap in one direction. It is thin, very little electricity passes at once, and the heating effect is small. But if the terminals between which the spark passes are connected up with some arrangement in which the electricity can be stored, a larger quantity will then pass at once, the spark will be fatter, hotter, alternating, disruptive, and therefore capable of clearing dirt out of the way. In Lodge's apparatus the current is supplied by an accumulator to an induction coil, the terminals of which are connected to the inner coating of a Leyden jar. The outer coating of the jar is connected with the sparking-plug. Fig. 60

shows diagrammatically the arrangement,[1] and Fig. 61 a section of the actual instrument. In Fig. 60 the two balls at A are adjusted so that the electricity flows into the jars until they

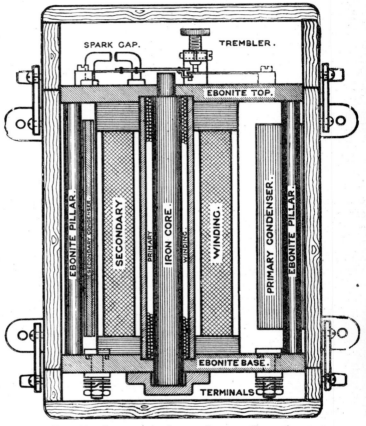

Fig. 61. SECTION OF LODGE IGNITER TYPE A
SHOWING THE CONSTRUCTION.

become, as it were, full, when they suddenly empty across the gap. At the same moment a discharge takes place at the sparking-plug. The spark lasts no longer than a millionth of a

[1] Two Leyden jars are shown here. In practice only one is used.

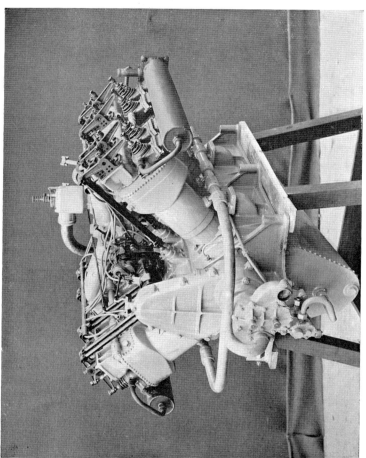

FIG. 63.—VEE TYPE OF MARINE MOTOR.

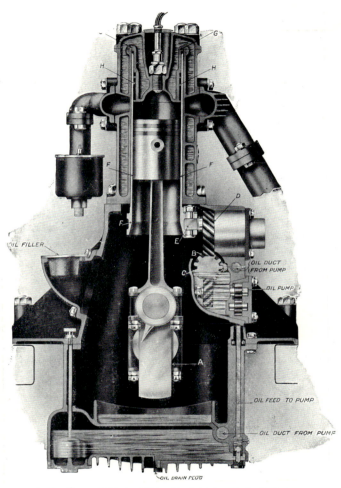

FIG. 64.—THE ARGYLL SINGLE-SLEEVE ENGINE.

To face page 69.

second, and so violent is it that water, oil, or dirt, though offering an easier path, do not deflect it. A spark can be obtained even when the plug is immersed in water.

The spark is timed by a cam motion. For engines with more than one cylinder a distributor must be employed, so that the explosion in each cylinder may be timed to take place at the right moment. This consists generally of a rotating disc with a metal stud, which makes contact with fixed studs in turn.

The petrol-engine is *par excellence* the engine for small powers, and it attracted attention from the first by reason of its extreme lightness. Apart, therefore, from its widespread employment for driving small machines, it is in locomotion—on rail an road, on sea, and through the air—that it has shown its greatest value. For these purposes it assumes varying forms, a few of which we are able by the courtesy of the makers to illustrate here. Fig. 62 is a 20 horse-power Wolseley engine for a motor-car. Two of the cylinders are shown in section, and this, together with the lettering and list of parts, will enable the construction and arrangement to be followed without difficulty. The fan on the right is for the purpose of drawing air through the coils of the radiator in which the water from the cylinder jackets is cooled. Another interesting type has the cylinders placed in pairs, each inclined equally to the vertical as in Fig. 63. This is a marine type of motor, and the method of construction economizes both space and weight ; on that account it has also been adopted for aeroplanes.

In the earlier petrol-engines the valves were almost universally of the poppet type or mushroom-shaped, and it is almost impossible to avoid a certain amount of noise when these are rising and falling nearly a thousand times a minute. Several engines have been designed, however, in which the ports are opened and closed by a sliding motion which is perfectly silent in action. One of the most interesting was the Argyll Single Sleeve engine, a beautiful section of which is shown in Fig. 64. A thin sleeve or tube pierced with holes corresponding to the ports is fitted between the piston and the cylinder. By means of toothed gearing and a small crank this is caused to move up and down and also to rotate, so that at the right moment the inlet or exhaust port is uncovered. The rotating motion renders it possible to use one sleeve only, and reduces the power that would be required merely to push the sleeve up and down between the surfaces.

One of the most remarkable engines ever designed is the Gnome, an external view of which is shown in Fig. 65. There are seven cylinders rigidly connected together and having pistons which operate on the same crank. The crank-shaft is fixed and the cylinders rotate on ball-bearings round it. The exhaust valve is placed at the end of each cylinder and is operated by a rod and lever worked from the main-shaft. Petrol and air are mixed in the carburettor and enter the space in the middle of the casing which contains the crank. Each cylinder receives its charge through a valve in the piston. The bearings and crank are oiled by forced lubrication.

This engine is extraordinarily light. The 100 horse-power size is the lightest yet made, and weighs only 220 lb. or 2·2 lb. for each horse-power developed. The cylinders with their fins are bored out of solid forged steel and are only ⅛ inch thick ; and the other parts are as light as it is possible to make them. The rotating cylinders act as a fly-wheel and give great steadiness of motion, while the rapid rotation through their air—1,000–1,200 revolutions per minute—keep them cool. In fact, it is a moot point whether they are not in this way kept too cool for the highest efficiency of working.

Considerable improvements were made in this engine in 1912, and it is now known as a monosoupape engine. As this name implies, the poppet-valve in the piston for admission of air and petrol has been abandoned, a series of ports, which are uncovered at the necessary moment by the piston, taking its place. At the conclusion of the exhaust-stroke the valve permitting the expulsion of the burnt gases remains open sufficiently long for pure air to be sucked in on the downward stroke of the piston. The fuel is then admitted towards the end of the stroke, through the orifices provided, in the form of a very rich mixture with air. Since this air merely serves as a " vehicle " for the petrol, a pump must be employed to force the latter from the tank. The air admitted by the exhaust port serves very effectively as a means of cooling. Another valuable feature rests in the fact that the supply of petrol can be throttled down until the engine makes only 200 revolutions per minute, thus allowing for a variation in speed, which is essential in aeroplanes engaged in military observations.

FIG. 65.—THE GNOME ENGINE FOR AEROPLANES.

FIG. 66.—A LARGE MARINE DIESEL ENGINE.

MEDIUM AND HEAVY OIL-ENGINES

The cost of light oil—petrol or gasoline—led to the attempt to design engines which would burn a cheaper fuel, and Priestman, of Hull, put such a one on the market in 1888. This used a medium oil of specific gravity about 0·8. It was followed in the early 'nineties by the Hornsby-Ackroyd engine, which involved a new principle of ignition. The chamber at the end of the cylinder into which the gases were compressed was heated at first by a lamp, and the compression raised the temperature to such a degree that the mixture exploded. This chamber had a number of thin vanes on its inner surface, which aided the passage of the heat from the metal to the mixed gases, and, after the chamber once became hot, successive explosions maintained the temperature, so that the heat of compression added to the heat of the chamber ignited the charge.

When gas-engines were first introduced they were of small size, and the cost of town gas prevented their entering into competition with steam-engines of large size. The use of cheap blast furnace gas, and of producer gas of which the by-products reduced the cost of the fuel, immediately made gas-engines serious rivals to steam-engines for use on land. Similarly both the petrol and medium oil-engine, though suitable for marine as well as land use, remained of small size owing to the cost of oil. History has now repeated itself, and an engine capable of using crude petroleum residues has once again revolutionized the production of power—this time both on land and sea.

The achievement is due to Dr. Rudolph Diesel, whose long series of experiments resulted in the design of an engine which has been one of the remarkable engineering successes of the past thirty years. It was first exhibited at the Paris Exhibition of 1900. Three years later an 80 horse-power engine was shown at Düsseldorf, but there was nothing to indicate that it would shortly enter the field of large powers, and invade the domain of marine propulsion. At Liége in 1905, however, an engine of 500 horse-power was shown, and by 1908 engines of 1,000 horse-power were running. To-day engines of 1,500–2,000 horse-power per cylinder are being constructed by a dozen firms in England and on the Continent.

Here is an engine requiring no boilers, capable of working with the cheapest oil fuel, which is easily stored, and occupies far less space than an equivalent amount of coal, and capable of undertaking all the ordinary duty that modern manufacture and transport impose.   It was stated in *Cassier's Magazine* (Special oil power number, 1911, p. 151) that if Diesel engines of 1,500 horse-power per cylinder were installed in the *Mauretania* the 70,000 horse-power of that vessel could be produced in one-fifth of the space occupied by the boilers and turbines.   The need for coal trimming and stoking would be abolished ; it would be possible to dispense with 192 stokers and 120 trimmers, or 312 men, whose wages amount to nearly £40,000 a year. With an equal weight of fuel it would be possible to steam four times the distance without taking in a fresh supply.   That the

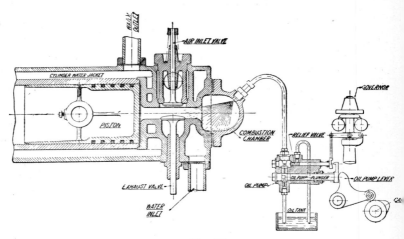

Fig. 67.

advantages were not exaggerated is obvious from the development of the Diesel engine for ships during the last ten years. In 1918 the world's tonnage of motor ships was about 530,000. In 1928 it was about 5,300,000.   There are now several motor ships with engines developing 20,000 horse-power, and much larger engines are proposed.

It will be interesting to examine the principle upon which the Diesel engine is based.   The efficiency of an internal-combustion engine depends largely upon the degree of compression.   But it will be recollected that when the compression of a mixture of oil or gas and air reaches a certain point ignition takes place, and there is

therefore a limit to the compression that can be employed. If,
for example, the compression in the Hornsby-Ackroyd engine
had been so high as to cause the explosion before the compression
stroke was completed, the engine would have stopped or reversed.
On this account higher compression can only be secured if the

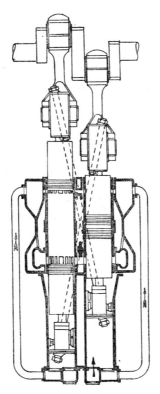

Fig. 68. FULLAGAR ENGINE.

fuel is forced into the cylinder at the commencement of the
explosion stroke.   An air-compressor driven by the engine itself
charges a steel vessel with air at 1,000 lb. per square inch, and
this air is used to inject the fuel into the cylinder.   This at such
a temperature from the previous compression that the mixture
burns smoothly and rapidly as it enters, expanding by the heat

of its own combustion, and producing a steady pressure upon the piston.   There is no explosion in the ordinary sense.

Owing to the high pressures employed the engine is very heavily built, and many firms now make what is called a semi-Diesel in which the principle is the same (see Fig. 67), but the initial pressures are lower.   Both two and four-stroke engines of Diesel and semi-Diesel types were made.   During the war, development has taken place in two directions.   The first is what is known as " solid injection ".   Instead of admitting the air and oil together, they are admitted separately.   The second is a return to the four-stroke engine, but with open cylinders and with two pistons, the explosion taking place between the latter.   The advantages of this method are three in number —lightness, less vibration, and more effective scavenging. Lightness is attained by the absence of cylinder covers, and there is less vibration because the explosion takes place between two moving pistons.   The most interesting engine of this type is the Fullagar, see Fig. 68, which was originally designed as a gas-engine but has been adapted by Messrs. Cammell Laird and Co. as a marine heavy oil-engine.   There are two cylinders and four pistons and the crossheads of the two upper pistons are connected by diagonal rods, shown by dotted lines, with the crossheads of the two lower ones.   It has been built to develop 1,000 horse-power, and is only about half the weight of an engine of similar power constructed on the older plan.

In reviewing what has been said about the internal combustion engine, it will be seen that its uses range over nearly the whole field for which power is ordinarily required.   The small oil or petrol engine is admirable for domestic purposes, such as driving a vacuum cleaner, or pumping water and driving a dynamo to light a house that is far from a town supply ; and to cut chaff, mow the lawns, drive milk separators and churns, thresh corn, in places where and under circumstances in which gas power is out of the question.   Large stationary engines for pumping and driving factories are in competition with gas-engines using town gas or having their own producer plant.   The crude oil-engine again is being used for electric light and power stations, for pumping in waterworks and docks, and for driving the machinery of factories and workshops.   In iron and steel works, however, the vast quantity of waste gases from the blast-furnaces and coke ovens renders oil-engines unnecessary. For the motor-car and the aeroplane the petrol engine alone stands, though benzene is now largely used as fuel. Improvements in steel manufacture have enabled it to be made so light and yet so powerful that within thirty years two new forms of locomotion have arisen—forms which were dreamt of for a century and are as yet only just emerging from a vigorous infancy. In this progress the usual order of events is to be observed. A new discovery or invention which ministers to comfort, convenience, or efficiency is at first available only for those who can afford the heavy capital outlay which its possession involves. But, sooner or later, according to the public service it can perform, this is brought by public supply within the reach of all whose needs outweigh their private resources. Thus, the private car has been followed by the commercial vehicle, the taxi, and the motor-'bus. But in the case of the aeroplane and the airship, it is possible, as we shall see later, that a public service may anticipate any large development of private enterprise.

In other forms of transport the internal-combustion engine is proving of equal value. The petrol-engine has been used in launches as long as it has been used in motor cars, and several ships have been already equipped with Diesel engines. Coal-burning under steam boilers is a wasteful process, but oil-power requires not only an adequate supply of fuel, but also the necessary depôts at ports of call from which a fresh supply can be obtained. Moreover, its cost does not encourage its use except in high-speed passenger ships and in ships of war.

Internal combustions have, however, one defect for marine purposes, which they share with the steam turbine : they are not reversible. They can be constructed to run in either direction, but reversal is not so simple and quick as in the case of the reciprocating steam engine. What is required is some form of transmitting device between the engine and the driving shaft which shall be capable of having its direction reversed while the main engine continues to run in the same direction. This can be accomplished in several ways which are outside the range of this chapter.

CHAPTER V

GENERATION AND TRANSMISSION OF ELECTRICITY

THE nineteenth century was without doubt the age of steam. The twentieth century will be the age in which he who can control the generation and distribution of electricity will hold the key to most of the operations of industry in his day. But the primary source of power will still be water, or steam, or the internal-combustion engine, and the countries in which water power or coal or petroleum is plentiful will command the markets of the world. For the real value of electricity lies in the ease with which it can be transmitted great distances by a thin wire over-head or underground. In this way, power can be produced at one centre with the economy that always attends operations on a large scale, and can be utilized in places where water is not available and where the hiss of steam or the cough of an internal-combustion engine would be impossible or undesirable.

MAGNETS AND ELECTRIC CURRENTS

In attempting to give some account of the way in which electricity is produced and distributed it is first necessary to recall a few elementary facts of electrical science. A magnet is a piece of steel possessing properties which cause it to take up, when freely suspended, a north and south direction. The end turning towards the north is called a north pole, and the opposite end a south pole. Between the poles in the space surrounding the magnet a force is exerted along definite curved lines, and these curves can be rendered visible by scattering iron filings on a paper laid over the magnet. These arrange themselves in definite directions which are indicated for different cases in Figs. 69–71. The presence of other magnetic substances in this "field of magnetic force" causes distortion ; the lines gather up and pass in greater number through soft iron than through any other substance. The curved lines are supposed to be continuous through the magnet, but the chief internal effect of importance to the electrical engineer is the permeability, which is measured by the ease with which the force is produced in the material.

Two magnets exert a force upon one another, according to the law that like poles repel and unlike poles attract. This is only another way of saying that the magnets act on one another in

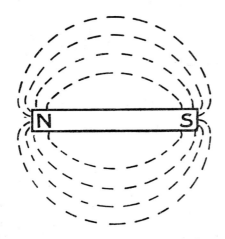

Fig. 69. MAGNETIC FIELD OF A BAR MAGNET.

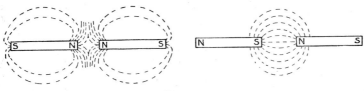

Fig. 70.

MAGNETIC FIELD OF A PAIR OF   MAGNETIC FIELD OF A PAIR OF
MAGNETS—LIKE POLES.   MAGNETS—UNLIKE POLES.

such a way that the lines of force of each may run in the same direction, and as far as possible through iron or steel all the way. The tendency for the force to pass through iron or steel creates a tendency for the iron or steel to move into such a position that the greatest amount of magnetisim is produced within it.

When steel is once magnetised it retains more or less unimpaired the properties which it has acquired. Soft iron, however, only possesses these properties so long as it remains in a magnetic field, acting in these circumstances as a temporary magnet. At the same time, since the earth is a magnet, most varieties of iron or steel have some residual magnetism.

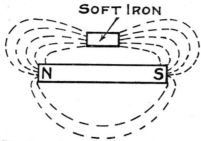

Fig. 71.   INFLUENCE OF A PIECE OF SOFT IRON ON A MAGNETIC FIELD.

Whenever therefore a piece of soft iron is placed in a magnetic field so that lines of force pass through, it acquires temporary polarity at the points where the lines enter and leave the material. The soft iron is then said to be magnetised by induction. The terms soft iron and steel have not now the significance that was implied twenty or thirty years ago. Soft iron is a somewhat rare material, and steel can now be obtained which exhibits a very wide range of " retentivity " or power of retaining magnetism, while some iron alloys are even non-magnetic. Other substances than iron and its alloys are capable of being magnetised, but not to anything like the same extent.

Now at this stage let it be taken for granted that an electric current can be produced and sent through a metal wire. This wire must form part of a continuous circuit or loop ; if it be broken at any point the current will cease to flow. It should be covered with cotton, silk, rubber, or one of the many substances which offer a large resistance to the passage of electricity, for the latter tends to cut across the shortest path.

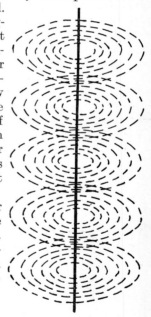

Fig. 72.   MAGNETIC FIELD OF A STRAIGHT WIRE CARRYING A CURRENT.

All metals and water may be regarded as conductors, and most other materials such as paper, wood, oil, and those mentioned above as non-conductors or insulators, though all insulators will break down under a sufficiently high electric stress. Magnetic force, on the other hand, is exerted freely through all these bodies, the only material which has any appreciable effect upon it being iron and its alloys.

A wire carrying an electric current possesses the same properties as a magnet, only in this case the curves which represent the direction of the force are circles whose planes are at right angles to the direction of the wire, as in Fig. 72. If

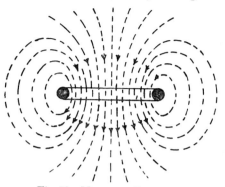

Fig. 73. MAGNETIC FIELD OF A SINGLE COIL.

the wire is coiled into a ring, Fig. 73, then one face tends to turn towards the north and the other towards the south pole, so that the plane of the ring becomes east and west. If the wire is coiled in a spiral so as to form a number of rings side by side, then the whole coil acts like a long magnet, and if freely suspended turns so that its axis points north and south,

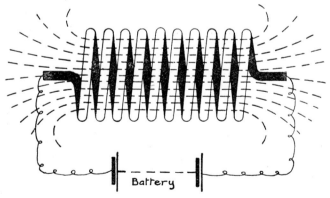

Fig. 74. A SOLENOID AND ITS MAGNETIC FIELD.

Fig. 74. Such a coil or " solenoid " is more powerful with the same current if it is provided with a soft-iron rod or core, which gathers up and concentrates, as it were, the magnetic force inside the coil. The effect of such a coil decreases with distance, in the same way as a magnet. If the current is increased the rings may be supposed to expand outwards, and if it is decreased or stopped they may be regarded as closing up until they disappear into the wire from which, apparently, they emerged.

The wire carrying a current is, in fact, surrounded by a magnetic field, and a small freely suspended magnet brought near to it tends to set itself at right angles to the wire. A galvanoscope or galvanometer consists of a coil of wire above or below or within which hangs a small magnetic needle. Each turn of the coil produces its own field and contributes its own force to turn the needle at right angles to the plane of the coil. If the current is reversed the movement of the needle is reversed, and if the coils are large and circular, and the needle small, an exact measurement of the force exerted can be made. This magnetic effect in turn affords a means of calculating the strength of the current in the coil.

THE MEASUREMENT OF ELECTRICITY

Just as it is impossible to obtain any clear idea about steam-engines without speaking of temperature and pressure, so it is necessary to describe an electric current in definite terms. The strength of the current, which governs the magnetic effect, is measured by the quantity of electricity which flows through any cross-section of the wire per unit of time,[1] and it is convenient to suppose that there is some force tending to drive electricity through the wire. This force is called *electro-motive force* or e.m.f., and is measured in *volts*, so called after the famous Italian physicist Volta. Different substances offer different degrees of *resistance* to the passage of electricity, and the resistance is measured in units called *ohms*, after another of the early workers. The strength of the current produced by an electro-motive force of 1 volt acting through a resistance of 1 ohm is called 1 *ampere*, after a celebrated Frenchman. If E is the number of volts, I

[1] The term "quantity" of electricity will acquire a clearer meaning after reading Chapter XX.

the number of amperes, and R the number of ohms resistance of a wire, then

$$I = \frac{E}{R}^1$$

Similarly the total quantity of electricity in a given time is measured in units, each equal to 1 volt multiplied by 1 ampere.

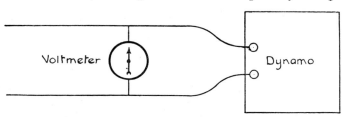

Fig. 75.   DIAGRAM SHOWING METHOD OF CONNECTING UP A VOLT-METER.

This unit is called a watt, and 1,000 watt-hours is a Board of Trade unit of electricity.

The simplest instruments for measuring electro-motive force and current are based on the galvanometer.   If the coil surrounding the magnetic needle consists of very thin wire it will have a high resistance and only a very small current will pass. When such an instrument is connected across the main wires leading from a dynamo as in Fig. 75 the electro-motive force acting on the coil will be the same as in the main wires but the current will only be a very small fraction of the main current. The deflection then will vary with the electro-motive force and the voltage indicated by the movements of the needle.   Such an instrument is called a voltmeter.

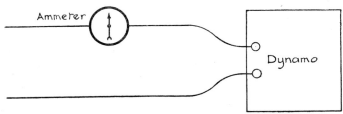

Fig. 76.   DIAGRAM SHOWING METHOD OF CONNECTING UP AN AMMETER.

[1] This formula used to be $C = \frac{E}{R}$ but I is now the international symbol for current.

G

An ampere-meter, or ammeter as it is generally called, may consist of a galvanometer with a coil of very thick wire connected in series as in Fig. 76. The electro-motive force is due to a fall of potential round the circuit. (If water instead of electricity were being considered the force tending to make water move from one point to another would be due to a difference of level.) The fall of potential depends upon the resistance, and with a low-resistance galvanometer will be very small. It will not absorb more than a fraction of the power in the circuit, but since all the current passes through it and the potential difference between its terminals will be very small its indications will vary with the strength of the current.

Other instruments depend upon the heating effect of a current, which is proportional to I^2R. If the resistance of a wire is practically the same for all temperatures, the amount of heat produced by a momentary current in it will depend only on the square of its strength. The heated wire will expand and, as the slack is taken up by a spring, will turn a pointer.

The total power, however, depends upon the voltage and the current strength, and an instrument for measuring the product of these is called a wattmeter. Space will only allow a brief statement of the principle upon which such an instrument is constructed. If a coil of wire suspended in a magnetic field carries a current it tends to set itself at right angles to the lines of force. This effect in a coil of high resistance will be proportional to the electro-motive force, and in a coil of low resistance to the strength of the current, so that the combined effect of two such coils (no iron being present) will measure the watts. If one is fixed and the other free to move, the motion of the latter will be a measure of the mutual attraction due to the current and the electro-motive force.

In generating stations the instruments record on rolls of paper the changes in the electro-motive force, strength of current, and power produced, so that very accurate calculations of the cost can be made. But this applies to the whole amount and, in order that each consumer shall be charged exactly for the quantity he uses, other instruments are required.

Behind the hall door or in some other out of the way corner of a house in which electricity is used for heating or lighting, is a small, compact instrument which the householder regards with a considerable amount of awe, not unmixed with suspicion.

For it is upon the indications of this apparatus that his quarterly bill for current is calculated, and as the amount is small or large so he is prepared to look upon it as an upright judge or as a biassed advocate of the people who placed it there. The instrument consists of an iron case with a window in front and having a set of dials something like a gas-meter. The internal arrangements are very considerable in different types, and it may perhaps be sufficient to indicate the principle upon which one of these performs its duty.

It should be noted at the outset that the pressure or " voltage " of a public supply is constant, or nearly so ; and as the quantity of electrical energy for which the consumer has to pay is measured by the product of the number of volts, the number of amperes, and the time, it is necessary to measure and record only the last two quantities. Suppose now a small electric motor is constructed so that at the voltage of supply a current of one ampere causes it to make one revolution per second, a current of two amperes two revolutions per second, and so on. Then the number of turns which the motor makes in any given time multiplied by the voltage of supply will give the total amount of electrical energy which has passed through it. Such a motor can easily be arranged to operate a set of dials so that readings are obtained directly in Board of Trade units of 1000 watt-hours.

ELECTRO-MAGNETIC INDUCTION

The principles upon which nearly all methods for the production and application of electrical power depend were discovered by Michael Faraday between 1830 and 1832. He showed that if a magnet be moved so as to approach a coil of wire the coil has a current of electricity produced within it. If the magnet is withdrawn a current in the opposite direction to the first is produced. In fact any movement of the magnet such that the lines of force in its field cut the wire causes a current, the direction of which depends upon the direction in which the line moves. As a wire conveying a current has a magnetic field it is capable of acting on another wire in the same way, either by actually moving either wire, or by causing the strength of the current to alter so that the lines move inwards or outwards, and thus cut the second wire in their motion.

If the reader can keep these actions and reactions in his mind he will have no difficulty in understanding how nearly all the apparatus employed in the practical applications of electricity work. He must always picture a conductor conveying a current as surrounded by a field of magnetic force, and he may, as a rule, assume that iron is used merely to concentrate this force and to give it direction. If either wire carrying a current, or a piece of iron, is free to move, the movement will take place in such a way that the greatest possible number of lines of force pass through the iron. For example, a coil of wire through which a current of electricity is passing (Fig. 74) will "suck-up" an iron rod until the latter protrudes equally at either end, and will exert considerable force in so doing. Two conductors carrying currents act upon one another by reason of the magnetic fields with which each is accompanied. And every electrical machine may be regarded as a magnetic machine in which the magnetism is produced by electric currents.

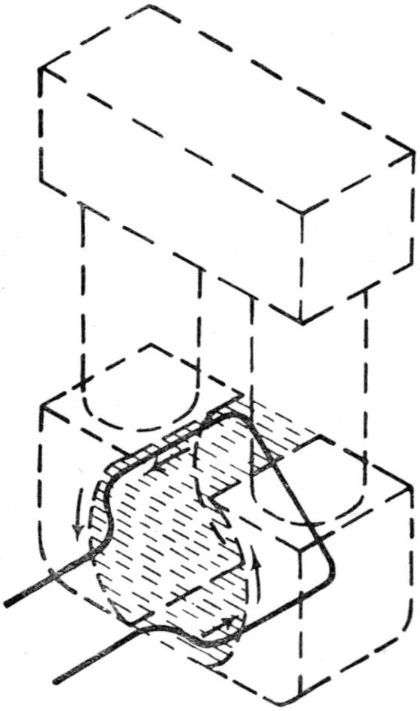

It is customary to express the strength of a magnetic field by the number of imaginary lines of force per square centimetre—measured at right angles to the direction of the force. Into the exact meaning of this it is not necessary to enter here, but it may be stated that the e.m.f. produced in a conductor is proportional to the number of lines cut per second ; that is, to the strength of the field and the velocity with which the conductor moves. And as the movement of conductors in a magnetic field is the method by which electricity is invariably generated for practical purposes, we may proceed to consider the construction and mode of working of generators, or dynamos as they are more usually called.

Fig. 77. Rectangular Coil Rotating between Poles of Field Magnets.

THE CONTINUOUS-CURRENT DYNAMO

Imagine a rectangle of wire to rotate between the poles of a magnet as in Fig. 77. As each side of the rectangle approaches the pole a current is induced in the direction shown by the arrows, and this increases until the centre of each pole is reached (when the greatest number of lines of force are cut per second), and then decreases until the rectangle has turned so that its plane is vertical. As the rotation continues, the side which at first passed the face of the south pole now passes the face of the north

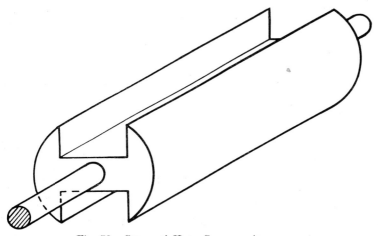

Fig. 78.   Siemens' H or Shuttle Armature.

pole, and *vice versa*. The induced current now is in the opposite direction to that indicated by the arrows.

A wire rectangle is too flimsy a thing to be rotated without some support, and a single coil can only cross the field twice in every revolution. So in the actual machine it is replaced by coils of similar shape, but of many turns, because each turn has a current induced in it, and the electro-motive force is proportional to the total length of wire cutting the magnetic field. In order to concentrate the magnetism of the poles in the gap between them, the rotating coil is mounted on an iron armature which, in the earlier machines, was of the form shown in Fig. 78. If each end of the wire is connected with a metal ring mounted on the shaft by means of a non-conducting disc, then the current can be drawn off as it is produced by brushes that make a sliding contact with the rings as in Fig. 79.

A continuous current is obtained by means of a commutator.

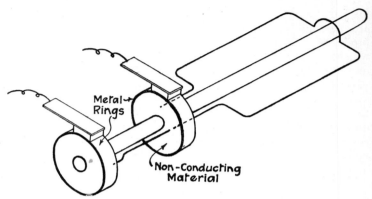

Fig. 79.   SLIP RINGS.

This consists of a metal ring or cylinder mounted on an insulating drum, and cut in halves at opposite ends of a diameter, as in Fig. 80.   Two brushes rest on this in such a position that when

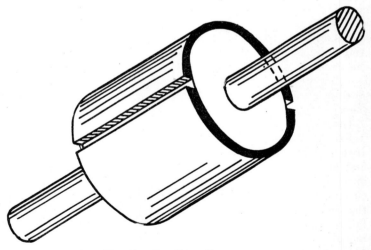

Fig. 80.   TWO-PART COMMUTATOR.

the current in the armature is just about to reverse the brushes change over from one-half to the other, thus rectifying the alternation in the armature.

In this arrangement the coils are only cutting the lines of force for part of the revolution, and while the cheeks of the shuttle are passing the pole-piece no electro-motive force is being produced.   Modern armatures are made in the form of a drum, which is built up of a number of thin sheet-iron discs, threaded on the shaft.   These are separated from one another by coats of varnish or thin paper to prevent induced currents flowing through and heating them, and they are stamped with teeth in the edge to form slots in which the armature coils can lie.   The coils are wound up separately and connected up with segments in the commutator.   The latter consists of copper bars mounted on a non-conducting sheath round the shaft, and separated from one another by strips of mica.

The brushes consist of carbon blocks held in a frame which can be rocked slightly backwards or forwards until there is the least sparking between the brushes and the commutator.   The reason for this adjustment is as follows.   The rotation of the iron armature causes distortion of the lines of force, so that instead of pursuing their usual path, they are dragged round slightly with the armature.   If the brushes do not take off the current from each pair of commutator bars when the voltage is at its highest value, there is a tendency for the current to jump across to the brush before or after the bars have passed under it. And since the voltage is highest when the coil is passing through the strongest part of the field, the brushes which would normally be exactly opposite the pole pieces have to be given a " lead " corresponding to the angle of distortion.   In a modern machine their position is mid-way between the poles.

The magnetism in the field magnets is produced by the machine itself.   Coils of wire are wound between the pole pieces, and these are connected up in one of the ways illustrated in Fig. 81.   In the first method the current from the armature divides, part going into the outer circuit and part round the field magnets.   If more current is taken in the outer circuit less goes into the field coils, the strength of the field is reduced, and the voltage of the machine falls.   This is called shunt winding.   In the other, or series wound machine, the same current flows round the outer circuit as the field coils, and when the former increases the field strength

increases and the dynamo rises to the occasion.   Machines which
have to bear varying loads have part of the winding in series and
part in shunt, with such a proportion that each compensates
the defects of the other.   They are then said to be compound-
wound.

The voltage of a dynamo depends upon the length of active
wire in the armature, that across the ends being ineffective, and
many turns of thin wire will produce a high voltage.   It also
depends upon the strength of the magnetic field.   But a long
length of thin wire has a high resistance and the current from

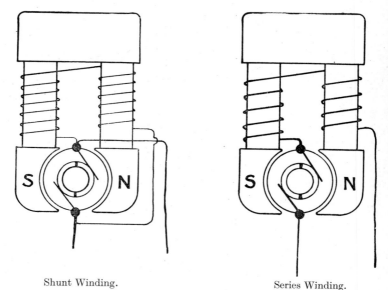

Shunt Winding.   Series Winding.

Fig. 81.   DIAGRAM SHOWING METHOD OF WINDING DYNAMOS.

such an armature will be relatively weak.   On the other hand, a
few turns of thick wire will produce a current of low voltage but,
as the resistance is also low, of considerable strength.   The
winding of the armature therefore determines the uses to which
a machine can be put, though, as will appear later, the character
of the current can be altered before use in any way that is desired
with considerable ease.

Dynamos to give continuous or direct current, or D.C. machines

FIG. 82.—ARMATURE OF D.C. GENERATOR.

To face page 88.

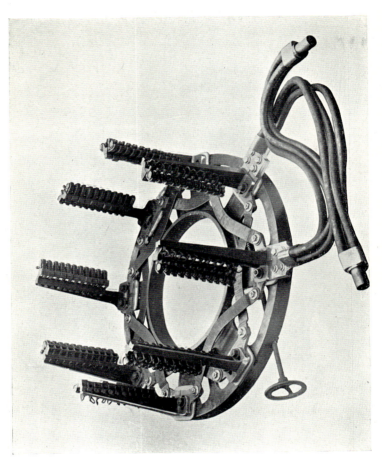

FIG. 83.—BRUSH GEAR OF D.C. GENERATOR.

To face page 89.

as they are called, are not usually constructed to give a very high voltage, and as high-tension direct currents have been used recently (e.g. in the wireless telegraph station at Clifden Bay) it will be interesting to describe how it is done. Suppose there are three machines each giving 1,000 volts, and imagine them connected so that the first terminal of the first dynamo is connected to the outer circuit, and the second to the first terminal of the second dynamo. The second terminal of this dynamo is connected to the first terminal of the third dynamo, and the second terminal of the third dynamo to the outer circuit. In this way the total electro-motive force becomes equal to the sum of the electro-motive forces of all three machines, or 3,000 volts. The attempt to secure the same result in a single machine would introduce difficulties of insulation.

The existence of only two poles which concentrate the lines of force in one direction leads to irregularities in the voltage which are very noticeable in lamps unless the machine is driven at very high speed ; and as a high speed involves special difficulties in construction to secure exact balance, the avoidance of vibration, and a sufficient lubrication of the bearings, all modern machines except very small ones, or machines to be driven by turbines, are made with four or more poles. These multipolar machines have the field magnets in the form of a ring with magnet cores projecting from the inner surface, and the coils are so wound that the poles are alternately north and south. Instead of cutting the field between a single pair of magnet poles in each revolution, the armature coils cut the field between two, three, or more pairs, and the successive impulses of electro-motive force occur more frequently. A lower speed is therefore possible, and the lamps do not flicker.

The arrangement will be understood from Figs. 82–3–4, which show the armature, brush gear, and the complete machine, of the type constructed by Messrs. Dick, Kerr & Co., of Preston. Fig. 83 shows clearly how the alternate brushes are connected up so that half of them collect current from coils passing north poles and half from south poles of the field magnets. Fig. 82 showed how the strips of copper forming the armature coils are placed in slots in the drum and kept in place by binding with steel pianoforte wire resting on mica bands. In a 2-pole machine each strip would pass completely round the armature, lying in slots diametrically opposite to one another ; but in the

8-pole machine it is only necessary to carry it through slots one-eighth of the circumference apart—that is, so that the two lengths shall be opposite N. and S. poles, and the currents in the two portions shall therefore flow in the same direction round the coils.

A word of explanation is perhaps necessary in regard to the " interpoles "—the small field magnets between each pair of main poles. These are so wound as to act in opposition to the pole behind, and thus to convert the gradual change of e.m.f. in the armature coils into a rather sudden one. In the absence of interpoles, which are fitted on most machines nowadays, the effect of a pole on an armature coil persists during the period when the armature coil is leaving the pole and when the corresponding commutator bars are leaving the brushes, and a certain amount of sparking occurs, which not only increases the wear of the commutator and brushes, but also causes loss of energy. With interpoles both these faults are avoided.

This particular machine is constructed to generate 700 kilowatts (a kilowatt is 1,000 watts) at 330 revolutions per minute. The e.m.f. is 500 to 525 bolts and the strength of the current 1,330 amperes, representing over 900 electrical horse-power.

THE ALTERNATING CURRENT DYNAMO

It has already been remarked that a dynamo or generator provided with slip rings instead of a commutator gives an alternating current, and for small machines this is the main constructional difference between them. Moreover, in machines giving continuous current in one direction it is obviously necessary that the commutator and the armature should rotate in order that the change of direction in the armature coils may be rectified. But if an alternating current is required the condition for its production is merely that conductors and lines of force should continually cut one another and this can be secured as easily by rotating the field magnets as by rotating the armature. In very large machines there is clearly an advantage in rotating the field magnets rather than the armature, because it is the armature which carries the main current; and when heavy currents at high pressure have to be taken from rings there is bound to be some loss owing to the imperfection of sliding contacts. On the other hand, a comparatively small low-tension continuous current is necessary to excite the field-magnets, and

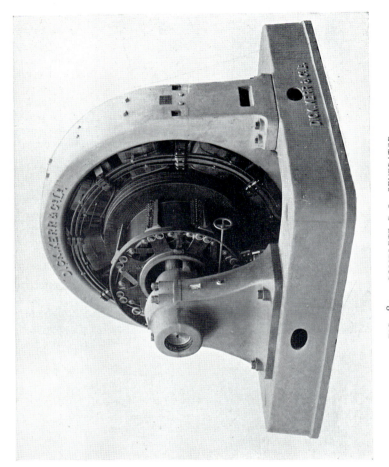

FIG. 84.—COMPLETE D.C. GENERATOR.

To face page 90.

FIG. 85.—LARGE A.C. GENERATOR.

To face page 91.

is conveyed through slip rings without any appreciable disadvantage. The current from the armature is then collected by cables attached to the fixed armature coils.

A modern alternating current generator consists of a rotating wheel, called a rotor, carrying on its rim a number of pole pieces, surrounded by coils of wire, which must be fed either with direct current from a separate dynamo, or from one mounted on the same shaft. Surrounding the rotor is a large ring-shaped frame, having coils fixed on its inner surface. The poles on the rotor are alternately north and south, and as it rotates the lines of force between successive poles cut the coils of the fixed armature or stator. For each pair of poles in the rotor passing the armature coils per second, there will be a complete alternation of the current, so that the periodicity or frequency will be given by the product of the number of pairs of poles and the number of revolutions per second. Thus suppose there are sixty poles and a corresponding number of coils, then at 1,200 revolutions per minute, or twenty per second, there will be 30 × 20 or 600 alternations per second.

The example just given assumes that the wire in the stator is wound continuously round successive poles, giving what is known as a single-phase current. If, however, the wire is in two portions wound round alternate poles, the current will be a maximum in one when it is at a minimum in the other, a condition known as 2-phase. More frequently there are three sets of stator coils, giving a 3-phase current.

Fig. 85 shows a 4-pole 3-phase machine giving 8,000 k.v.a. at 750 revolutions per minute, made by Dick, Kerr & Co. It is particularly interesting as showing the arrangements for ventilation. The outer casing of the stator is double, and air is drawn in at the bottom and expelled through an opening in the top. The rotor is kept cool by the operation of a fan fixed on the shaft. This and other features will be clear from Fig. 86. The four field coils with their laminated pole pieces will be clearly seen. In order to obtain the same e.m.f. at lower speeds, the number of poles would have to be increased, and the rotor constructed of larger diameter. The high speed at which turbines run reduces the number of poles which are necessary. Standard types of modern machines with two poles run at 1,500 revolutions per minute for 25 alternations per second, and at 3,000 revolutions per minute for 50 alternations per second. Machines with *four* poles run at 1,500 revolutions per minute for 50 alternations per second. The rotor then takes the form of a drum in which the coils are laid in slots on opposite sides, and is very similar in appearance to the old shuttle armature.

It will, perhaps, be interesting to notice the elaborate precautions which are taken in the winding of modern electrical machinery. In the coils of the armature shown in Fig. 82 the strip copper is wound on a model or " former " so as to be of the exact shape it will take in the machine. It is then cleaned and dried, the free ends wrapped with superfine linen tape, twice dipped in special varnish, and baked between each process. The portion which is to lie in the slot is then insulated with mica, parchment paper, and linen tape, dried in a vacuum, impregnated with oil and acid-resisting varnish, and baked for twelve hours. The slots are lined with leatheroid troughs to prevent abrasion. For peripheral speeds of more than 6,000 feet per minute the steel wire binding referred to on page 89 is insufficient, and the conductors are held in place by hard-wood wedges driven into dove-tails formed in the outer portion of the slots. The finished armature is baked for twelve hours and given a final spraying of air-drying, black varnish. It may be left to the reader to imagine the vast amount of patient investigation and experience which lie at the back of such a series of processes ; and these are quite apart from those concerned with the mechanical (as distinct from the electrical) features of a modern armature.

TRANSMISSION AND DISTRIBUTION

The size of the wire required to transmit electricity over a distance is determined by the resistance. This is proportional to the product of the resistance of the wire and the square of the current strength. On account of its high conductivity copper has been invariably used for the purpose, though aluminium is also used on account of its lightness. Its conductivity is less than that of copper, and its price per ton is a little higher, but volume for volume its weight is less than one-third.

As the power transmitted in a given time is measured by the product of the voltage and current, for large quantities either the voltage or the current, or both, must be high. But if the strength of the current is doubled, the losses due to resistance are

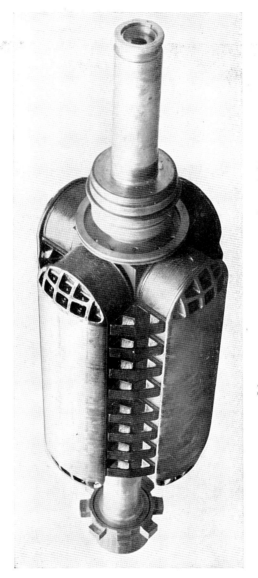

FIG. 86.—ARMATURE FOR LARGE A.C. GENERATOR.

To face page 92.

FIG. 87A.—TRANSFORMERS FOR RAISING VOLTAGE TO 132,000.

To face page 93.

multiplied by four, so it is usual to employ high voltages in transmission wherever possible in order to save the cost of copper. The determination of size, however, is not arbitrary, and several factors have to be taken into account.

While high voltages are desirable for transmission they are not always suitable for actual use, and one of the contrivances which adds so much to the adaptability of electrical power is the transformer. This consists of two separate coils of wire wound on a soft-iron core. When a current of electricity in one starts or increases in strength a current is produced in the *opposite* direction in the other. Similarly when the current in the one decreases in strength or stops a current is produced in the *same* direction in the other. As the " induced " currents in the second coil depend upon the number of lines of force cut by each wire per second, and the number of wires, the voltage depends upon the number of turns. And, since the power in the two coils is practically the same, it is possible to " step-up " from a current of low voltage and relatively great strength to one of high voltage and small strength, or *vice versâ*. The ordinary Rhumkorf or induction coil is a step-up transformer, and the coils used for converting the high-tension current from long-distance transmission lines to low-tension current for tramways and other purposes are step-down transformers. Both the entering and leaving, or primary and secondary, currents are alternating. The change from alternating to direct is effected by a motor generator, to be described later.

The transmission of electricity at high voltage necessitates special precautions, for not only is the tendency to leakage immeasurably greater, but a shock is highly dangerous. Consequently, though one of the earliest central stations in England— that at Deptford—produced an alternating high-tension current, nearly all public supply stations for town supply produced a strong current of moderately low voltage. A common practice now, however, is to generate electricity at 11,000 volts, transform up to 33,000 volts for transmission, and then transform down to 200–250 volts for supply. In the new grid or network of distribution now being developed in Great Britain, the voltage is raised to 132,000 for transmission and in the United States a transmission voltage of 220,000 volts is in use. This change has led to an enormous development in transformers and high tension switches.

In order to economize copper most electric light stations distribute electricity at 460 volts by means of what is known as the three-wire system. Suppose there are two dynamos connected up as shown diagrammatically in Fig. 87. If each dynamo is capable of producing current at 230 volts, the total electromotive force in the two outer wires will be 460 volts. The wires leading into the premises of each consumer, however, are always the inner and one outer, so that the electro-motive force inside the dwelling is only 230. If for power, 460 volts is used and the fittings must be of a special character.

If electricity has to be distributed across country over long distances, then there is an advantage in using high-tension currents. In England some of the electric railways considered in Chapter XIV are fed by a 6,000-volt system, and in Norway

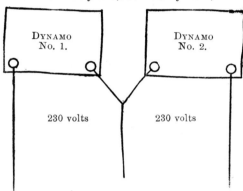

Fig. 87. Diagram to Illustrate Three-wire System.

the current used for the manufacture of nitrates by the process described in Chapter I has an electro-motive force of 10,000 volts. The Ontario Power Company, of Niagara Falls, sell their electricity to three subsidiary companies, one of which distributes it over 160 miles at 60,000 volts. The wires consist of aluminium $1\frac{1}{8}$ inches in diameter, and are carried on iron standards 55 feet high, with an average span of 500 feet. The over head cables are fixed to porcelain insulators which weigh 35 lb. each.

At Keokuk, on the Mississippi, the fall in the Des Moines Rapids has been utilized for a hydro-electric power station, from which electricity is distributed locally and as far as St. Louis, 144 miles away. For the long-distance transmission the current

FIG. 87B.—OIL SWITCHES FOR HEATING A 132,000 VOLT CIRCUIT.

To face page 94.

FIG. 87c.—132 K.V. OIL SWITCHES AND OVERHEAD STRUCTURE, OUTDOOR CONNECTION.

To face page 95.

is stepped up to 110,000 volts and the $\frac{5}{8}$ inch copper cable is carried on steel towers about 80 feet high and 800 feet apart. Instead of the bell-shaped insulator similar to those used for telegraph and telephone lines, the form adopted consists of seven porcelain discs, each ten inches diameter, and with corrugated or ribbed surfaces. These are strung together on a rod by means of malleable iron fittings cemented into them, and the cable conveying the current is suspended from the lower disc.

But a transmission system of still higher voltage is now in operation. The Pacific Light and Power Corporation erected two hydro-electric stations, about four miles apart, at Big Creek, seventy miles from Fresno in California. From there power was conducted to Los Angeles, 275 miles away, at a pressure of no less than 150,000 volts, where in addition to domestic and factory use it served also the Pacific Electric and Los Angeles railway systems. This installation not only had the highest voltage which had ever been employed on a commercial scale, but it is transmitted also over the longest distance which heavy currents of electricity have yet been conveyed either overhead or underground. To-day, however, heavy currents are being transmitted at 220,000 volts, and 132,000 volts has been selected for the electric grid which is now being developed in this country.

The employment of such enormous pressures has been rendered possible only by improvements in the designs and construction of switches and transformers, see Figs. 87A and 87B. When a circuit conveying a high-tension current is broken, the electricity tends to jump across, and the resistance that it encounters causes an " arc ", which may fuse the metal of which the switch is composed. It has been stated that at the Deptford Station when the main switches on the 10,000 volt circuit were broken, a man had to " beat out " the arc with a mat at the moment when the lever was thrown over !

An " arc " or tongue of flame formed by the current jumping a gap, is flexible, and is deflected by a magnet just in the same way that a flexible wire carrying a current would move so that its magnetic field corresponded to the one acting upon it. Switches are, therefore, often provided with magnets which prevent the arc becoming established by blowing it out as soon as it is formed. And all switches, whether for direct or alternating current, are operated partly by springs which cause the contact to be broken with extreme rapidity.

The switches used for very high tension are immersed in oil

which has a high insulating power, and are never operated directly by hand. They are opened and closed by electro-magnets, and the small switches which control them have only to deal with a weak low-tension current entirely independent of the main circuit. The high-tension switches were at first locked up out of sight and touch in a brick or concrete chamber, which was opened only for the purpose of occasional inspection and repairs. The more recent tendency, especially in the United States and Canada, is to have these switches out of doors. Fig. 87C shows outdoor connections and oil switches for a power station in South Africa.

MOTORS

The reader who understands the D.C. dynamo will have no difficulty in understanding the D.C. motor, for they are in all essential parts the same, and one machine can be used for either purpose. The magnetic effect of the coils on the field magnets and armature is concentrated between the pole pieces in such a way that when a current is sent through the machine attraction occurs between armature and field magnet, and the former turns. But as soon as the poles between which the attractive force is exerted come opposite to one another, the commutator arrives at a position in which the current is reversed. The armature pole is now of opposite polarity, and repulsion ensues. This is repeated for every pair of poles, and a continuous rotation of the armature is secured.

Like the dynamo, the D.C. motor may be shunt, series, or compound wound. The shunt motor develops very little power at low speeds with heavy loads, because most of the current goes into the armature and the magnetic field of the field magnets is therefore weak. As the armature rotates, however, it behaves as in a dynamo, and produces a back electric-motive force in its coils which is equivalent to a resistance, and drives more current round the field magnets. Such a machine is largely self-regulating.

In a series motor the same current passes round the armature and field magnets, and from the moment it is switched on there is a strong turning movement. This makes the series machine valuable for cases where large power at low speed is required, as in trams, lifts, and cranes, which have frequently to be stopped and restarted. Constancy of speed under varying loads

FIG. 88.—TRANSMISSION TOWERS.

To face page 96.

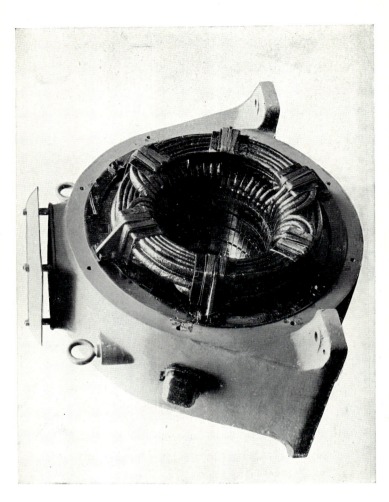

FIG. 89.—STATOR FIELD MAGNETS OF LARGE INDUCTION MOTOR.

To face page 97.

is secured by compound winding, in which the defect of each system is compensated by the other.

As the resistance of the armature coils is invariably lower than that of the field magnets, the greater proportion of the current tends to pass through them when the motor is at rest, and it is not until the machine has acquired some speed that the back voltage in the armature reduces this to a safe amount. In order to prevent the armature coils becoming overheated and "burnt out" all D.C. and some types of A.C. motors are provided with starters. These are boxes of resistance coils fitted with a switch, which as it is turned, causes the current to pass first through the whole length of resistance, then smaller parts of it in succession until the whole of the resistance is cut out and the current passes directly to the motor. The type of starter used on tramways will be described in Chapter XIV.

The problem of obtaining motion from an alternating current is a much more difficult one, but has been solved in three ways. If an alternating current is passed through the armature of an alternating current dynamo at rest the machine will not start, for the tendency to rotation in one half is balanced by the tendency in the other half. And even if the armature is rotating, the two opposite tendencies may be equal. But if the speed is such that each coil passes from one pole to another during half an alternation of the driving current, then the direction of the latter will be changed just in time to convert what would have been a repulsion into an attraction, and the machine will continue in motion at constant speed. Thus if the frequency is 100 per second, there will be 1,200 half-alternations per minute, and if there are four pairs of poles the speed must be $\dfrac{12,000}{8} = 1,500$ revolutions per minute. Such a machine is called a synchronous motor, and so long as it is not overloaded it will continue to run at constant speed.

The second type is called an induction motor. Suppose there are two or three sets of coils in pairs in the stator, and suppose them to be fed with 2-phase or 3-phase current, so that the maximum magnetic effect is produced successively all round the ring. Any conductor, such as a copper bar, will follow the coils as each one is successively excited, because of its tendency to move to the strongest part of the field. The rotor consists, therefore, of a number of copper bars, each end of which is fixed

H

to a copper ring, forming a sort of squirrel cage from which this type of rotor takes its name. If the initial load is not high, this motor is self-starting, and as no current passes into the armature conductors no slip rings are required.   If such a motor is required to deal with heavy initial loads, the coils are wound on the rotor and current is led into them by means of slip rings and carbon brushes. When the machine is fairly started the brushes may be lifted from the rings. Figs. 89 and 90 show the stator field magnets and armature of a 315 horse-power induction motor constructed by Dick, Kerr and Co.

The third type of motor is used for railway work, and takes single-phase current. It is similar in principle to the direct-current machine, but whereas in that case the commutator reverses the direction of the current in the armature, in this case the reversal takes place by the alternating current in the field magnets. The armature coils are short-circuited by connecting opposite brushes in pairs, and the brushes are fixed so that each set of coils is closed at the time when the field-magnet poles exert the most powerful turning effect. This possesses all the merits of the D.C. series machine, with the additional advantage that the current to work it can be conveyed cheaply over great distances by relatively thin wires.

We are now in a position to understand how an alternating current is converted into a direct one. If the current is sent through the armature of a synchronous motor, entering by slip rings, the machine will, as has been observed, rotate at constant speed.   If, moreover, the armature-shaft be fitted with a commutator to which the other ends of the coils are attached, the armature will turn at the correct rate to enable the commutator to deliver to the brushes a direct current. A Rotary Converter of this kind usually has fitted on the same shaft a small induction motor, which serves to start it and run it up until it is in step with the alternating current which drives it. The field magnets are fed by current from the D.C. end of the machine.

ELECTRICAL STORAGE

The value of electrical power is enormously increased by the fact that it can be stored. This is accomplished in cells which are distinguished from those used in the generation of electricity on a small scale for electric bells, telephones, and experimental work, by being called storage cells, or secondary cells, or more

FIG. 90.—ROTOR FOR LARGE INDUCTION MOTOR.

To face page 98.

 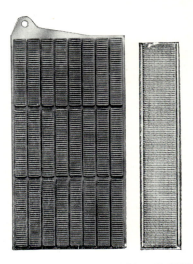

FIG. 91.—SINGLE POSITIVE
  TUBE AND PLATE FOR
  EDISON CELL.

FIG. 92.—SINGLE NEGATIVE ELE-
  MENT AND PLATE FOR EDISON
  CELL.

 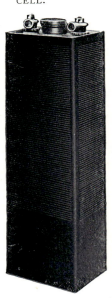

FIG. 93.—SET OF PLATES AND COMPLETE EDISON CELL.

To face page 99.

generally accumulators. A number of cells constitutes a battery. The battery can be fixed up permanently, or enclosed in a box, taken to a generating station, charged, and then taken away to the place where the electricity is to be used. This is just as simple as taking a piece of clockwork to be wound up and then removing it to another place to drive machines.

There are, broadly speaking, three types of secondary cells in use, two of which are very similar, and depend upon a discovery made by Planté more than fifty years ago. He found that when a current of electricity was passed for some time through a cell containing two lead plates immersed in dilute sulphuric acid a current could afterwards be obtained in the reverse direction. At first, not a great deal of the electricity could be stored in this way, but by repeatedly charging and discharging the cells, the plates became capable of taking up and retaining, as it were, a greater quantity. The total amount of electricity put into the cell can in no case be recovered, and under ordinary conditions there is a 20 per cent or 30 per cent loss. The storage is due to a somewhat complicated chemical change, or rather series of changes, in which one of the plates is converted into lead peroxide, PbO_2, a chocolate-coloured substance. The immersion of the plates in the acid may be assumed to lead to a thin layer of lead sulphate, $PbSO_4$, being formed on both plates. The passage of the current causes the separation of hydrogen at the negative plate, and this reduces the lead sulphate to lead, which it leaves in a spongy condition. At the other plate, an oxidizing action occurs, and the lead plate is oxidized to the chocolate peroxide. The liquid in the cell becomes denser, showing the presence of more sulphuric acid than before the passage of the current.

The tendency for the spongy lead to be oxidized and the peroxide to be reduced causes a current to flow in the opposite direction to the charging current when the plates are connected to an external circuit. The e.m.f. is always about 2 volts—or 2·2 volts for a fully charged cell, falling a little as the cell is discharged. With lead plates the active materials are at first produced only in a thin surface layer, but repeated charging and discharging increases the depth and enables the duration of charge and discharge to be increased. The process, however, is very tedious, and the active material is liable to break off in flakes from the surface of a flat plate.

An improvement was effected by Faure, who employed an alloy of lead and antimony cast in the form of a grid and pressed into the interstices of the grid a paste of red lead, Pb_3O_4, and sulphuric acid. This secured at once a greater amount of active material, and reduced the time required for "forming" the plates. Numerous other modifications have been introduced in order to secure a greater sponginess throughout, so that chemical action can proceed more readily and affect a great quantity of material. Thus, while the percentage occupied by the pores in an early type of Planté plate was only 25, that of a modern chloride cell reaches 60 or 70.

It will, perhaps, be not without interest to indicate one way in which this has been achieved. The Chloride Company make their positive plates in the shape of perforated slabs with spiral rosettes of thin lead strip pressed into the holes. These are "formed" by the original Planté process. The negative plates contain pellets of a fused mixture of lead and zinc chlorides in the perforations. On passing a current of electricity through these plates in the cell, the zinc is removed, and the lead is reduced to the spongy condition.

Lead accumulators require a great deal of care in management. They must not be charged or discharged too rapidly, or the active materials tend to become displaced. The amount of liquid necessary, and the use of the heavy metal lead, renders them of great weight. The corrosive character of the acid requires the cells to be made of glass, ebonite, or, for very small ones, celluloid. But the many attempts which have been made to discover a lighter metal that would serve the purpose, and a method of construction that would withstand hard usage, have resulted in a single success which may now be described.

The Edison Accumulator was invented about 1904, and though it made slow progress at first, improvements in manufacture have secured for it during the last few years a considerable reputation. Mr. Edison employs iron and nickel instead of lead, and a solution of caustic potash instead of sulphuric acid. The construction of the plates involves a degree of mechanical ingenuity thoroughly in keeping with the standard of the twentieth century. The positive plate consists of a nest of nickel-steel tubes, each one of which is formed by a perforated nickel steel strip wound in spiral. These are packed tightly with alternate layers of nickel oxide and flakes of metallic nickel.

Fig. 91 illustrates the single tube and a complete plate. The negative plate is also of steel and contains oxide of iron pressed into a number of lozenge-shaped pockets, see Fig. 92. The liquid and the immersed plates are contained in closed-ribbed nickel-steel box, Fig. 93. The cell contains water with 21 per cent of caustic potash, and the solution requires a little water occasionally. The chemical changes which take place have not been fully worked out, but the net effect is supposed to be that the oxygen of the nickel oxide is transferred to the iron plate, rendering that more highly oxidized. But ordinary chemical analysis reveals no difference in the composition of either plate charged or uncharged.

The ability of this cell to withstand hard wear may be gauged by the fact that some were tested by being lifted and dropped 2,000,000 times, the process being continued for twenty-two days and nights, without any mechanical defect arising, and with a loss of only one-quarter per cent in efficiency. On one occasion a fire in a garage boiled out all the liquid, and yet, when filled, the battery is said to have been as good as ever. Moreover, the Company quote cases in which cells in regular practice are charged for short periods at five times the normal rate—a proceeding that would, in lead accumulators, cause buckling of the plates and displacement of the active material.

The e.m.f. is only about 1·2 volts, so that more cells are required for a given voltage than in the case of ordinary storage batteries.

Accumulators are used very frequently in electrical power stations for dealing with variations of load, charging being accomplished when very little current is required in the mains. They are also used to a limited, but, in the case of Edison cells, to an increasing extent, for traction. It will be clear that if a tram-car could carry its own generators with it an enormous capital outlay in overhead and underground equipment would be saved, but hitherto the weight and fragility have stood in the way. Since two Edison cells are required to give the same voltage as one lead cell there is very little saving in weight by their use. But cars have been equipped in this way in many towns, and are said to be doing good service. The necessity for proximity to a charging station when the storage battery runs down will militate against its extensive use for motor-cars, but very considerable progress is being made in connection with tradesmen's vans, which have a definite round, and which always return to the same place. Small portable types are used to a large extent in displacement of primary batteries—for miners', policemen's, and domestic lamps, for surgical and medical and dental work ; for motor-cars, submarines, railway and other signalling, for electric clocks, and for a score of other purposes.

The present type of Edison cell was placed on the market in 1908 ; in 1909 the rate of production reached 500 A-4 type cells per day, and in 1913 it had grown to six times that amount. About forty American railway companies have adopted the battery for train lighting, and a number of main lines and light railways are stated to be using these accumulators for motive power. The Ford motor-car was in 1914 equipped with self-starting and lighting apparatus driven by Edison cells, and this alone involved the manufacture of no less than 240,000 sets of 6-volt batteries per annum. A more recent type of cell, which is made in England, also employs nickel and iron in alkaline solution. It is known as the NIFE accumulator.

Meantime, the older forms of lead accumulator continue to be improved, and several other types somewhat similar to the Edison are being tested. Whether the use of self-propelled vehicles carrying their own store of electricity on ordinary roads becomes general or not, there are scores of other directions in which accumulators will find a ready application. Perhaps some of us will see the time when rails will be torn up or left to rust in their concrete beds, when the tall poles with their overhead wires will be pulled down, and when motor-cars, vans, and buses will be charged up at night ready for their next day's journey.

So much for a very brief outline of the way in which electricity is produced, distributed, and stored. Later on some attempt will be made to describe how it is employed in manufacture for domestic purposes, in wireless telegraphy, and in locomotion on land and sea. But all these applications are merely encouraging signs of what it can and will do in the future. Wherever there is a cheap source of power such as water, or easily mined coal, or a plentiful supply of oil, there electricity can be generated and distributed in thousands of horse-power along a web of overhead or underground wires, which can be tapped at any point in their length and used to drive machines silently, effectively, and with a grim purpose that overcomes all obstacles. Already, as we know, it illuminates the night, drives mills and factories, railways, trams, and steamships, coats baser metals with copper, nickel, silver, and gold, makes the blocks by which many of the illustrations in this book are produced, manufactures some of the nitrogen compounds upon which the wheat supply of the world will ultimately depend, and in the Kjellin furnace excites the particles of cold steel until they glow and flow like water.

But, above all, what cannot fail to strike even the casual observer most forcibly, is its cleanliness. True, most electric generating stations produce smoke, but this is due to inefficient methods of using coal. If, as has been suggested, central stations were established on or near the coalfields, and the coal were converted into gas and used for driving gas engines or for raising steam, then all the heat and light and power which the world requires could be obtained with an infinitesimal fraction of the smoke and grime which hang like a pall over every industrial town.

" And," as Mr. Milnes has said, " when once the problem is fully solved, when once power shall be conveyed by wire or, possibly, by wireless induction, from any source to an application, then the factory town is doomed. And when our productive centres are no longer squalid with dirt, when the mill is planted on the hill-side, when the web is woven and the tracery designed where light is bright and Nature beautiful, then beneath the touch of unsoiled hands a fairer fabric may issue from our looms than has ever yet delighted the daughters of men. Then shall pride in the results of toil—toil's best reward—be once more the portion of the worker ; then shall cleanliness of work beget cleanliness of home, and therewith cleanliness of life, of speech, of thought, wherein is the perfection of man's manliness. And production, taking on somewhat of the true creative character, may again hold out to the craftsman some share in the God-like privilege of gazing on the work of his own hands, and seeing that it is good."

CHAPTER VI

HISTORY loses much of its dramatic force by its inability to tell us who produced fire for the first time and whether he burnt his fingers with it. If a facile pen driven by a vivid imagination could have described the looks of astonishment and awe on the faces of whose who witnessed the birth of artificial light and heat, it would have given a picture of an event more important to the future of mankind than all the petty wars and scholastic controversies with which the books are filled. The wonder which it created lingered for many centuries ; for long after its value in extracting metals was known, it continued to enter into the most sacred rites of religious observance.

Though the principal use of fire was, and still is, the winning of metals without which few of the tools and appliances of the modern world could be made, the production of light has had a very important effect in enabling man to overcome the disadvantage of circumstance, and it marks one of the most clearly defined steps from savagery to civilization. The admonition of the proverb to rise with the lark and go to bed with the sparrow, though enjoying the warrant of history, would interfere seriously with the customs of the twentieth century, in which the Daylight Saving Bill was, for many years, classified with the annual records of the Sea Serpent.

For many centuries such light as the world required was furnished by the vegetable wick fed with animal or occasionally mineral oil, and it is only a little more than a hundred years since lighting by coal-gas was introduced. During the last fifty years coal-gas has been supplemented by paraffin and petroleum, which had the advantage of portability, and the final method of lighting by flame arose with the calcium carbide and acetylene industry in 1894. The electric light was known in the laboratory from the time of Davy in the early years of the nineteenth century, but until cheaper methods of producing electricity than by the use of primary batteries had been invented no commercial application was possible. But since 1879 when the first installation of Jablochoff candles was exhibited progress has been rapid.

It is a little curious that the discovery which has enabled gas to maintain its position was made incidentally in an attempt to improve electric lamps. In 1884 Dr. Auer von Welsbach was trying to impregnate the fine carbon threads used in incandescent lamps with one of the oxides which have the property of glowing brilliantly when heated to a high temperature, and he found that the temperature of an ordinary gas flame fed with air on the principle of Bunsen's burner, was sufficient for the purpose. These mantles were made by soaking a ramie fabric in an emulsion or thin paste of oxide of cerium with not less than 1 per cent nor more than 2 per cent of oxide of thorium, and then drying them. When they are suspended over a flame the cotton is burnt off, leaving a delicate and fragile framework of the mixed oxides.

At first the mantles were sold before the gauze had been burnt away, or they would have shaken to pieces in travelling. But it is rather important that this burning should be thorough, and the consumer's burner did not always do this effectually. The mantles, therefore, are thoroughly burnt in the factory, and then dipped into collodion—which consists of nitro-cellulose or gun-cotton dissolved in alcohol and ether. On drying, the alcohol and ether evaporate, leaving a thin film of nitro-cellulose which holds the framework together. When a new mantle is lighted this film burns off with a lurid flame, leaving the skeleton behind.

The Act of Parliament under which gas companies worked until recently required them to supply gas of a specified illuminating power when burnt in a burner giving a luminous flame. But the luminosity of an incandescent mantle depends entirely upon the heating power of the gas. All the light comes from the mantle, and the " candle-power " of the gas used with these mantles is a term which has no meaning whatever. More-over, a large quantity of gas is used for cooking, for warming rooms, and in gas-engines, for none of which is illuminating power a measure of its value. Indeed, it is stated on good authority that over 80 per cent of the gas manufactured is used for purposes in which the heating effect is the chief criterion of its value. If people were charged a price for butter which varied only with its colour, the world would laugh, and the persons who decreed the conditions of sale would be regarded as merely encumbering the

earth. And the case was really as bad as this. For gas of high
calorific value can be made more economically than gas of high
illuminating power, and gas consumers were paying more than
they need have done—paying, in fact, for something they did
not want and having no guarantee that what they did want was
supplied to them.

While it is not intended here to enter into the merits of gas
lighting versus the electric light, it may be remarked that some
investigations by Dr. Rideal tend to show that the former is
actually the healthier of the two. The flame keeps up a continual
circulation of the air and creates a hot layer some 12° C. higher
in temperature near the ceiling, which passes through the porous
plaster and effects ventilation where it would hardly have been
suspected. The reader will probably have noticed in a room
lighted by gas, and with a plaster ceiling covered only by the
roof, that the parts under the rafters remain white while the
spaces between are discoloured. This discoloration is the result
of the air filtering through and leaving in the surface pores the
fine particles of dirt that it contains. With the electric light,
on the other hand, the air is said to be relatively stagnant, and
to become vitiated more rapidly in cases of overcrowding.
This, however, does not apply to the new gas-filled lamps, which
radiate a considerable amount of heat.

There are, however, special conveniences attached to the use
of electricity, and we shall proceed to examine some of the
principal items of recent progress. It will be convenient to deal
with incandescent and arc lamps separately.

ARC LAMPS

The arc lamp developed out of a discovery by Sir Humphry
Davy in 1808. He was passing the current from a battery of
many cells giving an electro-motive force of 2,000 volts, through
two copper rods, and he found that when they were separated
by a small amount the electricity sprang across in a sort of
flame. The heat caused the flame to rise and form a curve,
and from this the name " arc " is derived. The metal rods were
soon replaced by those of carbon, the ends of which glow
brilliantly and give far more light at lower cost than could be
obtained from any common metal.

If the current is direct, one of the carbons, called the positive,
has its end pitted or worn into a hollow or crater, and this is

the hottest and brightest portion. The other, or negative, carbon is worn to a point, but at only half the rate of the positive rod. An alternating current makes each pole positive and negative in turn, and causes them to wear away at equal rates. If the alternations are less than 40 or 50 per second, the alterations in brightness are distinctly noticeable. There is a simple method of ascertaining whether an arc is fed by alternating current. If a walking-stick be whirled round, it will move some little distance between successive passages of the current in the lamp, and will thus be alternately in light and darkness. With a direct current the illumination will be constant, and the whirling stick will appear as a continuous blurred disc.

Before proceeding to consider modern developments there are two features to be noticed. One is that the vaporization of carbon which takes place in the arc develops a back electromotive force in opposition to the current that produces it. This amounts to 35 or 40 volts, and no pressure less than this will keep it alight, even though the ordinary distance apart of the carbons is 2 mm., or about one-twelfth of an inch. As the actual resistance of the arc is low when it has once started, only about 5 or 10 volts over this back e.m.f. is necessary to maintain it, so that the usual pressure on direct-current arc-lamp circuits is 45 to 50 volts. A resistance coil is attached to each lamp to make the arc a small portion of the total resistance and thus maintain a steadier current.

The second feature is the regulating device which keeps the carbons at a distance apart as they burn away. One of the carbons is fixed and the other, in one type, is attached to an iron rod which passes up the centre of a coil of wire which forms a solenoid. Current flowing through the coil causes it to suck the rod up and thus separate the carbons. The coil is wound with two wires, one in series and one in shunt. When the current is cut off, the carbons come into contact ready for starting again. As the current is switched on, the series coil separates the carbons, but, if it draws them too far apart, the current passes through the shunt, which is wound in the opposite direction, and forces them together again.

In modern lamps, one or both of the carbons is invariably made with a softer core which consists of powdered charcoal and potassium silicate compressed into a rod. In an ordinary open D.C. lamp the positive carbon is cored and the other solid. The

burning away of the rods is one of the principal disadvantages,
and forms no inconsiderable portion of the cost of maintenance
—not only in the cost of the carbons themselves, but also in
the labour of replacing them.

A decided economy is effected by the use of closed lamps in
which the arc is surrounded by a globe pierced with small holes
so that the circulation of the air is impeded and the carbons
last longer. Such a lamp needs attention only after 90 to 100
hours. When direct current is used both carbons are solid, but
with alternating current it is usual to have one cored and one
solid.

An ordinary open arc requires about 1·4 watts per candle-
power, or just about the same as a vacuum type metal filament
lamp ; but with the best quality of carbons the consumption

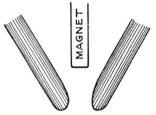

Fig. 94. Arrangement of Carbons in Flame Arc Lamp.

Fig. 95. Arrangement
of Carbons in Or-
dinary Arc.

may be as low as 1·1 watts per candle. The closed arc is less
efficient, requiring 2·3 watts for the same light, but this is
compensated for by the saving in carbons and labour.

The newest development, however, is in lamps which produce
a flame. In 1898–1900 Brewer used calcium fluoride in the
carbons and this, being more easily converted into vapour
than carbon, increased the length of the arc to 20 millimetres
or $\frac{4}{5}$ of an inch. Later, carbons with a large core containing
potassium silicate and a fluoride were used, the colour of the arc
depending upon the particular fluoride employed. With calcium
fluoride a good yellow light with excellent fog-piercing properties
can be obtained for 0·4 watt per candle-power. Cerium fluoride
gives a white and strontium fluoride a red light, but these require
0·7 watt per candle. For street lighting, therefore, the yellow
flame was invariably used, while for matching colours indoors

either the white open lamp or, better, an enclosed arc, was alone suitable. Almost the only use of the red flame was stated by Mr. Maurice Soloman [1] to be in butchers' shops, where its ruddy glow enhanced the colour of the meat !

Apart from its economy, the flame arc had an additional advantage in that both carbons are arranged to point downwards in a Vee, Fig. 94, and the cone of shadow produced by the lower carbon in an ordinary lamp, Fig. 95, was avoided. The tendency of the flame to creep upwards was checked by an electro-magnet between the carbons, which repels the flame so that it formed a downward curve. These lamps were only made in large sizes. They were never less than 1,000 candle-power, and were frequently two or three times as powerful. Arc lamps have been very generally replaced by the more powerful filament lamps which are now available.

A new arc lamp, Figs. 96–7, of novel construction, practical value, and great scientific interest was produced in the laboratories of the Edison and Swan Electric Co. in 1915. It is enclosed in a glass bulb which contains an inert gas at a pressure of about five inches of mercury, and is furnished with three terminals. Two of the terminals lead to a spiral of tungsten wire, called the atomizer, and the third to a bead of tungsten on the end of a wire of such a length that the bead is close to the spiral. A resistance box is used with the lamp and the connections are so arranged that when the current is switched on it passes through the spiral and makes it red-hot. The switch is then thrown over so that the current passes between the bead and the spiral when the bead immediately glows with a brilliant white light. In the 100 candle-power lamp the bead is only 2 millimetres in diameter, so that it forms an admirable point-source (hence the name " pointolite " lamp) for experiments in optics.

The scientific interest of the lamp lies in the fact that heating causes an emission of " electrons " (which are explained in Chapter XX) and the conductivity of the gas in the globe is so increased that an arc between the bead and the spiral is possible. This is of interest again in connection with the thermionic valves used in wireless telegraphy and telephony. In order to render it more effective, the wire of the spiral is coated with certain metallic oxides which emit electrons freely when heated.

In the 500 candle-power lamp, Fig. 97, there are four terminals

[1] *Science Progress.*

and a square plate as well as a bead of tungsten. There are here three stages. First the atomizer is heated, next the arc is formed between the spiral and the bead, finally the arc is produced between the bead and the disc. The difficulties of manufacture in stepping from 100 to 500 candle-power required a year to overcome them. But in 1919 lamps of 1,000 candle-power were being made, and one of more than 4,000 candle-power had been constructed. The 1,000 candle-power lamp has a globe only six inches in diameter. A voltage of at least 50 is required, though lamps requiring only 30 volts are made. The efficiency of the small size is 0·65 watts per candle, but the larger ones take less. The brilliancy of the bead in the 100 candle-power lamp is 12,000 candle-power per square inch, and 16,000 candle-power per square inch in the 500 candle-power, but one of the former has been run up to 60,000 candle-power per square inch, at which point the globule is practically molten.

INCANDESCENT ELECTRIC LAMPS

The first practical incandescent lamp was the successor of numerous attempts to produce light by passing a current through a fine metal wire. De Moleyn suggested enclosing the wire in a glass globe from which all the air had been exhausted, and an American named Starr co-operated with an Englishman named King to construct a lamp with a slender rod of carbon. The final success was achieved by Thomas Alva Edison and James Wilson Swan, of Newcastle—another fertile English and American combination.

The filament or fine thread was originally a fibre of bamboo which was carbonized by heating in a closed vessel with charcoal. This was cemented to the ends of two pieces of platinum wire which were fused in one end of the glass globe, and served to convey the current to and from the filament. After these wires are sealed in, the globe is exhausted by connecting the other end to an air-pump. When the required degree of vacuum has been obtained the bulb is sealed up. The use of platinum for leading-in wires is based upon the fact that it has the same rate of expansion as glass, and the joint will not, therefore, crack on cooling.

The filaments of modern carbon lamps are not bamboo. Cotton-wool is made into a paste with zinc chloride, which

FIG. 96.—THE POINTOLITE LAMP, 100 CANDLE-POWER.

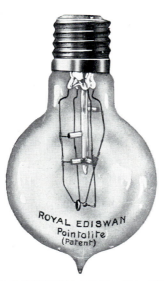

FIG. 97.—THE POINTOLITE LAMP, 500 CANDLE-POWER.

To face page 110.

FIG. 99.—THE GAS-FILLED
LAMP.

FIG. 101.—THE QUARTZ MER-
CURY LAMP.

FIG. 100.—THE MERCURY VAPOUR LAMP.

To face page 111.

dissolves it, and is forced through a fine hole into a mixture of alcohol and hydrochloric acid. The liquid jet is thus converted into a tough thread, which is dried, cut into suitable lengths, and wound upon carbon blocks of the shape it is intended to assume when complete. The blocks are packed in powdered charcoal and heated to a high temperature, which makes the filaments hard, black, and shining. They are joined to the platinum leads by holding the filament and wires in contact and dipping in benzine while a current is passed. The rise in temperature at the bad contacts causes decomposition of the benzine, and the deposition of carbon round the joint. The filament is then made uniform by heating in an atmosphere of benzine vapour. The thinner portions become overheated and carbon is again deposited. The rest of the process has been described in connection with bamboo filaments.

The Ediswan and similar lamps held the field for nearly twenty years, but during the past twenty-five some formidable rivals have appeared. The first of these was the Nernst lamp, introduced in 1898. It consists of a thin rod of the same metallic oxides as are employed in the incandescent gas mantle, but includes complications not met with in any other type of incandescent electric lamp. The rod offers a very high resistance in the cold, and the current when it enters the lamp passes first through a platinum wire spiral which is coiled round it. The heat from this raises the temperature sufficiently to cause the rod to glow. But as the temperature rises, the resistance falls considerably and less current is required. The excess is disposed of by connecting a piece of iron wire to the two ends of the rod, through which the unnecessary current is shunted. This lamp has now been superseded.

The year 1905 saw a revival of metal filament lamps, for which platinum had been found unsuitable, owing to its low melting-point twenty years before. The first was the Osmium lamp, but the wire is so brittle at ordinary temperatures that it was soon replaced by an alloy of osmium and tungsten, called osram. This wire, again, is not flexible, and is made in short horseshoe-shaped threads, which are joined end to end in series. About the same time Siemens and Halske brought out the Tantalum lamp, and this was from the first a genuine success. As an instance of the unrecorded tragedy which often lies behind discovery and invention, the writer may mention that he knew a

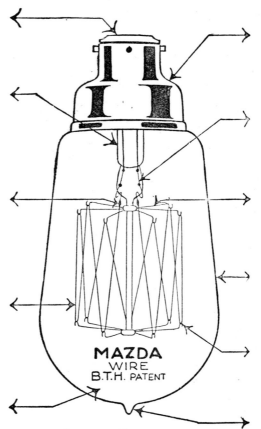

Fig. 98. The Points of Excellence in a Glow Lamp.

1. Vitrite glass button forming moisture-proof cap.
2. Accurately sized, machine-made "stem" and "mount" for filament. Ensures accurate centring af filament in bulb and provides a strong support.
3. Sleeve joint between filament and leading-in wires. Note:—Only *two joints*, not ten or twelve as with the old pressed filament. Joints are not rigid, thereby avoiding breakage at these points.
4. Drawn-wire filament of one continuous length of drawn tungsten wire. Ensures uniformity of filament and superior strength and durability.
5 First-class vacuum uniformly effected by the chemical method of exhaust. Reduces blackening—maintains candle-power—ensures reliability by prevention of arcs or explosions.
6. Uniform and strong cap symmetrically centred and firmly fitted on bulb. Secured with special waterproof cement, which prevents loose caps.
7. Platinum wires securely welded to leading-in wires with weld knot clamped into the glass, relieving the platinum from any strain.
8. Spider supports of most suitable material—provide flexibility for filament at turns. A great improvement over the old rigid supports.
9. Uniform bulb blown in moulds and gauged to exact and uniform size, carefully selected free from flaws and blemishes.
10. Well-rounded loops at the turns of the filament. Avoids the old rigid and sharp joints and the tendency to breakage and short-circuiting of the filament.
11. Small, strong, securely made tip—reduces tip breakage, which destroys many lamps.

prospector who was rendered a helpless cripple by rheumatism acquired on a journey which resulted in a rich find of tantalum ore just before its value for incandescent lamps was proved.

A disadvantage of the Tantalum lamp is the great length of wire which is necessary in order to offer sufficient resistance to the current, and numerous efforts were made by lamp manufacturers to discover another material. The method of obtaining wire is to draw a thin rod through a series of conical holes of gradually decreasing size in a hard plate, or through similar holes in diamonds. The high melting point of tungsten was in its favour, but the difficulty was to obtain a drawn wire sufficiently thin and flexible to permit of a considerable length being coiled inside a small globe. But, as Moissan had stated, as a result of his researches with the electric furnace in 1892 (see Chapter IX) that malleable tungsten could be obtained, persistent efforts were made in that direction, and have been crowned with success.

According to an account in the *Electrician* of 26th August, 1913, Messrs. Siemens and Halske first tried an alloy of nickel and tungsten containing 6 per cent of the former metal. This was rolled into a rod 1 mm. in diameter, then drawn through steel and diamond dies, and the nickel distilled off in a vacuum. Later they were successful by a process described in their English Patent of 1907. Compressed blocks of tungsten powder were raised to a white heat in a vacuum, when the particles " sintered " together by semi-fusion.

A year earlier, in 1906, the General Electric Company of America took out a patent under which tungsten powder, tungsten oxide, and glucose were compressed and squirted through a hole into the form of rods 5 mm. in diameter and 20 mm. long. These were heated to 1,000° C. in a vacuum to decompose the glucose and oxide, and then to a point just below the temperature of fusion of the metal. The rods were then rolled and drawn white hot, the heat for the latter process being supplied by an electric current passing diametrically through the wire at the die. To-day the filament is made of a tungsten wire drawn through a diamond die, and it is suspended on molybdenum supports. The old carbon filament lamp has now disappeared.

About 600,000,000 filament lamps are now made annually, and the process of manufacture has been enormously improved. They are shaped by machinery from glass tubing. This tubing is drawn continuously night and day and the General Electric Company alone produce about 10,000 miles a year. The exhaustion has to be very thorough to remove oxygen and water vapour. It is accomplished by a pump of the Gaede type (see p. 396), and the objectionable gases are removed by coating the filament with phosphorus. When the filament is heated this volatilizes, and the excess condenses on the glass, where it forms an imperceptible film. As a further example of the number of details which are considered in the design and manufacture of such a familiar object, Fig. 98, which embodies the claims made by the British Thomson-Houston Company in respect of their Mazda lamp, may be studied.

A new type of lamp which has carried the evolution of lighting a stage further, was placed on the market about fourteen years ago. It is known as the "half-watt" lamp,[1] and is a metal filament lamp which invades the field previously commanded by the arc. The filament is of tungsten and is closely wound. The globe is filled with nitrogen at about two-thirds the pressure of the atmosphere. Usually if a metal filament lamp is not sufficiently exhausted, it becomes very hot when in use ; but this defect has been overcome in the new lamp by making the globe unusually large and keeping the closely wound filament as low as possible (Fig. 99). One lamp of this type of 60,000 candle-power has been constructed. These have replaced arc lamps for outdoor lighting and the illumination of large buildings for, apart from first cost, they effect an enormous reduction of labour in maintenance.

The effectiveness of filament lamps for different purposes varies with the form of the incandescent thread. For use in a lantern a light of the highest possible intensity, concentrated in the smallest possible area, is required. The Nernst lamp came very near to this ideal because the glow was distributed along a fairly short, thick line. In long filament lamps the wire is sometimes made with the coils very close together to secure a suitable effect. But for the ordinary lighting of rooms the light should be distributed as much as possible. The ordinary metal filament lamps suffer from the fact that the filaments are end on to the lower part of the bulb ; they are excellent for side lighting, but if used for top lighting should be provided with a globe to

[1] The term half-watt expresses hopes which have not been realized in small sizes and the term gas-filled lamp is now generally used.

FIG. 103.—NEW TYPE ELECTRIC FIRE.

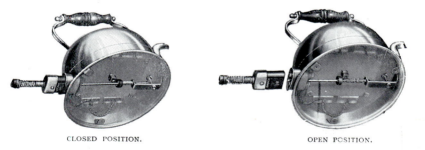

CLOSED POSITION.   OPEN POSITION.

FIG. 104.—AUTOMATIC SAFETY-KETTLE.

To face page 114.

FIG. 105.—AN ELECTRIC BOILING PLATE : CONSTRUCTIO[?] WHEN CONVERTED.

FIG. 105A.—AN ELECTRIC BOILING PLATE : CONSTRUCTIO[?] WITH RESISTANCE SPIRA[?] ENCLOSED IN POSITION.

FIG. 105B.—AN ELECTRIC BOILING PLATE COMPLETE.

To face page 115.

distribute the light. While all globes cut off some light, they serve a useful purpose in destroying the glare which results from an intense light concentrated in a thin wire. The ideal method of lighting is by diffusion from the walls and ceiling of the room, and this can be achieved most successfully by the use of gas-filled lamps.

VACUUM-TUBE LIGHTING

An interesting lamp which has been introduced in recent years, is the Cooper-Hewitt Mercury Vapour lamp. If a high-tension current of electricity is passed through a partially exhausted glass tube, the air or other gas within glows with a soft light, the colour depending on the gas or gases present. In the Cooper-Hewitt lamp, a tube has an iron electrode at one end and mercury in a small reservoir at the other. The current is switched on and the tube tipped so that a stream of mercury reaches from the reservoir to the iron electrode. When the level of the tube is restored the mercury flows back, the circuit is broken, and the arc which is formed is immediately converted into a greenish glow which fills the whole tube. Two forms are made, one in which the tipping of the tube is effected automatically as soon as the current is switched on, and the other in which the tipping is effected by hand. The former is illustrated in Fig. 100. The efficiency of this lamp has recently been increased by the use of quartz instead of glass for the tube. Quartz is rock crystal—hard, strong, and requiring the oxyhydrogen blowpipe to fuse it. The amount of light produced is greater as the temperature and pressure of the mercury vapour is increased, and higher temperatures and pressures are produced than would be possible in glass tubes. As quartz is extremely transparent to ultra-violet rays, a much greater proportion of these are emitted, and though this is not good for the eye it is admirable for photographic work. On the other hand, the light is whiter, has less of the greenish colour which is so characteristic when glass tubes are used and the lamp itself is shorter (see Fig. 101).

A special manufacturing difficulty arose in fixing the electrodes. For this purpose it is necessary to use a material which expands on heating at the same rate as the material of the tube. For glass tubes platinum possesses the requisite property, and the iron electrode was attached to the platinum before it was fused in. But quartz expands at only one-twentieth of the rate of platinum, and to have employed this metal would have resulted in fracture of the tube on every occasion.

The problem was solved by the use of " invar ", an alloy of steel and nickel, discovered by M. Guillaume, which has a rate of expansion very little greater than that of quartz. Unfortunately invar undergoes an alteration of properties at a red heat, and cannot therefore be fused in. The method adopted is to grind a tapered rod of invar into a conical hole in the tube and to fix this in with cement.

The most beautiful and effective system of lighting, however, is that devised by Mr. Moore, an American. A high-tension current of electricity is passed through a tube containing a gas at low pressure, and the whole tube is filled with a glow, the colour of which depends upon the nature of the gas. With air the colour is rosy red, with nitrogen yellowish red or golden, with carbon dioxide it is white, and with neon a brilliant orange red. The tube is $1\frac{3}{4}$ inches or more in diameter, and it may be of any length up to 200 or 300 feet. An excellent example is to be seen in the escalator tunnels at the Liverpool Street Station of the Central London Railway, to the manager of which the author is indebted for particulars. The tube is fed by a 3-phase alternating current of 17,500 volts, and the total length is 274·8 feet. The candle-power is stated to be 55 per yard, and the consumption is from 1·3 to 1·7 watts per candle.

For internal lighting the tube is arranged on the ceiling and gives an illumination which is admirably distributed. With an ordinary source of light the intensity varies inversely as the square of the distance, becoming one-quarter at twice the distance, and one-ninth at three times the distance. In the case of the Moore light, however, the intensity varies inversely as the simple distance, being one-half at twice and one-third at three times the distance from the tube. The greatest disadvantage is the high voltage required—not less than 5,000 as a rule, and this at present effectively prevents its adoption except in certain circumstances.

The contrivance is not quite so simple as it appears at first sight. In addition to a transformer to give the high-tension current from the low-tension supply there is a valve for automatically adjusting the vacuum. This is a very ingenious device, and is rendered necessary by the fact that the discharge tends to increase the vacuum and a small quantity of air must

FIG. 106.—AN ELECTRIC COOKER.

To face page 116.

FIG. 106A.—AN ELECTRIC BAKING OVEN : SHOWING INTERNAL ARRANGEMENT.

To face page 117.

be admitted. A narrow branch tube from the lighting tube, see Fig. 102, is bent twice at right angles so that the open end is upwards and surrounded by a small bath of mercury. The end of this branch is closed by a carbon plug which is sufficiently porous to allow air to flow through it, but is sealed when the plug is covered with mercury. The alteration of the level of the latter is effected by raising or lowering a glass tube, the low end of which dips into it. The upper end of this tube contains a bundle of iron wires and is surrounded by a coil of wire so connected that as resistance of the gas in the main tube varies, the tube is sucked out of or pushed into the mercury. For about one second in every sixty the tube breathes in this way, and the proper degree of vacuum for the greatest efficiency is obtained.

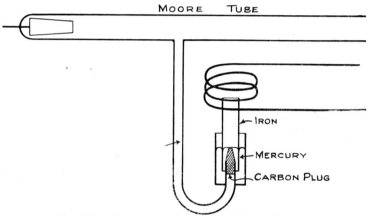

Fig. 102. REGULATING DEVICE FOR MOORE TUBE.

In few matters is the steady growth in efficiency so strikingly shown as in the improvements in electric lighting during the last twenty years. The old arc lamp with solid carbons required, as we have seen, about 1½ watts for each candle-power and the enclosed arc still requires 2·3. Even allowing for the cost of labour and renewal of carbons, this was able to compete with gas for outdoor lighting where a powerful illuminant was required. The yellow flame arc takes only 0·4 watt per candle, and is therefore three and a half times as efficient, and the red and white flame arcs are twice as efficient as the older lamp.

The old carbon filament required about 4 watts per candle, and was much more expensive than gas. But the Osram lamp, taking about 1·3 watts, brought domestic lighting by electricity below the cost of gas for the first time, and this has been still further reduced by the gas-filled lamp. The cheapest light of all, however, is that from the Westinghouse Cooper-Hewitt Mercury Vapour lamp in a quartz tube, which only requires about 0·2 watt per candle.

If a watt could be converted wholly into luminous energy it would produce no less than 56 candle-power. Against this the quartz mercury vapour lamp can only show an efficiency of 10 per cent, and the most effective type of filament lamp an efficiency of 5 per cent. Most forms of lamp give not only light, but heat ; what is needed is a relatively cold light, but so far no method of making a solid substance glow without heating it (except very faintly as described in Chapter XIX) has yet been discovered.

In a vacuum tube the particles are caused to glow at a relatively low temperature, and if some method could be devised of so exciting the atoms (or electrons) that the bulk of their energy gave rise to luminous waves, the problem would be nearer solution. The Moore light and mercury vapour lamp are the nearest approaches to the desired end, but both have disadvantages for domestic lighting.

In this connection it is perhaps worth while recalling some experiments performed by Nikola Tesla about 1892. By the use of current alternating with extreme rapidity he electrified the space between two metal plates some distance apart, so that vacuum tubes placed between them glowed brightly. These experiments were referred to in a humorous speech by the late Professor W. E. Ayrton, in which he made a suggestion to the effect that street lamps might in future be abolished and a vacuum-tube walking-stick serve to light the way. Continuing in the spirit of prophecy, he looked forward to the time when fires and smoke would disappear ; when a man would be able to bask in the rays of the electric field, recline on the graceful curve of an equipotential surface, and rest his feet upon a fender composed of horizontal lines of force.

ELECTRIC HEATING

When a current of electricity passes through a conductor heat is produced, and the greater the resistance offered to the passage of the current the greater is the proportion of the electricity which is converted into heat. The arc and filament lamps, which have already been described, are illustrations of this fact, though in those cases the heat is desired only in order to raise bodies to the temperature at which they produce the greatest amount of light, and the heat formed at the same time is so much waste.

But while the problem of obtaining a greater amount of light from a given amount of electrical energy has so far proved a matter of difficulty, there is no trouble in converting a large quantity of electricity into heat. Moreover, there is an absence of smell, smoke, and ash inseparable from coal and almost inseparable from oil, together with a possibility of regulation and adjustment that renders heating by electricity of particular value. For as the heat produced in any part of the electric circuit is proportioned to the square of the current, the resistance, and the time, the mere movement of a switch which throws extra resistances into or out of the circuit, will regulate the temperature to a nicety impossible with flame.

Attention will be drawn to the use of the electric arc in welding, in Chapter VIII, and the whole of Chapter IX is devoted to the great range of manufacturing processes in which the electric furnace is now employed. Consideration will therefore be confined in this section to some of the domestic applications.

The material in which the heat is produced may either be a thin wire or strip of metal having a high resistance, or a fabric composed of metal and asbestos, or a thin metallic film deposited upon a strip of mica, or rods of carbon which have been coated with a preparation that prevents oxidation. Of the many types available for heating rooms we have selected one manufactured and recently put on the market by Messrs. Ferranti, Ltd., see Fig. 103. It consists of a rod of " globar ", a specially prepared carborundum, mounted in the axis of a parabolic trough of accurate form and with a mirror-like surface. This surface is chromium-plated, and the parabola reflects the beat upwards and downwards, while the length of the rod, about 18 inches, secures a wide angle of radiation horizontally. The hot rod is

of very small dimensions, compared with the reflector, and
68 per cent of the heat produced is radiated into the room
It is sent out in various forms, some of which are portable,
so that they can be moved to different parts of the room. The
one we have chosen to illustrate has a tile setting, which gives
it some similarity to a fireplace, but cleaner and neater looking
than fireplaces designed to burn solid fuel.

Some of the most ingenious electrical heating devices are
intended for use in the kitchen. Fig. 104 shows the Metro-Vick
automatic safety-kettle. If an electric kettle boils dry the heater
becomes too hot and may burn itself out. This involves the
expense of a new heating element. But in the kettle illustrated the
electricity is disconnected by ejection of the heater. This action
depends upon the fact that the bottom of the kettle is domed,
and the heater is placed just below it. When the water boils
away beyond a certain level, the temperature of the dry centre
of the dome rises, expands upwards, and releases a spring plunger,
which forces the connector from its socket. By means of an
additional attachment, consisting of a mercury switch which
cuts off part of the current when tilted, this kettle can be so
arranged that it will not boil over.

The production of heating appliances for use in the kitchen has
called for a considerable amount of experiment. The conductor
in which the heat is produced should be protected from liquids
or solids that may fall upon the stove or hot-plate. It must
be adequately insulated so that there is no possibility of the cook
receiving a shock. The heat must pass readily from the hot
wire, ribbon, or rod to the vessel or food which is to be heated or
cooked. How these conditions are secured is well illustrated by
the construction of the Cosmos type of boiling-plate (Figs.
104–5). This consists of a cast-iron disc with a spiral groove
on the under surface. A flexible magnesium " tube ", through
which passes a nickel-chromium wire spiral, is laid in the groove.
When this is treated with steam in a closed vessel at a temperature
not exceeding 300° C. the magnesium is converted into crystalline
magnesium hydroxide, which swells out and fills the groove.
The magnesium hydroxide is a sufficiently good electrical
insulator to prevent an electric current in the spiral escaping
to the iron plate, and yet a sufficiently good conductor of heat
to allow the heat to flow through from the spiral. The space
inside the heating coil is filled with alundum cement, the groove

is covered with an asbestos plate, and that again by a cast-iron plate which forms the under-surface of the heater.

A plate of this kind, starting all cold, will raise 2 pints of water to the boiling-point in $8\frac{1}{2}$ minutes, or 5 pints through the same range of temperature in $14\frac{3}{4}$ minutes. When the plate has once become hot 4 pints of water can be boiled in less than 8 minutes.

Heating devices modified according to the position they are to occupy and the purpose they are to serve, are used in the electric cooker shown in Fig. 106. One great advantage claimed for a cooker of this kind is that as there are no products of combustion in the oven, very little air circulates through it, and the loss of weight of the food is very small. By other methods of cooking this loss lies between 25 and $33\frac{1}{2}$ per cent, whereas in the electric oven the loss is never more than 10 per cent. An ordinary lunch or dinner for four to six persons can be cooked for an average cost of $3\frac{1}{2}d$. when the cost of electrical energy is a penny per unit.

A larger oven, used by bakers and confectioners, is shown in Fig. 106B. The loaves or confectionery are placed on trays which hang from chains passing over toothed wheels. The wheels turn at such a rate that a tray passes the oven door, on the right of the illustration, once each minute. This oven has six trays. With eight trays each 73 inches long by 17 inches wide the baking space is 70 square feet. There are no hot spots where the cakes are blackened or cold spots from which they emerge pale and anæmic. Every cake passes through hot and cold spots in turn, so that each one is cooked to the same attractive tint. For every £1 worth of small goods baked the cost of the electricity in a particular case worked out at 2·35 pence.

Among the numerous domestic heating appliances to be seen to-day in the shop windows are electrically heated irons for the laundry. This, again, is a case where careful regulation of the temperature is desirable to avoid scorching. Thousands of people to-day use an electric toaster on the breakfast-table, and find that the pleasure of really hot fresh toast is well worth the sixteenth of a penny per slice which the process costs them. If this method were general—and it might easily become so— the miniature trident fork will one day occupy an honoured place in a museum of antiquities

On a larger scale are the electric or radiant heat baths which are so beneficial for rheumatism in the joints. The patient lies on a padded couch and is covered with a padded lid, with only his head and face visible. At various points this shell has openings in which are fitted electric radiators, which pour their genial rays upon the distressed limbs and cause that copious perspiration which eases the pain and lessens the stiffness. Such baths are included in the electrical equipment of the great White Star liner *Olympic*. Down in the engine-room the engines are throbbing with steam raised by coal, and the radiations which they are sending to the patient are the sunbeams which fell upon the earth in past ages, have been stored up for countless years, and are now liberated for his benefit.

We cannot close this chapter without emphasizing again the fact that the final achievement of applied science is cheapness. The more efficient metal filament lamp, if used merely to replace the carbon lamps, would have hit the electric supply companies hard. But, as a matter of fact, cheapness increases the number of consumers, and the whole effect of scientific discovery and invention is to enlarge the comforts and conveniences of a greater number of people. At first, some commodity may be scarce and expensive. Then the engineer, the chemist, the scientific manufacturer bring their minds to bear upon its production, and from a luxury to be enjoyed only by the rich it becomes almost a necessity within reach of the very poor. In this way fine linen and silk, lace, many kinds of food, the electric light, comfort and speed of travel, and a host of other results of invention are enjoyed by people who, in the absence of discoverers and inventors, would have regarded them with hopeless longing. And if this progress has not lightened burdens, nor lessened misery and want, it is not the fault of scientific man, but the cussedness of human nature and the failure of people to realize the trend and meaning of the age in which they live.

CHAPTER VII

It is a trite saying that we live in a mechanical age. Every operation ordinarily performed by man that can be carried out by a machine, is handed over to the care of whirling wheels, rocking levers, and rolling teeth. The numerous electric laundries, machine bakeries, and penny-in-the-slot machines add their evidence to the bicycle and the motor-car. And the applications of machinery in manufacture and daily life are becoming so numerous that the smaller steps in progress escape observation. For this is an age less of crude and obvious progress than of delicate refinements. The newspapers frequently announce that an increase in the price of raw material has prevented some company paying a dividend, or has raised the price of some manufactured article. But it is rare to see the announcement that an increased dividend or a decreased price is due to some small piece of ingenuity, some secret wrested from Nature, or some trick performed with Nature's laws.

While it would be impossible in a book of this size to notice a tithe of the contrivances by which time and labour are saved, and greater accuracy secured, in modern workshops, it would be equally undesirable to ignore altogether the general progress which has been made during the last thirty years. But it will be clear that the examples must be chosen because of the generality of their application, the striking character of the scientific principles involved or the results achieved, or the extent to which they represent the magnitude and power of human effort.

Let us therefore consider first some ways in which time and energy are economized in the workshop.

THE TRANSMISSION OF POWER

Most workshops are equipped with long lines of overhead shafting from which the machines below are driven by belts and pulleys. This has to rotate continuously whether one or fifty machines are working : the power required to drive it and the wear and tear are practically the same for one machine as for all. The belts require attention, and if one breaks the machine is idle. Should the main belt give way the whole of the work comes to a standstill. This applies not only to engineering work-shops, but also to all factories where machinery is employed. For instance, in the textile factories it has hitherto been the custom to effect the main drive from the engine by ropes working in grooved pulleys, and to drive the machinery from the main shafting by leather belting. A glance into many shops reveals an overhead mass of whirling wheels and a veritable forest of belts.

Fig. 107. A Ring Oiling Bearing.

The practice is rapidly gaining ground of using electrical-power, and driving each machine by an independent electro-motor. When the machine is not working no current is used, and at any time only so much is consumed as is necessary for the work in hand. The cumbrous method of altering the speed of a machine by a belt and stepped or cone pulleys is then unnecessary, the mere adjustment of a lever being sufficient to alter the speed. During the past ten years or so the textile factories of Lancashire, which had hitherto held aloof, have begun to adopt the method and thousands of small electric motors of from $\frac{1}{2}$ to 1 horse-power have been installed.

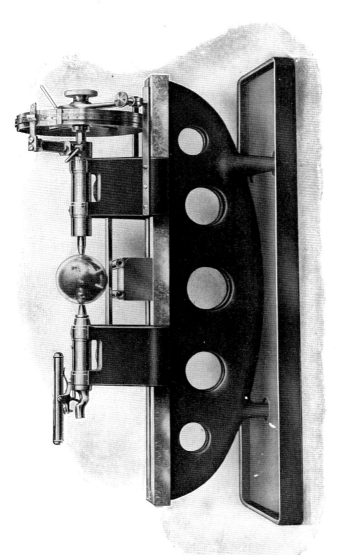

FIG. 108.—NEWALL MEASURING MACHINE FOR TESTING ACCURACY OF BALLS.

To face page 124.

FIG. 109.—BALL-BEARING FOR OVERHEAD SHAFTING.

To face page 125.

There is a manifest advantage in a large works where the various shops are spread over a wide area. For obvious reasons it is desirable to keep all boilers as near together as possible, and where steam-engines are used the concentration of the boilers leads to great waste of heat in the pipes conveying the steam to the engines. In a scattered works of this kind gas-engines can be used with advantage, but the better plan is to use steam or gas-engines to drive dynamos in a central power station, and to distribute the electricity along slender wires to the various departments.

Another source of expense in connection with overhead shafting is the maintenance and repair of the bearings. The older types required frequent adjustment and renewal, and had slung below them unsightly drip tins to catch the oil which leaked from the ends. The modern shaft bearing, of which one example is shown in Fig. 107, is not only constructed of better materials, but is capable of closer adaptation to the line of shafting. It is adjustable in a vertical direction to allow for differences in level or settling of its supports, and its horizontal direction can be altered to coincide with the axis of the shaft. Moreover, it is fitted with an automatic arrangement known as a ring oiler, which not only keeps the shaft lubricated, but prevents drip and loss of oil. This consists of a ring somewhat larger than the shaft—say, half an inch—the lower portion of which dips in oil contained in a circular trough formed on the ends of the bearing. The ring rotates with the shaft, carrying with it a film of oil, part of which is deposited on the upper side. This film spreads over the surface of the shaft and keeps the surface lubricated.

Examination of a bearing for an engine or heavy machine will show that it consists of two parts—a cast-iron frame and a brass or gun-metal bush, which is cut in half so that it can be adjusted for wear. In many cases the bush will be found to have wide grooves cut along its inner surface in the direction of the axis, and filled with a white metal. This white metal is poured in in molten condition when the shaft is in place, and the brasses adjusted so as to clasp it loosely.

A little consideration will show that it is extremely difficult to get a perfect fit between a heavy bearing and shaft, and any departure from true alignment will lead to excessive friction and wear at certain points. The white metal is one of many alloys on the market, called anti-friction metals. A typical antifriction metal, etched and examined with the help of a microscope, will be found to consist of hard crystals embedded in a softer matrix. These hard portions resist wear, while their soft bedding enables the metal to adjust itself to pressure. By casting it in grooves after the shaft is in position it soon adjusts itself to the surface so that the pressure and wear are evenly distributed. Moreover, it is at all times easily replaced at far less cost than would be required to replace worn brasses.

None of the bearings described overcome the difficulty that however well they may be lubricated the rubbing absorbs a considerable amount of energy. As the rolling of two surfaces over one another is very much easier than sliding, bearings are made which consist of a ring of case-hardened steel rollers mounted in a circular frame surrounding the shaft. These offer an extraordinarily small resistance to pure rotation, but if there is any end movement this involves sliding and its attendant disadvantages. The most flexible bearing, however, is one consisting of one or more rings of steel balls running in a groove or race in the body of the bearing. Ball bearings have long been used for bicycles, and they found early application in such machines as required high speeds for small loads, or slow speeds and heavy loads. During the last few years, however, they have been applied to all kinds of light and heavy machinery at all speeds. The Hoffmann Manufacturing Company of Chelmsford make the balls of case-hardened steel from $\frac{1}{16}$ inch to 3 inches diameter, and in any one size the balls do not differ from perfect spheres or from one another by $\frac{1}{10000}$ of an inch. The works run night and day throughout the year and produce over a million balls every twenty-four hours. The machine in which they are gauged before being sent out is shown in Fig. 108.

A beautiful example of a hanging bracket made by this company is shown in Fig. 109. This is capable of adjustment vertically, and the single ring of balls allows of adaptation to the line of shafting. The bearings can swing slightly, but this can be prevented when necessary by the set-screws shown at the right hand of the elevation. In a series of tests conducted at the National Physical Laboratory one of these bearings was compared with two others—one an ordinary hanger with needle lubricator, and the other fitted with ring oiler. The results are set out in the table given below.

CAST PHOSPHOR BRONZE.
PRESSED STEEL.

FIG. 110.—BALL CAGES FOR SKEFKO BEARING.

FIG. 111.—LARGE BALL-BEARINGS FOR TAKING END THRUSTS.

To face page 126.

FIG. 112.—RENOLD ROLLER AND BLOCK CHAINS.

FIG. 113.—RENOLD SILENT CHAIN.

To face page 127.

	Standard Plain Hanger.		"Hoffman" Ball Bearing Hanger (Patent).
	With Needle Lubricator.	With Oil Ring Lubricator.	
Starting Effort .	830 in lbs.	770 in lbs.	9·9 in lbs.

Revs. per minute.	Coefficient of Friction.		
80	·015	·015	·0013
130	·014	·014	·0014
250	·016	·013	·0015
500	·017	·012	·0016

Another interesting form of ball bearing is made by the Skeffko Ball Bearing Company, of Luton. The balls are of steel hardened throughout and are contained in a cage (Fig. 110) of pressed steel or phosphor bronze so shaped that two rows of balls occupy very little more space than one row. A very large bearing arranged to take up an end thrust in a shaft 18 inches in diameter is illustrated in Fig. 111. Comparison with the man standing behind will give some idea of the size. Both types of ball bearing illustrated are employed on agricultural and textile machinery, paper-making, steam and gas-engine governors, petrol motors, small marine propellers, motor and other wagons, tramway axles, electrical machines, and almost every type of machine that is made, with an average saving of nearly 20 per cent in the power required.

In transmitting motion to a machine, or from one part of a machine to another, toothed wheels are frequently employed. The wheels may be made of cast iron or steel, and the teeth cast or cut in the same material, or made of wood, raw hide, or other material which reduces noise and shock. The aim in designing wheel teeth is to secure rolling between the teeth in contact, and there are several beautiful devices for shaping the surfaces. As a mechanism, toothed wheels are older far than the steam-engine, so we shall say nothing further about them here beyond the remark that they have recently come into use for ship propulsion. Attention is drawn elsewhere in the book to the fact that the Hon. (now Sir) C. A. Parsons succeeded in cutting gearing that transmits 98 per cent of the power supplied to it, and was very nearly noiseless in action. The method by which a relative absence of noise has been secured is interesting.

The method of cutting the teeth was to fix the blank wheel to a table which was rotated by a worm. As the table rotates, a cutter carves out the spaces between the teeth. Any small error in the machine was found to recur at regular intervals, so that it accumulated at certain parts of the wheel being formed. The Hon. (now Sir) C. A. Parsons overcame this difficulty by fixing the blank wheel upon a second table which had a " creep " of about 1 per cent over the first. The main table was then rotated about 1 per cent slower and the inaccuracies inherent in the gearing of the machine were distributed evenly over the new wheel.

It will be clear that in the case of belt driving, if there is to be much difference between the speeds of two shafts, a considerable distance between them is absolutely necessary, or such a small portion of the rim of the smaller pulley will be gripped by the belt that much slipping will occur. On the other hand, toothed wheels become unnecessarily large when the shafts are far apart, and they are liable to be noisy. The method first used extensively on bicycles, in which a chain passes over two toothed wheels or sprockets, is much more elastic in regard to distance, is free from any possibility of slip, and can be made to work at least as silently as any other device for transmitting power. While chains are still used in enormous quantities for cycles and motor-driven vehicles, they are rapidly gaining ground in workshops and factories, not only for small, but for large powers. Twenty years ago chains to transmit 50 horse-power were rare ; to-day they are made to transmit 500. They are used to communicate power from motors to overhead shafting, and from overhead shafting to machines of all kinds—lathes, drilling, planing, and shaping machines, drop hammers, textile, wood-working, and printing machines, pumps and blowers. Moreover, they are used to transmit motion from one part of the machine to another, for regulating the feed, and driving the pump which supplies lubricant to the cutting tool.

The type selected for description is that made by Hans Renold, Limited, of Manchester, who consider that by the use of chains instead of belts in their own workshop they save more than £600 a year. Among the advantages which are claimed for this, in common with other makes, are absence of slip and more regular feed, saving in power and in wages of attendant, greater output, less wear and tear on machines and tools, a saving of space, less noise than toothed gearing, and longer life than belting under unfavourable conditions.

The three types made are shown in Figs. 112 and 113, while Fig. 114 shows how the silent chain engages the teeth of the sprocket wheel. The silent chain will run at a speed of 1,250 feet per minute, the roller chain at from 400 to 900 feet per minute,

and the block chain at from 200 to 500 feet per minute, but these speeds are frequently exceeded. The chains are made to transmit from $\frac{1}{4}$ horse-power to 500 horse-power in the first case and 100 horse-power in the other two.

The extent to which this method of driving machinery is increasing will be understood from the statement that apart from those intended for cycles and motor-driven vehicles, this one firm sold sufficient to transmit over 40,000 horse-power in 1909 and over 50,000 horse-power in the first *six months* of 1912.

Apart from actually transmitting power these chains are used for several other purposes, which it will be interesting to mention here. A special roller chain is employed to convey the table of a printing machine to and fro at

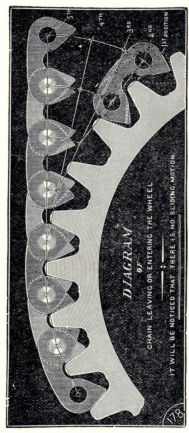

Fig. 114. Diagram to show Action of Silent Chain.

high speeds; another with specially shaped links to hold and carry type; a third with blocks having raised numbers between

K

the links for numbering articles ; and a fourth, with links so
constructed that it will bear compression, is used for ramming
home the shells in breech-loading artillery. But probably
two of the most interesting examples are illustrated in Figs 115
and 116. The first of these is a mortising machine and the
illustration shows how the chain, which carries cutters on every
link, is made to rotate round a frame, while it is pushed endwise
into the wood. The front portion of the block is removed to
show the shape of the mortise. Compared with the tedious
process involving the brace and bit, hammer and chisel, the
machine is marvellous in the speed and accuracy with which
it performs the operation.

The second figure shows the ingenious coal-cutting machine
invented by Mr. Austen Hopkinson. Here the problem is to
undercut the seam of coal so that it can be more easily removed
by blasting or the pick. It consists of a block chain passing
round two large sprocket wheels. The blocks are specially
designed to carry tool holders, and can easily be detached from
the chain for renewal. As the wheels revolve, the cutters rip
out the coal in the same way as the teeth of a saw.

There are perhaps a few cases in which the elasticity given by
a belt—and more particularly the ease with which it is thrown
off—render it more desirable than a chain drive. An engineer
of the writer's acquaintance tried a chain on a coal-breaking
machine, and found it so effective that when, as he put it, a
curbstone got amongst the coal, some damage was done.
Formerly such an occurrence merely threw the belt off and saved
the crusher from injury. Where chain drives are employed for
a pulsating load such as pumps, special spring sprockets are
used. These have a rim separate from the boss, held in place
by springs which allow of a little play between the two. There
is no doubt that under suitable conditions chain driving is a real
economy, saving power, increasing speed, and raising the output
of the machines.

An extremely interesting and effective means of transmitting
power with a variable speed which is entirely under control,
is illustrated in Figs. 117 and 118. It consists of two hydraulic
units, one acting as a pump and the other as a motor. They
may be in the same casing or some distance apart and connected
only by the pipe which conveys the fluid—generally oil—from
one to the other, while obviously the two shafts may be at any

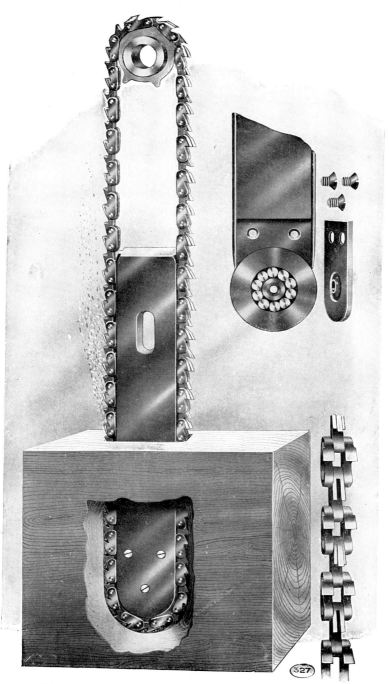

FIG. 115.—MORTISING MACHINE WITH CHAIN CUTTER.

To face page 130.

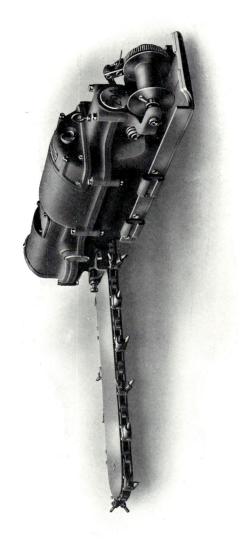

FIG. 116.—THE HOPKINSON COAL-CUTTING MACHINE.

To face page 131.

angle. Both pump and motor consist of a number of barrels
mounted round the shafts. Each barrel has a bucket piston and
a rod which presses at one end upon the bottom of the bucket and
at the other upon an inclined disc. The speed of the driven shaft
depends upon the rate at which oil is supplied to its motor. The
driving shaft rotates at constant speed but the rate at which oil
is pumped is varied by altering inclination of the disc, and thus
varying the stroke of the pistons. The speed of the driven shaft
may, therefore, vary from that of the driving shaft in one direction
down through zero to a similar speed in the reverse direction.

From Fig. 118 it will be observed that the pump and motor
are set back to back and that the centre valve-plate between them
admits oil from one to the other. Only a small quantity of oil
is required and that is used over and over again. On the extreme
left will be seen the arrangement for altering the inclination
of the disc and thus varying the speed. This and similar contrivances were largely used during the war for the elevation and
training of big guns. In times of peace it finds its greatest use
in the steering gear of ships ; for capstans, hoists, winches and
cranes ; for printing machinery and hydraulic presses. It has
been made to operate a swing bridge, to drive machine tools, and
for many other purposes.

While the foregoing devices for transmitting energy are
improvements on old and well-established methods, the one next
to be described is based on an entirely new principle. Everyone
knows that sound is propagated through a medium, solid, liquid,
or gas, as a wave-motion. Particles in the immediate neighbourhood of a sounding body are set in motion, and the motion
travels outwards in ever-widening circles like the waves formed
when a stone is thrown into a pond. Each particle in a wave
swings backwards and forwards about its original position, causing
alternate compressions and rarefactions. It is the motion and
not the particles, which reaches the drum of the ear and cause
the sensations which we recognize as speech, music, laughter,
or noise.

The speed of sound in any medium depends upon the density
of that medium. In air at the ordinary temperature it is about
1,200 feet a second ; in water it is nearly 5,000 feet a second, and
in elastic solids it is much more rapid. In the denser medium the
motion travels farther than in one less dense. Thus a faint
scratch on the end of rod of wood, or metal can be heard quite

easily by a person who places his ear at the other end. The distances at which sound can be heard under water led to the development of submarine signalling described in Chapter XVI and was proved of immense importance in the detection of submarines during the war.

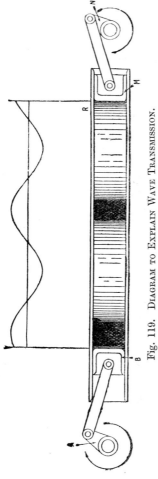

Fig. 119. Diagram to Explain Wave Transmission.

About fifteen years ago, M. Constantinesco, a Roumanian, discovered that if water was contained in a pipe, and a series of impulses, corresponding in frequency to waves of sound, was applied by means of a piston at one end, the motion could be communicated to a piston at the other. The arrangement is shown diagrammatically in Fig. 119, and though there are a number of details which have to be taken into consideration, the principle upon which the apparatus works will be clear. It was used during the war for timing the fire of machine guns mounted on aeroplanes, and firing between the blades of a propeller; and 30,000 of these appliances were made by W. H. Dorman & Co., of Stafford, for the British and Allied Forces before the Armistice. It is now being applied by the same firm to rock-drills, riveters, and other tools.

Fig. 120 shows an external view of a generator. There are two pistons which are driven from the shaft in the centre. The two spheres are known as capacity vessels and their purpose is to absorb surplus energy which, under certain circumstances, may be produced. The speed is such that 40 impulses per second or 2,400 per minute are given to the water, and these are trans-

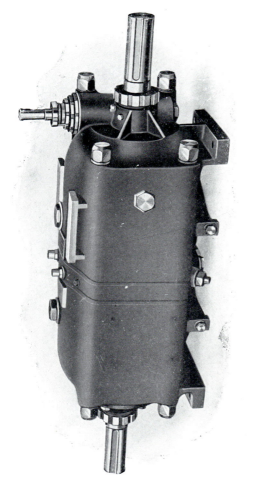

FIG. 117.—VARIABLE SPEED OIL TRANSMISSION GEAR.

To face page 132.

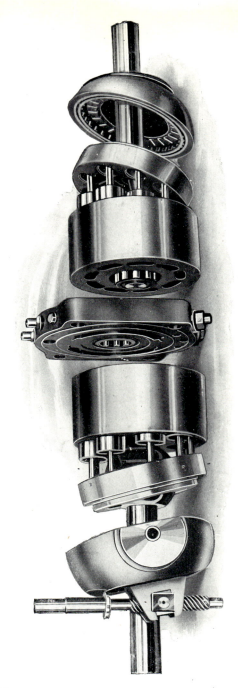

FIG. 118.—VARIABLE SPEED OIL TRANSMISSION GEAR : DISMANTLED TO SHOW CONSTRUCTION.

To face page 133.

mitted to the tool through a " Flexstel " pipe, the construction
of which will be clear from Fig. 121.   The drill may be mounted
on a cradle or other form of stand.   Water passes to the drill
point by a tube through the hammer, and the amount can be
regulated by a screw cock.   No fine dust is produced.   The drill
is rotated at each stroke by a small independent wave motor
operated from the main supply.

   Up to the present the most promising development appears
to be in rock-drills for mines and quarries, but it is also adapted
for any purpose for which compressed air has hitherto been
applied.   The transmission of energy by compressed air is so
wasteful that the new system ought to prove a considerable
advantage.

   Not infrequently a piece of machinery or a length of shafting
needs to be started or stopped immediately, and for this purpose
a clutch is required.   Thus, suppose a line of shafting is in two
portions, one connected with a source of power and the other
to a machine ; then the object is to have some sort of connection
which can be made or released at a moment's notice.   This is
often accomplished by fixing a disc with teeth on its face at the
end of the power-shaft, and providing a similar disc with recesses
in place of teeth at the end of the driven shaft.   The second
disc is capable of sliding along, but must turn with the shaft
upon which it is mounted.   In this case the engagement of the
two discs cannot be effected without jerk, and is almost impossible
at high speeds.   A common plan is to replace the first disc with a
drum open at the end and with the inner surface of the rim
bevelled.   The other disc has the outer surface of its rim bevelled
to fit, and when the two bevelled or coned surfaces are pressed
together the one turns the other by friction.   As a general
rule the smaller or inner coned surface is covered with leather or
other material which gives a more gradual bite or purchase.

   The most perfect type of friction clutch, however, is that
designed by Professor H. S. Hele-Shaw and illustrated in Figs. 122
and 123.   Between the drums on the driving and driven shafts
is placed a number of rings of pressed steel and phosphor bronze,
the two metals being used alternately.   The former have lugs on
their inner edges, and the latter lugs on their outer edges, and
both have a Vee groove running all round them.   The rings with
lugs inside slide over a steel drum fixed to the driving shaft, and
those with lugs outside fit inside a hollow cylinder which slides

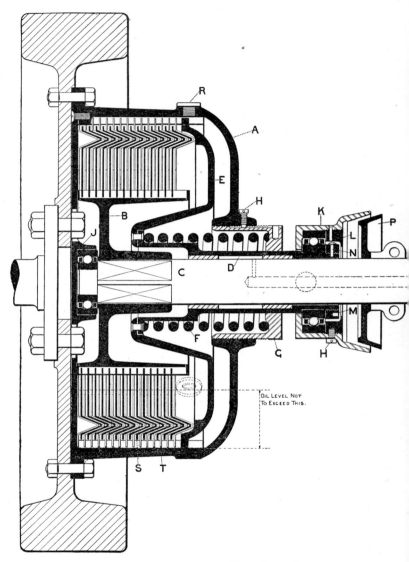

Fig. 122. SECTION OF THE HELE-SHAW CLUTCH.

page 134]

FIG. 120.—WAVE GENERATOR.

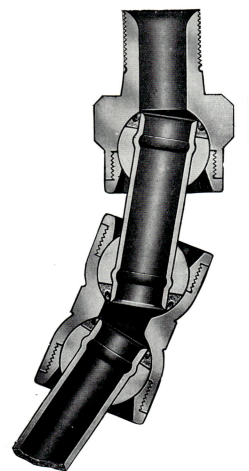

FIG. 121. —FLEXSTEL PIPE FOR WAVE TRANSMISSION.

along the driven shaft. The rings are immersed in a bath of oil, and as they are pressed together the oil between them is gradually squeezed out, the friction increases very slowly until the plates become locked together, and the motion is transmitted from one shaft to the other.

Very similar clutches are made with flat discs or rings, but this particular type has all the advantage of gradual action possessed by cone clutches by reason of the Vee groove in the rings. The friction, which is liable in ordinary cone clutches to cause overheating, is in this one spread over so large a surface as to render the heat produced negligible. Moreover, the Vee grooves always retain some oil, and therefore ensure perfect lubrication. The spring shown in Fig. 122 prevents the pressure between the rings being greater than is necessary to transmit the power.

The Hele-Shaw clutch is very largely used on motor vehicles, and in this connection its perfect lubrication enables it to be used to regulate the speed. In some of the London newspaper offices it is used to control the huge printing machines, and the great masses of metal and rolls of paper moving at high speed in these form a severe test of its efficiency. It is used by leading makers of motor-fire-engines, for the steam pinnaces of H.M.S. *Dreadnought,* and in numerous other instances where reliability and uniformity of action are a *sine qua non.*

MACHINE TOOLS

Anyone who has been through an engineering workshop will realize that the machines it contains can be classified in groups according as the tool or the work moves. In the drilling and shaping machines holes are bored, or a plane surface is made on a piece of material which is fixed rigidly to the table of the machine. In the lathe, the boring machine, and the planing machine the work as a rule moves and the tool is fixed—perhaps it should be explained that a boring mill is a lathe without a back centre, the object being bolted to a horizontal or vertical face plate. There are two or three interesting scientific principles involved in the use of these machines, a better understanding of which has had an important effect on recent design. So long as the cut is continuous and in the same direction it is a matter of very little consequence whether the work or the tool moves, and there are generally advantages in having a fixed tool. The lathe for the external surfaces of long objects and the horizontal or vertical boring mill for internal machining and facing short objects are not likely to change. In the planing machine, however, the object moves backwards and forwards and—originally—the cut was made only one way. A saving of time was effected by fixing the cutting tool in a reversible socket in which it was rotated automatically at the end of each stroke, thus cutting in both directions. But if the object is at all heavy a good deal of energy is wasted in starting, stopping, and reversing the direction of the table, and many machines are now made in which the object is fixed and the tool holder travels backwards and forwards. But one tool cutting at once will not satisfy the modern demand for speed, and frequently two tools are set to work at once, one taking a roughing and the other an intermediate or finishing cut. This demand is partly responsible for the development of the modern milling machine. In this the tool is a hard steel wheel with teeth shaped with the correct angle for cutting, while the work moves backwards and forwards beneath it. In one sense this disobeys the rule given above, in which any reciprocating motion should be given to the lighter part. But the milling cutter moves relatively fast and the work slowly and with few reversals. The finish from a milling tool is very much smoother than that from an ordinary tool, because the large number of teeth following one another closely are wide enough to permit of overlapping. A smooth surface instead of a series of channels is formed. As an example of the work done in this way Fig. 124 shows the Acme screw thread, and the cutter by which it is chased at one operation.

Perhaps no change is greater in workshops than the wide application of grinding. The most accurate work is now performed by a carborundum wheel which, spinning round at a high speed, tears off the metal with its thousand points and creates showers of sparks in its passage over the surface. From a tool used in the fettling shop for removing roughly the surplus metal on castings, the grinding machine has within twenty-five years become an instrument of precision, to which has been entrusted the most accurate workmanship that modern manufacture demands. From the thin pins that hold together the links of a bicycle chain to the smoothing of an armour plate, the engineer depends upon the machine which is familiar to all through the itinerant knife-grinder.

FIG. 123.—PAIR OF RINGS OF HELE-SHAW CLUTCH.

FIG. 124.—ACME THREAD AND CUTTER.

To face page 136.

FIG. 125.—A LARGE BORING MILL.

To face page 137.

The efficiency of all the machines which depend upon a steel cutter has been enormously increased since 1900. In Chapter VIII the discovery of high-speed tool steel by Messrs. Taylor and White, of the Bethlehem Steel Company, will be described. This steel enables the speed of overhead shafting to be increased from 90 to 250 revolutions per minute, and raises the amount of metal which can be torn off per hour from 30 to 137 lb. Since then a large number of other special tool steels have been produced, some of which owe their properties to the presence of vanadium, which has a most powerful influence upon the steel with which it is alloyed. The result is that work which formerly took weeks is now executed in days.

As an example of a large modern machine tool we illustrate in Fig. 125 a boring mill made by Messrs. Richards, of Broadheath. Ordinarily the machine will deal with a casting or forging 20 feet diameter and 10 feet high, but by moving backwards the uprights that carry the tool-bridge work of 24 feet diameter can be bored or faced. The machine is of massive proportions and is capable of taking very heavy cuts with high-speed tool steel. The table is 15 feet diameter and rests upon an annular surface of white metal. It has teeth round the edge and is driven through gearing by a 50 horse-power motor. The speed may be varied from 0·238 of a revolution to 10·27 revolutions per minute. A 10 horse-power motor serves to raise or lower the tool-bridge, and provides the quick motion for setting the tools. Another 10 horse-power motor moves the uprights along the side beds. The tool holders are balanced by a patent spring contrivance at the upper ends, instead of the older arrangement of chains and balance weights. No less than twelve rates of feed are provided, ranging from 0·0301 inch to 1 inch per revolution of the table. The whole machine weighs about 50 tons.

Apart from size and accuracy the greatest advance has been made in automatic machine tools. The material is fed in at one end and a whole series of operations are performed upon it without any attention from the man. In fact so little attention is required that a man or boy can take charge of five or six machines. However complicated these may appear to the uninitiated they are in reality very simple. They have been developed step by step from the original machine in which every movement was effected by hand. First one motion was rendered automatic, then a second, then a third, and so on, until the machine can do everything but pick up material from the floor. Thus in some grinding machines the plate upon which the object would usually be fixed by bolts and slips is a pole of a magnet, and the movement of a switch holds the work in position as the carborundum wheel passes over the surface. In this way some half dozen objects may be gripped at once, and when the process is complete they are instantaneously released by a single movement of the hand.

A common type of automatic machine is one in which steel rod is fed in at one end and is converted into small cheese-headed screws, with slotted heads, in its passage. The separate tools required in the process are mounted on small carriages which move up to the end of the rotating rod and retreat when they have done their work. As the last of these cuts off the screw it is seized by a pair of steel fingers and transferred to a vice which grips it firmly while a steel saw mounted on a carriage advances and cuts the groove in the head for the screw driver.

THE TRANSPORT OF MATERIAL

Perhaps no part of modern works or factory equipment is more remarkable than that which transports the material from one place to another. All shops in which heavy articles are dealt with have an overhead travelling crane, which moves up and down and from side to side, picking up here and depositing there huge weights that a dozen men would be unable to move. While in most of these the object is slung by chains to a hook, the latter is sometimes replaced by a powerful magnet, which picks up a ton or more of iron as easily as a toy magnet picks up iron tacks. It is quite startling to see one of these magnets lowered on to a heap of pig iron and made to pick up a dozen or more of pigs by a slight movement of the crane-driver's arm.

One method adopted both in factories and yards is to employ an overhead railway with a single rail. From this is suspended by two wheels a small cage containing an electromotor, a man, and a winch. The cage picks up material and conveys it expeditiously to its destination. A good example is to be seen at the Victoria Station, Manchester, and very complicated arrangements are to be found for moving barrels in modern breweries. Occasionally a similar method is combined with a hoist so that several floors can be served.

Where there is plenty of floor space travelling belts are frequently employed. Thus at Messrs. Lever Bros., of Port Sunlight, as fast as the bars of soap are stamped they are placed on endless belts which convey them to the packing shop. There the band passes between two or more pairs of tables at each of which are girls who seize the bars as they pass, wrap them in paper with almost incredible swiftness, and hand them to other girls who pack them in boxes. The boxes are nailed up, placed on another band and transported—partly underground—to the wharf. Here, as they emerge in a continuous stream from a tunnel, they are piled up on a platform hanging from a crane, and slung into barges for shipment to all parts of the world

Broad, heavy belts or bands of this kind are used for all kinds of material—perhaps on the largest scale for coal and corn. In these cases they are usually called conveyors. The band is not always level, and often proceeds up and down hill, but some means has then to be taken to prevent the material slipping down. It often forms part of a machine. Thus in an ordinary threshing machine the corn falls from the ear on to a belt which is violently shaken from side to side, while a blast of air from a fan passes over in the opposite direction to that in which the corn is being conveyed. The shaking causes the lighter husk to rise to the surface, whence it is blown away, while the corn is carried forward on the belt and tipped into a sack. A similar method is used in gold-mining machinery. The crushed quartz, among which are fine particles of gold, is washed on to an india-rubber band which also has a shaking motion. The heavy gold remains on the belt while the coarser but lighter quartz is brought to the surface and washed away.

ELECTRIC WELDING

A neat process for joining two pieces of iron or steel, which has been in use to some extent since 1886, but has been developed considerably during the last few years, is that of welding by electricity. The usual process as carried on in the shops is as follows : the two pieces to be joined are connected up with a source of electricity (from a dynamo or public supply acting through a transformer) giving a strong current at low voltage. The ends grasped in sliding holders are then pressed together, and being rough they touch only at a few points. The resistance at the junction is therefore much greater than at other parts of the circuit, the ends are raised to the softening point, and an excellent joint is formed. A slight bulge round the joint owing to the force employed in pressing the soft ends of the rods together, is removed by subsequent hammering, which is beneficial in other ways. This process was devised by Professor Elihu Thomson, and a current of from 50,000 to 100,000 amperes at from 1 to 5 volts is used. The use of massive clamps prevents any other portion of the apparatus than the bar under treatment being overheated.

A modified form of apparatus enables quite thin strips or rods—not more than $\frac{3}{4}$ inch diameter—to be welded ; and copper, brass, and practically all metals and alloys can be joined in this way. In the case of iron and steel it is necessary to keep the temperature below that at which the metal fuses, and for this reason considerable pressure must be used. For other metals the pressure need only be sufficient to bring them into contact as the extreme ends fuse, when the current is immediately cut off.

There are, however, other methods which are useful not only in the workshop, but also in the shipyard and on outdoor repair work generally. In one a flame arc is formed between two inclined carbon poles and blown forward on to the joint to be welded as the ends of the carbons are moved along over the surface. Another method is to make the object to be patched or repaired one pole, and to move a single pole over the defective portion. In these cases a rod of soft iron is often used and small dabs of fused iron are plastered along the joint, and afterwards well hammered to render the joint solid.

The metal electrode process, while possessing many advantages, was not altogether free from objection, the most serious of which was the tendency of the molten metal to become oxidized. This was overcome by covering the pole with a flux which melts at the same temperature as the metal, and which flows over and protects the joint. The result is that the range of work successfully attempted during the last four years has enormously increased. In 1918 the Admiralty launched at Richborough a barge 125 feet long, 225 tons displacement, electric welded instead of riveted joints. Still more recently Messrs. Cammel Laird & Co., of Birkenhead, have launched the 500 ton all-welded coasting-vessel shown in Fig. 126, while larger vessels built on this plan are contemplated in the near future. The value of

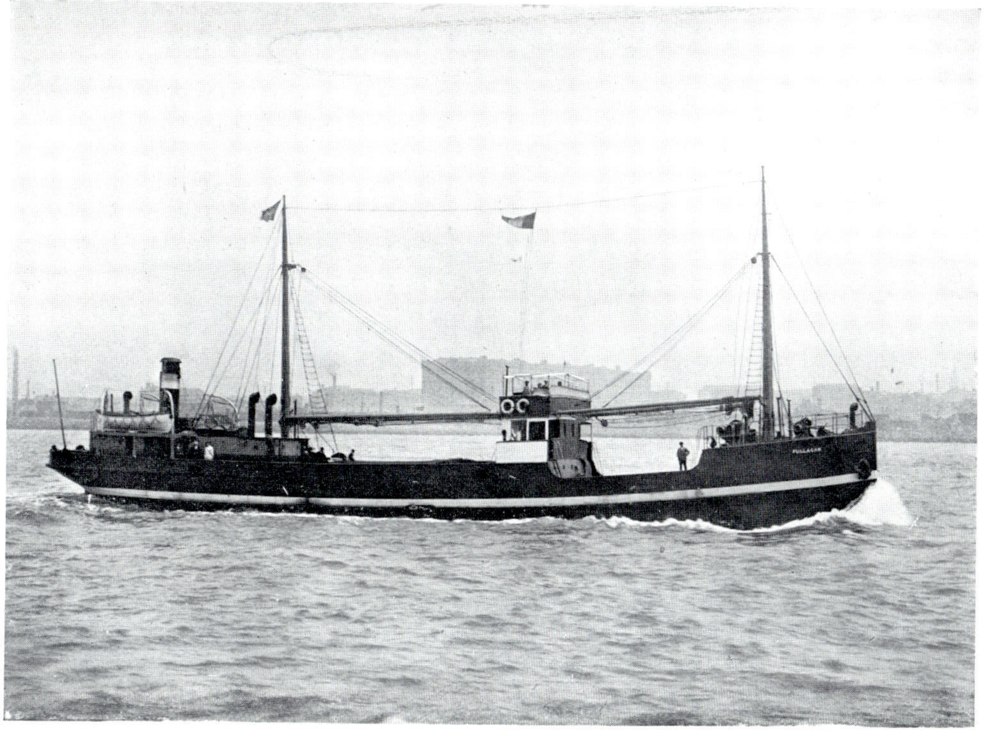

FIG. 126.—AN ELECTRICALLY WELDED SHIP: M.S. "FULLAGAR."

(M.S. Motor Ship.)

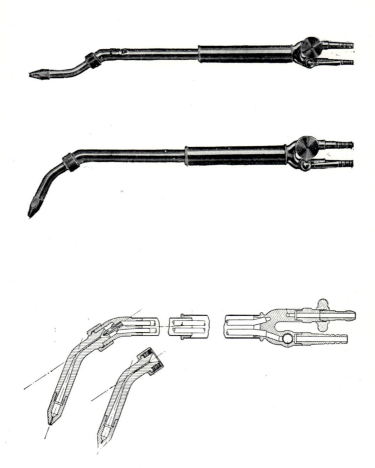

FIG. 127.—CONCENTRIC BLOWPIPE FOR OXY-ACETYLENE
CUTTING.

To face page 141.

the process lies in the elimination of marking-out, punching and caulking, a saving of from 5 to 10 per cent of plating, or, say, 100 tons in a 3,000 ton ship, greater strength than a riveted joint, and a reduction of labour costs by about one-half.

The extent to which electric welding is now employed would hardly be credited by those outside the workshops in which a wide variety of metal work processes is carried on. Thus electric cables, steel band saws, tyres for wheels, tramway rails, bicycle parts, steel tubing, coils of piping for refrigeration (see Chapter X) and many other articles are jointed by this process, in addition to repairs on railways and tramways, in shipyards, in boiler shops, and many other kinds of work.

CASTING AND WELDING BY THERMITE

Another portable process that has a very wide range of application is the use of thermite, though probably it belongs more particularly to the foundry. It was invented by Dr. Goldschmidt, and depends upon the fact that when powdered aluminium is mixed with a metallic oxide and ignited, it burns with a very high temperature—about 3,500° C.—removing the oxygen from the metallic oxide and liberating the metal in a molten condition. As this temperature is more than sufficient to melt every known metal the process can be used to make small castings of the rarer or more refractory metals and alloys.   For this purpose a quantity of powdered aluminium is mixed with the necessary proportion of the metallic oxide or oxides, in a crucible, and a fuse of some material which ignites more easily than aluminium, which requires a temperature of 700° C., is placed on the top. When the fuse is ignited the whole mass flares up, and in a minute or two the metal is ready for pouring.

This process has been largely used as well as the electric one for welding together the end of tramway rails. In an ordinary railway track it is necessary for each length of rail to be independent of and separated from the next one, by an amount which will allow for expansion in hot weather. But as the rails for electric tramways are used to convey the current, they must be in continuous metallic connection. Formerly this was accomplished by connecting each rail with the next one by a metal strip bolted on the side below the rail head. A tramway rail embedded in concrete and paving, however, is less likely to buckle at high temperatures than an exposed rail, and it is now the custom to weld the ends of the rails together. For this purpose a small crucible containing the powdered aluminium and iron oxide is fixed on a tripod stand over the rail joint. The paving is removed at this point and a mould is made round the rails. The fuse is fixed, ignited, and in a minute or so after the flare a hinged bottom to the crucible is allowed to fall, and the metal pours into the mould below. The latter is afterwards broken away, and the protruding metal ground away to the level of the rail head. There are few towns in which this process is not employed when the track is being relaid, but as the repairs are generally carried out during the night it is rarely seen by respectable people other than those engaged on night work.

OXYACETYLENE WELDING AND CUTTING

Striking as are the results of the processes described they are in some circumstances eclipsed by a new agent which has been placed at the disposal of the engineer. This is acetylene gas, to which reference is also made in Chapter IX. Formerly the hottest flame obtainable in a blowpipe was produced by a mixture of oxygen and hydrogen, which gives a temperature of about 2,000° C. But hydrogen never was cheap, and acetylene is—at any rate relatively so. Moreover, a mixture of oxygen and acetylene produces a temperature of 2,400° C., and is therefore 20 per cent hotter than the oxyhydrogen flame. And when after Moissan's discoveries in connection with the electric furnace calcium carbide, which in contact with water generates acetylene, came to be manufactured in quantity, engineers and metal workers availed themselves of the new process. For welding purposes the parts to be joined are heated with the flame and are then brought into contact and hammered. Or if a patch is being put on or corner joint made in thin sheet, the metal is heated and dabbed with the end of a thin soft-iron rod, much in the same way as the plumber uses a stick of solder, or any of us use a stick of sealing wax.

Nearly all the ordinary processes of welding can be carried out by this method, and a great many pieces of work which would be spoilt by being placed in the smith's fire are easily dealt with. Not only has it provided an alternative method of jointing in many well-established forms of construction, but it has aided in no uncertain way that enormous development of

FIG. 128.—CUTTING A SEMICIRCULAR PLATE WITH OXY-ACETYLENE BLOWPIPE.

To face page 142.

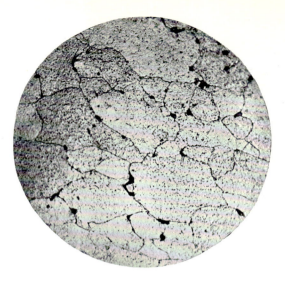

FIG. 129.—NEARLY PURE IRON.

THE FINE DARK LINES SHOW THE OUTLINE OF THE CRYSTALS,
AND THE SPECKS ARE SLAG.

FIG. 130.—MILD STEEL.

FERRITE (LIGHT) AND PEARLITE (DARK) SHOWING THE DIRECTION
OF ROLLING.

To face page 143.

mechanical practice which has taken place during the last fifteen years. In this and in other ways workshop practice is being revolutionized.

But if oxyacetylene welding is an example of progress, oxyacetylene cutting is a far more startling one. If a jet of oxygen gas is allowed to play upon red-hot iron, the metal burns in the gas with brilliant scintillations. The oxide which is formed melts at a lower temperature than the metal and is blown away almost as rapidly as it is formed. The most effective type of blow-pipe for this purpose is the concentric one illustrated in Fig. 127. From the diagrams it will be seen that the oxyacetylene flame is produced at the mouth of the space between the inner and outer tubes, and oxygen is blown through the middle of it. When such a jet is moved over the surface of sheet iron it cuts a hole clean through. The usual workshop methods for cutting are shearing and sawing. To the former there is a limit of thickness—more than $1\frac{1}{2}$ inches is rarely attacked—and the latter is slow. If a large hole has to be made in the middle of a sheet of metal it must either be bored out or a number of holes drilled round the margin and the piece chipped out with hammer and chisel. These operations are carried out with far greater ease by the oxyacetylene jet, and with astonishing rapidity. An elliptical manhole—say 16 inches by 10 inches—in a 1-inch boiler plate only requires four or five minutes, and an armour plate 6 inches thick can be cut clean through at the rate of a yard in ten minutes. Fig. 128 shows a large thick rectangular plate being cut to semicircular form by a jet mounted on the end of a radial arm.

The extreme portability of the apparatus—the acetylene and oxygen are contained in steel cylinders—renders it of particular value for repair work. One of the most interesting examples of its recent use was on a large passenger vessel—the *Commonwealth*—which had had her bows stove in and stem twisted in a collision. The stem and damaged plates were cut out by the oxyacetylene blowpipe, and a new bow was fixed within three weeks from the vessel entering dry dock.

It ought perhaps to be stated that an oxyhydrogen jet with excess of oxygen, or with oxygen driven through it, will serve the same purpose, though the temperature is lower.[1] For it is

[1] Oxyhydrogen jets are used by some Sheffield firms for cutting armour plate.

an interesting fact that while the cost of hydrogen prevented its employment for commercial purposes, it has long been appreciated and used by burglars for effecting an entry into steel safes in search of plunder. Such criminals could adopt a method which the profits of legitimate industry were too small to justify.

CHAPTER VIII

THE world's production of steel in 1927 was more than 99 million tons. If this amount were rolled into a flat bar 6 inches wide and half-an inch in thickness, it could be wound 160 times round the earth. Rolled into a plate 2 inches in thickness and floated on the ocean, it would form a pathway 165 feet wide from Liverpool to New York.

The period from 1890 to the Great War is one of the most remarkable in the history of the iron trade. Great Britain had increased her production by 15 per cent, Germany had become the second largest producer of iron in the world, and the United States had added an amount equal to the whole production of Great Britain—a production that had taken 130 years of solid progress to achieve. In 1912, Great Britain poured out of her furnaces over 10,000,000 tons of pig iron, Germany nearly 13,000,000, and the United States more than 25,000,000. The great American development has been partly due to the rich deposits of ore on the shores of Lake Superior, which are easily mined and produce a good quality of metal. Though discovered in 1845 the difficulties of transport made it impossible for the iron-masters of Pennsylvania to use them for thirty years. It was not until the Sault Ste. Marie and other canals brought the great lakes into navigable intercommunication that these vast stores of raw material became available. Since the war the balance of production has been completely altered. In 1927 the U.S.A. produced over 36 million tons, the German production was the same as in 1912, and Great Britain's contribution had fallen to 7,350,000 tons. Nearly half the steel produced annually comes from American furnaces.

THE MANUFACTURE OF IRON

During the last forty years the changes which have taken place in the manufacture of iron and steel have been mainly in the direction of improvements in quality, increase of yield, and economy of fuel. In order to understand how these have been effected it will be necessary to recall briefly how iron is reduced from its ore. From very early times, until the middle of the eighteenth century, iron ores were smelted in masonry furnaces with charcoal, and the necessary temperature was attained by blowing in air with a bellows—often worked by a water-wheel. After this period the use of water for blowing gave way to the steam-engine, which became a satisfactory source of power in the hands of James Watt in 1769. Coke began to replace charcoal in 1735.

The ores of iron are usually oxide of iron mixed with *gangue* or earthy matter. The processes are rather complicated, and several reactions between the air, oxide, earthy matter, and fuel proceed simultaneously in different parts of the furnace. Limestone is added to form with the gangue an easily fusible slag, which floats on the surface of the molten metal. The slag is tapped off occasionally and is conveyed to the slag tip—some varieties are used for repairing roads. The metal is run into sand moulds about 3 feet long, 4 inches wide, and 4 inches deep, and the resulting castings are called pig iron. A supply of ore, fuel, and flux (limestone) is fed in continuously at the top of the furnace, and iron may be produced daily for months or years.

Greater economy was secured by the invention of Neilson in 1829, by which the air was heated before being blown into the furnace. For over thirty years this air required a separate supply of fuel to raise the temperature of the iron pipes through which it was passed. In 1863 Sir William Siemens introduced the regenerative principle by which the hot gases from the furnace were led into one of two brick chambers filled with bricks so arranged as to leave open spaces or *chequers*. When this chamber was hot the gases were diverted through the other, and the air from the blowing engine was passed through the hot one. The chambers were therefore engaged alternately in storing up the heat and giving it up again to the blast.

THE ECONOMY OF FUEL

Within the last twenty years economy in the production of iron has been further effected in two ways. One is an additional method of utilizing the hot gases from the top of the furnace.

The following table shows their composition in two cases—where coke and raw coal are being used :—

	Coke. %	Raw Coal. %
Carbon monoxide (CO)	25	28·0
Carbon dioxide (CO_2)	12	8·6
Nitrogen (N)	59	53·5
Hydrogen (H)	2	5·5
Methane or Marsh-gas (CH_2)	2	4·4

The fuel most generally used is coke, but raw coal is employed in the west of Scotland, and a mixture of coke and raw coal in South Staffordshire. Charcoal, the original fuel, is still used in North America, Sweden, and Styria, where timber is plentiful ; the gases evolved have a similar composition to those obtained from coke. In all cases there is a sufficient proportion of carbon monoxide, hydrogen, and methane to render the mixture inflammable. For each ton of coal charged into the furnace 130,000 cubic feet of gas are produced, so there is clearly a vast source of power going to waste. In 1892 B. H. Thwaites suggested that this gas should be utilized in gas-engines, and the plan was put into operation by the Glasgow Iron Company in 1895. It is estimated that the blast-furnaces of this country yield sufficient gas to produce 750,000 horse-power if used in gas-engines, and a number of other iron-masters have followed the lead of the Glasgow firm. In Germany and elsewhere, however, progress has been much more rapid. By 1906, within ten years of the first installation, there were no fewer than 349 gas-engines developing 385,000 horse-power, and the majority of them were using blast-furnace gas. In the same year the United States Steel Corporation decided to instal similar engines to develop 150,000 horse-power,—representing 10 per cent of the total power required.

A further supply of gas is obtainable from the coke ovens. In this case, as well as where raw coal is used in the blast-furnace, it is becoming customary to collect the tar and ammonia, which are valuable by-products. The ammonia is converted into ammonium sulphate and sold as a manure, for which purpose it is worth £12 a ton. The tar is distilled and used for oil fuel, disinfectants, and other purposes in the same way as the tar from town gasworks.

A more recent economy relates to the removal of moisture from the blast. Ordinary air invariably contains vapour water and the amount varies from day to day. Assuming an ounce of water in every 50 cubic feet, and a blast of 40,000 cubic feet per minute, the amount of moisture entering the furnace would be 300 gallons per hour! While the presence of even the minimum quantity of water in the air may be objectionable, the variation is still more so, because it causes the furnace to work irregularly and renders it difficult to secure a uniform quality of iron.

Now the amount of moisture that air can retain depends upon its temperature. For every temperature there is a definite percentage of water vapour which it can hold. When this percentage is reached the air is said to be saturated, and any reduction of temperature results in the precipitation of some of the moisture. Hence by strongly cooling the air practically all the water can be thrown out.

In 1904 Mr. Gayley, of the Carnegie Steel Company, Etna, Pennsylvania, carried out some tests with a blast-furnace operated by ordinary air and by air which had been deprived of its moisture by cooling. The cooled air reduced the consumption of coke by 20 per cent, increased the yield of iron by 25 per cent, and effected a net saving of 150 horse-power. The machinery employed to dry the gas consisted of an ammonia compression plant (see Chapter X) which would have been capable of making 225 tons of ice in twenty-four hours.

In 1911 Mr. Gayley reported that, as a result of six years' working, the average saving of fuel had been 10 per cent and the average increase of yield had been 12 per cent in one furnace; while in another furnace the figures were 7·5 per cent and 23 per cent. Again, in the Warwick furnace at Pottsdown in the same State, the result of reducing the moisture in the blast from 9 grams to 3·5 grams cubic metre was a saving of 21 per cent of fuel and an increased output of 23 per cent on 750 tons of iron. Lastly, Guest, Keen & Nettlefold, of Cardiff, report a saving of from 13·4 per cent to 18·4 per cent of fuel, and a gain in output of from 14·1 per cent to 26·4 per cent.

The value of the method appears to depend to some extent on the temperature at which the furnace is normally worked, and it does not necessarily follow that it will give much advantage in a dry climate or under all conditions. It is claimed by those who have used the process successfully that the more regular working of the furnace is in itself almost a justification. Professor Josef Erhenwerth calculated that the difference in the amount of fuel required to produce 25 cwt. of iron in summer and winter owing to the difference in the amount of moisture should be 1 cwt. In actual practice it turned out to be $\frac{3}{4}$ cwt. There is no doubt that the cost of installing refrigerating plant is against its more general adoption. An estimate for the equipment for six furnaces at Skinningrove is said to have been £70,000. Nevertheless, the results which have been achieved furnish a remarkable example of the interdependence of industry. No man concerned with industrial development can afford to ignore the progress which is being made in spheres widely separated from his own.

THE NATURE OF STEEL

Iron exhibits a marked variation in properties according to the amount of carbon it contains. Pure iron is a chemical curiosity, produced in a very small quantity in the laboratory for the purpose of research. The chief difficulty of obtaining it is the readiness with which it combines with carbon at the temperature of a furnace—in fact, carbon permeates iron even below its melting point. It is this property of combining with carbon that gives the metal its wide range of utility. So long as the percentage of carbon is small the iron is soft and easily bent, and when two pieces are made hot and then pressed together they unite—the process is known as *welding*. It melts at a very high temperature (about 1,600° C.) and passes through a pasty condition in which it can be rolled, beaten, or pressed into a variety of forms.

If the percentage of carbon is increased the metal becomes harder, stronger, and more elastic. With a still higher percentage it becomes more brittle and less tough, and the melting-point is lowered, so that it becomes liquid at about 1,100° C. Iron containing less than 0·1 per cent of carbon is called wrought iron, with from 0·1 per cent to 2·5 per cent it is called steel, and with more than 2·5 per cent it is called cast iron.

Cast iron is classified according to the fracture and is termed white, grey, or mottled. The grey or mottled appearance is due to the separation of carbon in the form of graphite. It is fairly elastic without possessing any great tensile strength and easy

to work with machine or hand tools.   But it melts suddenly and
cannot be forged or rolled.

Leaving out for a moment the properties of steel it is evident
that here are two varieties of the same metal, differing ostensibly
only in the carbon content, which are adapted to a wide range
of workmanship and purpose.   If tensile strength is required,
wrought iron can be used, for although the cost is high there
are few shapes which cannot be produced by the smith.   But
if tensile strength is relatively unimportant there is practically
no form which cannot be obtained by moulding.   In all cases
involving intricacy of outline or hollow spaces, casting is a far
cheaper process than forging.   Moreover, since the addition or
subtraction of carbon converts the one into the other, an article
can be made by the cheaper process and then converted into
wrought iron by removal of the carbon, a removal which can be
effected below the temperature of fusion.

In this process the castings are packed in iron boxes with
hæmatite iron ore (Fe_2O_3) and heated in a furnace for from
five to twelve days.   The oxygen in the hæmatite converts the
carbon of the castings into carbon monoxide, which passes away,
and the castings are found to have the appearance of wrought
iron, without, however, the fibrous structure and consequent
strength which is induced by rolling.   Many parts of agricultural
implements are made of so-called *malleable* castings.

Now consider steel.   With all percentages of carbon it has a
higher tensile strength and is more elastic than wrought iron ;
and it is always less brittle than cast iron.   It can be forged and
welded, but the process is more difficult as the percentage of
carbon rises.   It can be melted and cast, but with more difficulty
as the percentage of carbon decreases.   In these respects it
resembles both wrought and cast iron, but it possesses one
property which distinguishes it from either.   If it is raised to a
high temperature and cooled quickly it becomes intensely hard.
Moreover, if it is heated again to a lower temperature and then
cooled quickly a degree of hardness is obtained which depends
upon the temperature of the second heating.   This process is
called *tempering*, and it is the fact that steel can be tempered
which makes it so useful for tools, because the necessary degree
of hardness can be obtained without undesirable brittleness.

It may fairly be said that while its cheapness and wide range
of application have, apart, from its inherent qualities, retained

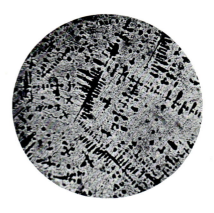

FIG. 131.——WHITE PIG-IRON.

THE BLACK CRYSTALS ARE AUSTENITE
WHICH HAVE CHANGED INTO PEAR-
LITE DURING COOLING, WITHOUT
LOSING FORM.

FIG. 132.——GREY PIG-IRON.

BLACK FLAKES OF GRAPHITE, IN GROUND
MASS OF PEARLITE, WITH ONE PATCH
OF PHOSPHIDE.

FIG. 133.——MARTENSITE IN
QUENCHED STEEL.

To face page 150.

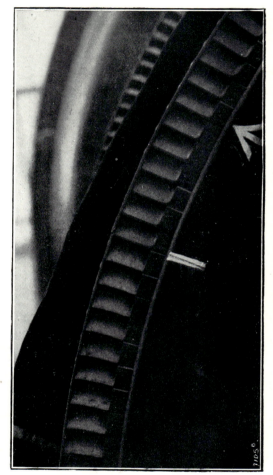

FIG. 134.—RESISTANCE OF STAINLESS STEEL TO EROSION.

To face page 151.

cast iron in favour, the last fifty years have seen the replacement to a very large extent of wrought iron by steel. It was Sir Henry Bessemer who first showed how steel could be produced quickly and cheaply, and his process was well described in the earlier volume. It consists essentially in burning out the carbon and other impurities in cast iron by forcing air through the molten metal and then adding sufficient spiegeleisen (an alloy of manganese and iron with a high percentage of carbon) to produce steel of the desired quality. It is worthy of note that a vote arranged by the *Scientific American* in 1896, as to the invention which had proved of the greatest benefit to mankind, resulted in favour of Bessemer's process for the manufacture of steel.

In recent years the proportion of steel manufactured by the Siemens-Martin process has increased, and is generally preferred. A longer time is required, but for that reason the process can be more closely watched, and any desired grade can be obtained with greater certainty.

There are two other processes for the manufacture of high-class tool steels. One—the *cementation* process—is very similar to that already described for the production of malleable castings. But in this case it is the addition and not the removal of carbon which is effected. A pure variety of Swedish iron is packed in boxes with charcoal and heated for from eight to eleven days at a temperature of 1,000° C. When unpacked the bars have a blistered appearance—hence the name *blister steel*. They are broken and sorted by men who have learned to distinguish the character of the metal from the fracture, then reheated in piles and hammered into bars. It should be observed that the absorption of carbon has been effected at a temperature *below* the melting-point of the metal.

The length of time required for the process just described is leading to its disuse, and the next process, by which *crucible cast steel* is made, is much quicker. Wrought iron is mixed with charcoal and melted in a fireclay crucible. In the course of a few hours—usually about four—the iron will have dissolved the carbon, and can be cast into moulds, The ingots can then be rolled or pressed into the desired form. Much of the special steel which is now so important is made by melting the ingredients in an electric furnace as described in Chapter IX.

Many investigations have been undertaken to ascertain the

relation between chemical composition, internal structure, and mechanical properties of iron and steel. The fact that steel of the same composition can exist in varying degrees of hardness shows that the percentage of carbon alone is insufficient to determine its properties, and that much depends upon its thermal history, i.e. to what temperature it has been heated and how it has been cooled.

Let us first consider the information that can be obtained from chemical analysis. Grey and mottled cast iron consist of a white, hard substance mixed with graphite. When the metal is dissolved in hydrochloric acid this graphite is unaltered, but the gas which comes off is not pure hydrogen, but hydrogen containing some hydrocarbons, or bodies consisting of hydrogen and carbon. If cast iron containing not too much carbon is melted and run into metal moulds it becomes " chilled " and then consists of white cast iron. On dissolving this in hydrochloric acid there is no residue of graphite, and the carbon is all evolved in the form of hydrocarbon gas. Hydrochloric acid has no action on free carbon, and could only give the hydrocarbon gas if the carbon were present in the form of a compound with the iron. It is evident, therefore, that there are two forms in which carbon exists in cast iron—free and combined. Moreover, as the hardening property of steel is intimately connected with the presence of carbon, it is evident that when the percentage of it lies between certain limits there is some special variety of compound which is still to be explained.

Our knowledge of the structure and constitution of metals and alloys has been enormously extended in recent years by the aid of the microscope. The surface of the metal is polished— a process which causes the harder constituents to stand out in relief ; or it is treated with acids or other reagents which attack and destroy some portions and reveal those which were otherwise indistinguishable. As different constituents become capable of identification names have been given to them, and, though the whole question is still in the throes of acute controversy, a few of the more firmly established facts and theories may be given.

It has been known for many years that when a piece of iron is allowed to cool down from 1,000° C. or thereabouts there is a point at which cooling suddenly ceases. The wire, rod, or strip glows brightly and undergoes a change of volume. Below this point the metal is magnetic ; above it is non-magnetic. If the metal is heated, e.g. by an electric current, instead of cooled, the same phenomena are observed. The change was explained by saying that iron existed in two forms, one stable only at a bright red heat and the other at ordinary temperatures, and that the point of recalescence, as it is called, was the point at which the one became converted wholly into the other. More careful study with improved instruments for measuring temperature has shown that there are two points of recalescence, and it is concluded, therefore, that there are *three* forms of iron—*allotropic* forms is the scientific term. These are called α-iron or ferrite, β-iron, and γ-iron.

The readiness with which carbon dissolves in iron and the marked effect which it has on the properties suggests that one or more compounds of the two elements are formed. At least one of these is recognized both by chemical analysis and under the microscope. It has the formula Fe_3C, and has been given the distinguishing name of *cementite*. A solid solution of cementite in γ-iron is called *austenite* ; a similar solid solution of cementite in α-iron is called *martensite*. During slow cooling ferrite and cementite separate in microscopic layers and the hardness of the latter gives rise to a pearly appearance on polishing. Hence the name *pearlite* for this mixture. More rapid cooling causes separation in granules, and the mixture is then termed *sorbite*.

The appearances under the microscope are illustrated in Figs. 129–33. White cast iron invariably contains crystals of austenite which have become changed into pearlite on cooling, though if the percentage of carbon is more than 4·3 crystals of cementite will be formed independently. Grey cast iron contains graphite, pearlite, and either ferrite or cementite according as the carbon is higher or lower than the amount required to form austenite. The constitution of steel is far too complicated a matter to be pursued further in the space available.

The theory of the constitution of steel has been the subject of an enormous amount of controversy, but whatever their explanation, the facts which have been discovered have been of incalculable value in enabling the steel-maker and the engineer to understand and make allowance for the peculiarities of the material upon which so much depends. The safety of such a structure as the Forth Bridge—which even to-day stands as

one of the great engineering achievements of the world—rests
not only upon the proportions of the different members and
upon the number, distribution, and soundness of the rivets, but
to an equal extent upon the microscopic structure of the steel.
In the days when it was built this was unknown or but dimly
recognized, and the engineers had to depend upon the behaviour
of test pieces and to take care to put in girders, ties, and struts not
large enough as they felt, but too large. Even chemical analysis
—now supplemented extensively by the microscope—was
not in use in all works. Up to the beginning of the present
century there were steel works known to the writer in which
no chemist was regularly employed, and in which the simple
necessary tests were carried out by workmen under the super-
vision of the manager, but with very little knowledge of what
they were doing. The temperature of the furnaces, now known
to be so important, was never measured. To-day steel-makers
know and control the temperatures to a nicety. At a recent
meeting of the Institution of Mechanical Engineers, Sir Robert
Hadfield stated that from 3,000 to 5,000 measurements of temperature were made in his works per week. And when these
furnaces have yielded up their burden, samples of the steel are
bent, stretched, and broken in the testing machine ; analysed in
the chemical laboratory ; etched or polished, and examined under
the microscope, which reveals their innermost secrets ; and
before the metal goes to the engineer to be entrusted with
delicate duty in some machine, or to play its part in a great
structure, its every peculiarity is known. The very molecules
have told their tale !

SPECIAL STEELS

The properties of steel are profoundly modified by the presence
of other elements than carbon, and by the actual amount of
each. In some cases the effect is good, in others bad, and during
the last twenty-five or thirty years an enormous amount of
work has been done in investiagting the effects. The chief
properties of steel which are important from an engineering point
of view are tenacity, ductility, and hardness. The first of these
determines the resistance to breakage by pulling at each end,
the second determines the ease with which it can be rolled into
plates, drawn into wire or bent, and the third determines the
resistance to wear by rubbing surfaces. A fourth property, at

present not fully investigated, is the resistance to corrosion by air, water, and other fluids.

One of the earliest substances to be alloyed with iron was manganese. With low carbon steels a small amount increases tenacity and decreases ductility. The axles and tyres of wheels may have up to 1 per cent but a steel containing even 0·6 per cent would be quite unsuitable for boilers and structural work. In steels containing more carbon the percentage should not rise above 0·3. This effect continues with all types of steel until with 5 per cent to 7 per cent of manganese and 0·5 per cent carbon the metal is brittle. Sir Robert Hadfield has shown, as a result of an investigation which occupied ten years, that a cast-steel bar containing from 8 per cent to 20 per cent of manganese can be bent considerably without fracture, and his manganese steel is useful, but very hard and difficult to work in the cold. It is extensively employed for the points of tramway rails and in other cases where an extremely hard, and yet not brittle, metal is required.

Perhaps the most widely used alloys of steel are those containing chromium, nickel, and tungsten. Chromium and nickel both increase the toughness, and are used largely for armour plate and projectiles. Tungsten steel for tools was introduced by Mushet in the middle of last century. It is a self-hardening steel which does not require to be suddenly quenched in order to temper it. The use of these metals has been extended in recent years by improved processes of manufacture. Thus Dr. Ludwig Mond's processes for the production of nickel led to a considerable increase in the supply of that metal. But the most remarkable progress owes its origin to Moissan's work with the electric furnace (Chapter IX). Not only chromium and tungsten, but titanium, molybdenum, and vanadium were then obtained for the first time in quantity and in a high degree of purity, and the electric furnace is now very generally used for preparing rich alloys of these elements with iron to add to steel.

Tungsten or molybdenum is contained in the new high-speed tool steel. The original Mushet steel, which contained tungsten and was self-hardening, would cut hard steel at the rate of 8 to 10 feet per minute, and soft steel at the rate of 10 to 15 feet per minute for heavy cuts, and 20 to 25 feet per minute for light finishing cuts. Similarly a milling tool would cut at 30 to 40 feet per minute. In 1900 Messrs. Taylor and White discovered steels that would work satisfactorily at a low red heat. It has generally been supposed that if Mushet steel was heated above cherry redness (815° C. to 845° C.) it was spoiled. But they found that if it was heated to the point when the metal began to crumble when touched (1040° C. to 1100° C.), and then allowed to cool steadily, its hardness and toughness are increased to an extraordinary degree. There are now a number of varieties on the market, and, curiously, some of them are not steel in the proper sense of the word because they contain no iron. In one, with which the writer is familiar, nickel is the principal constituent. But so long as a tool will do the work that is required of it, no one will quarrel about the name.

Of other substances used for alloying with steel, vanadium and silicon may be mentioned. The former is contained in some of the tool steels to which reference has been made. As it combines with nitrogen to form a nitrate it is possible that its value lies in the removal of dissolved gases and the production of an ingot free from minute cavities. For example, titanium, which is an element of similar properties, is sometimes added to molten iron in order to produce especially close-grained and sound castings. It is also used in steel manufacture, but does not appear to have such a powerful effect as vanadium, 0·5 per cent of which, in a particular case increased the tenacity by over 50 per cent steel, with 0·35 per cent of silicon is used for springs.

One of the most interesting discoveries of recent years is that of stainless steel by H. Brearley in the Brown-Firth Research Laboratories, Sheffield. An investigation carried out in connection with erosion in gun barrels resulted in the production of a steel which was unaffected by atmospheric moisture. Highly polished pieces exposed to the air of the laboratory for several weeks retained their brightness under conditions in which ordinary steel rapidly became rusty. The new alloy contains from 12 per cent to 14 per cent of chromium and about 0·3 per cent of carbon, and some difficulties had to be overcome before it could be hardened and worked successfully in the forge. A much higher temperature is required for tempering. It can be hardened by heating it to between 950° C. and 1,000° C. and quenching it in oil or water, and softened by heating to between 850° C. and 870° C., and allowing to cool very slowly. For machining it is in its best condition when heated to between 750° C. and 800° C. and allowed to cool in air. These facts illustrate not only the delicacy of the operations and accuracy of the measurements in a modern steel-works, but the amount of experimental work which must have been necessary in order to secure the highest utility from the material.

Stainless steel is used for cutlery, surgical instruments, springs, dental instruments, golf-club heads, finger-plates for doors, furnishings for stoves and grates, hollow-ware, scientific apparatus, and motor-car parts. It is also applied to the blades of steam turbines—not so much for its freedom from rusting as for its resistance to erosion. The result of a prolonged test which was carried out by Messrs. Thos. Firth and Sons, of Sheffield, on a 2,000 horse-power turbine in their works is shown in Fig. 134. The resistance of the stainless steel to the action of the steam is obvious.

The effect which the production of these special steels has had upon industry cannot be over-rated. Combined with improved methods of casting, forging, and working they are largely responsible for the development of the motor-car, and the aeroplane. The main problem which required solution was a sufficiently powerful engine of small weight; so long as cast iron was the only available material for the framework and cylinders, the power could not be increased without increasing the weight. Modern discoveries in the manufacture of steel and of aluminium have enabled engineers to construct engines weighing less than half as much as would have been possible thirty years ago.

FORGING

Since steel can either be forged or cast and run directly into mould from the furnace in which it is prepared, there is little to say about the latter process. For small articles of simple form, cast steel is an admirable material; but there are two difficulties in securing large castings. One is the tendency of the metal to give up dissolved gases, forming blow-holes, and the other is the tendency for the constituents—particularly such impurities as sulphur and phosphorus—to be unequally distributed throughout the mass. Nevertheless, large castings are obtained, weighing as much as 60 or 70 tons—the stern casting of the *Mauretania* for example. It is usual to add a small quantity of aluminium to the metal before pouring into the mould, and this is said to produce a casting freer from cavities and of more uniform composition. Another method is to apply hydraulic pressure to the metal directly it is poured into the mould. This same process has been used by Mr. Talbot in an attempt to prevent segregation. A large ingot, cooling from the outside, retains a liquid centre for a considerable time, and the tendency is for certain of the constituents to concentrate in this central liquid core, so that there is lack of uniformity in quality. The blow-holes and objectionable segregation occur in the upper portion of the ingot, and in gun manufacture this portion, to the extent of nearly one-half, is rejected. A sounder ingot throughout is therefore very desirable if it can be obtained. The process consists in applying lateral pressure by means of a hydraulic press while the steel is solidifying. Only the largest size ingot has been experimented with, because a small one cools before it can be removed to the press. With a 23-inch ingot there is only about 15 to 20 minutes in which to carry out the operation.

Not only is it necessary for structural purposes to have steel as uniform in quality as possible, but the development of the steam turbine has thrown a new and greater responsibility upon the steel manufacturer. The rotor of a turbine is a heavy mass of metal weighing several tons, spinning round at 1,000 revolutions per minute. As explained in the chapter on the steam-engine, enormous forces are brought into play if the centre of mass is not at the geometrical centre. For such work as this forged steel is often used and great expense is uncurred in machining that would be unnecessary if cast steel could be relied upon.

The fact that steel can be both cast and forged has brought together two groups of operations which were formerly quite separate : the foundry and forge.

Forging is the process of hammering, pressing, bending, and jointing metals while they are in a hot, pasty condition. In that respect it differs from wire drawing, spinning, and stamping in which the metal is worked in the cold, though a good deal of heat may be produced by friction. Metals which melt at a sharply defined temperature are incapable of being welded in the ordinary sense, though processes of jointing them will be described later ; wrought iron and steel are the most important welding metals. So long as the surfaces are hot enough, and free from a coating of oxide, mere pressure will suffice to form a joint. In order to secure the necessary cleanliness a small quantity of borax or something similar, which melts and forms a

FIG. 135.—AFTER PROPELLER BRACKETS OF WHITE STAR LINER "BRITANNIC." TWO CASTINGS. TOTAL WEIGHT, ABOUT 100 TONS.

To face page 158.

FIG. 136.—A PRODUCT OF A MODERN FORGE.

protecting covering, is dusted over the surface. Any oxide which has been formed is dissolved in this film and is squeezed out by pressure. A welded joint is improved by subsequent hammering.

But welding is only one of the minor processes of forge work. It is mainly a manual process, and with the large masses of steel that are manipulated nowadays manual processes have little scope. The modern forge that has grown out of the mediæval smithy is a huge structure that hums with machinery and reverberates with the thud of steam-hammers. Readers of the earlier volume will be familiar with the form, power, and delicacy of the latter tool. With it the metal may be subjected to the lightest tap or to a blow that shakes the very ground upon which the building stands. It is difficult to conceive of a tool with such a wide range of utility passing out of use, and probably no forge will ever be without one. But a century is a long time for any mechanical device to exist in its original form, and even to-day steel-makers are expressing a preference for the hydraulic press, first applied by Sir Joseph Whitworth in the middle of last century, and the rolling mill invented by Cort in 1783.

Perhaps the most interesting process allied to the use of the steam-hammer is drop-forging. The anvil contains the lower half of a die or mould, of the shape which the metal is required to assume. The upper half of the die is affixed to the under side of a heavy block of steel that can be moved up or down between guides. By means of a clutch this block can be drawn up to a height of from 2 to 10 feet and allowed to fall. If, when it is raised, a piece of hot metal is placed in the lower die, the falling weight smashes it at one blow into the required form. In this way many parts of a modern motor-car are constructed. Where the change of form is only slight the metal need not be heated, a single blow in the cold being sufficient for the purpose.

As an example of modern forge practice we may consider the formation of a large, seamless steel tube. The illustration, Fig. 136, shows such a tube, which was produced by the Darlington Forge Company. It is 9 feet inside diameter, 7 feet 8 inches long, and weighs 26 tons. The steel was first cast into a solid ingot, then a small circular hole was cut in the centre, and the mass of metal expanded in the hydraulic press to the required size for machining. The reader of an arithmetical turn of mind may be interested in calculating the thickness of the metal from the dimensions given above and the fact that 1 cubic inch of steel weighs 0·26 lb.

Forging is at the best a crude process, and the machine shop has always to be requisitioned to finish off the handiwork of the smith. An allowance must always be made for the amount to be removed in the lathe or planing machine, and as these are, or were until recently, relatively slow, it was customary for the smith to work as closely as he could to final dimensions. The invention of high speed tool steel, however, has made machining a cheaper process than accurate forging, and has considerably modified forge practice. It now costs less to run a lathe or planer than to maintain the furnace and steam-hammer. An interesting example is given in Harbord's *Metallurgy of Steel*. A 6-inch crank-shaft with the cranks at right angles would formerly have taken a long time to forge, and would have left the smith with only a thin skin of metal for the machine hand to remove. Nowadays it would be cut out of a slab like Fig. 137, the shaded portions being removed by a band saw in the cold. The cranks are then in

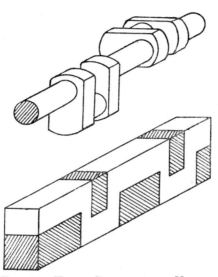

Fig. 137.   How a Crank-shaft is Made.

one plane. The shaft would be heated and twisted until the cranks were at right angles, and the portions of the shaft between roughly rounded by a light steam-hammer. The surplus metal would then be removed in a machine at the rate of 130 lb. per hour.

Of the other many interesting developments of the smith's art, space will not permit description. The ignorance as to the composition of iron that made Bessemer's process a commercial failure for the first four years has been dispelled, and the metallurgist no longer works by rule of thumb. Not only the effect of composition, but also the effect of previous history, on the properties of steel is now known with a degree of accuracy that would have astonished the ironfounder of thirty years ago. The Engineering Standards Committee have laid down exact speci- fications of the material to be used for various structural purposes, and the fiery furnaces, obedient to the intellectual control of man, pour out 500 tons of iron or 50 tons of steel per day, with a composition that can be calculated beforehand to one part in a thousand. The new century has presented man with materials that raise engineering from a primitive art to an exact science ; for he has learned to look beyond the mere naked-eye appearance and the approximate results of a crude test. Armed with the microscope he penetrates the hidden molecular society of which the steel is composed, and assures himself of the presence of that harmony which brings strength, of the absence of that discord which brings weakness and ultimate disunion.

M

CHAPTER IX

THE ELECTRIC FURNACE AND ITS APPLICATIONS

In 1892 Henri Moissan, of the Sorbonne, in Paris, commenced a series of investigations on the electric furnace, and thereby sowed the seed of industries which now utilize several million horse-power. Not content with startling the world and causing a flutter in feminine minds by making real diamonds by an artificial process, he discovered a number of new substances, and laid the foundation of manufactures which have exercised a profound influence on economic development. Other substances of great industrial value, which were scarce because they resisted the high temperature of the blast-furnace, became available in quantity and in a high degree of purity. New fields were opened for industrial enterprise, a fresh impetus was given to water as a source of power, and great factories sprang up around Niagara, in the Alps, and on the steep hill-slopes of Norway and Sweden. The hum of machinery arose once more amongst the mountains.

Apart from the use of heat as a source of power, the value of a high temperature depends upon three facts. Firstly, most bodies melt, and can therefore be moulded and cast into any desired form; secondly, the fluid condition renders mixtures more intimate and facilitates chemical change; thirdly, many substances are resolved into their elements and many new compounds are formed.

When Moissan began his experiments probably the highest temperature employed in industrial operation was about 2,000° C. Consider what this means. Water melts at 0° C. and boils at 100° C. Tin melts at 235° C., lead at 330° C., and zinc at 420° C., all below red heat. Then among the more commonly occurring metals there is a gap until we reach those of the coinage—silver, gold, and copper, which melt at 945° C., 1,035° C., and 1,050° C. Higher up the ladder of temperature cast iron melts at 1,000° C., pure wrought iron at 1,600° C., and platinum at 1,770° C.

The temperature of a blast furnace probably does not exceed 1,600° C. at the hottest point, and to get a higher temperature it is necessary to use gaseous mixtures in which the particles come into more intimate contact with one another. Readers of the earlier volume will remember that in the puddling process the iron became pasty as the carbon was removed, while in the Siemens regenerative furnace, heated by gas, the metal remains perfectly liquid at the end of the operation. In the Bunsen flame, fed with the proper mixture of gas and air, a temperature of 1870° C. is attainable ; 2,000° C. is produced by a mixture of oxygen and hydrogen ; and 2,400° C. by a mixture of oxygen and acetylene.

But in none of these cases can the actual temperature of the flame be communicated to the substance on which it plays. By an inexorable law of Nature heat flows from a high temperature to a low—downhill and not uphill—and some of it is lost in the transfer. A large amount is carried away by the waste gases, some is lost by radiation, and some escapes by conduction and convection. All these losses can be reduced by enclosing the flame in a casing which offers considerable resistance to the passage of heat, and this is the main advantage of a furnace over an open fire or flame. The heat is produced by the energy with which substances enter into combination, and the temperature will rise as more and more heat is generated, until the amount lost in a given time is equal to that produced in the same interval.

Moreover, when the temperature reaches a certain point the very substances whose formation produces the heat begin to decompose ; the combination is no longer possible, and the change which led to the evolution of heat is reversed. There is therefore a limit to the temperature obtainable in this way, which is independent of the losses by waste gases, radiation, conduction, and convection.

Now the conversion of electricity into heat is based upon a different principle. The passage of a current through a conductor is invariably attended by the production of heat, and a corresponding amount of electricity disappears. The greater the resistance which the conductor offers, the greater is the amount of heat produced. If, therefore, a copper wire, which is a good conductor, is replaced for a short distance by a wire of the same material but of smaller diameter, or by material of lower conductivity, a much greater amount of heat is produced there than at any other part of the length through which the electricity flows.

Again, the heating effect of a current is proportional to the square of its length. A current of 2 amperes produces four

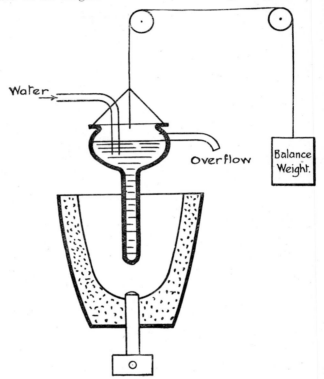

Fig. 138. DIAGRAM OF THE ORIGINAL ELECTRIC FURNACE INVENTED BY SIR WILLIAM SIEMENS.

times the amount of heat that a current of 1 ampere will produce ; a current of 3 amperes nine times, a current of 4 amperes sixteen times, and so on. If, therefore, a powerful current is conducted to a furnace by heavy copper cables and is then required to pass through loosely packed material of low conductivity and high resistance, this material may be raised to a white heat. Such an arrangement is called a resistance furnace because the heat is produced by the high resistance of the material with which the furnace is charged. On this principle furnaces were constructed by Sir William Siemens in 1879 and Cowles in 1886. The Siemens furnace is shown in Fig. 138. It consisted of a carbon crucible which was attached to one wire (or lead) of the source of supply, and a carbon rod [1] dipping into the material which was connected with the other wire. In this way a pound of iron was melted in an hour, and many other experiments were made which showed that the discovery was of great value. Siemens also constructed an arc furnace, which is shown in Fig. 139. Cowles' furnace was used on a commercial scale for preparing alloys of aluminium and copper.

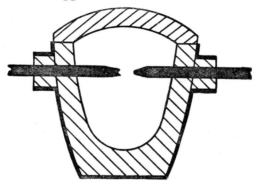

Fig. 139.   SIEMENS' ARC FURNACE.

The disadvantages of these furnaces for accurate laboratory experiments are the varying resistance due to the closeness or otherwise of the packing, and the alteration in composition during the process. Moissan therefore employed a different type, called an " arc " furnace, of which one form is shown in Fig. 140. The current is led in by two carbon rods which meet just over the substance under experiment, and are then separated. The points in contact are raised to a white heat, and when they are separated the electricity bridges the gap so formed and produces the highest temperature which has hitherto been obtained.

Moissan's furnace body consisted of blocks of lime enclosing a cavity in which the substance to be heated was placed. The

[1] The diagram shows a water-cooled metal electrode which was sometimes used instead of a carbon rod.

cavity was covered with a block of lime to prevent loss of heat
or access of air.   The heat was thus concentrated in a small
enclosed space, and the non-conducting property of the lime
served to prevent loss.   Thus in one experiment the cover was
3 cms. ($1\frac{1}{4}$ inches) thick, and yet when the current had been
switched on for ten minutes, and the under surface was melting,
it could be lifted by hand.   Magnesia will withstand a higher
temperature than lime and has the advantage that it is the only
oxide that is not reduced by carbon at the temperature of the
furnace ; but it conducts heat more readily.   When it was

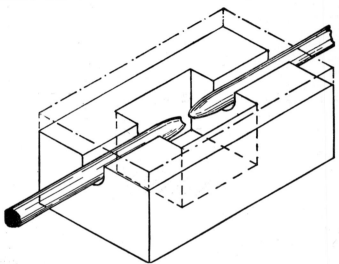

Fig. 140.   Moissan's Electric Furnace.

necessary to use it, thin plates were employed as a lining alternately with plates of graphite.   In some of the experiments on
a larger scale the furnace consisted of blocks of limestone, which
were speedily converted into lime.   The absence of any materials
other than lime or magnesia or carbon rendered it possible to
prepare substances of a high degree of purity.

In the earlier experiments the electricity was supplied by a
4 horse-power gas-engine and dynamo, which gave about 40
amperes at 55 volts.   Later a 45 horse-power steam-engine
driving a dynamo giving 440 amperes at 80 volts was used.

Finally, 100, 150, 300 horse-power was concentrated in the form of heat in the small enclosure containing the substance under examination. The temperature attained is impossible to measure accurately and difficult to estimate. But it certainly reached 3,500° C., and probably 4,000° C. would not be an exaggeration.

Under the influence of the enormous concentration of the higher powers employed the limestone gave off torrents of carbon dioxide, then the lime began to melt, and for experiments requiring the highest attainable temperature the lining of magnesia and graphite had to be used. The glowing crater could not be observed with the naked eye, and dark glasses had to be worn.

All the metals that melt below 1,000° C. or 1,200° C. boil in the temperature of Moissan's furnace. Thus in five minutes 103 grams of copper lost 26 grams, and flames of luminous copper vapour half a yard long streamed out of the holes through which passed the carbon rods. The fact that gold is volatile at temperatures near its melting point has long been known, and special precautions have to be taken in assaying the precious metal to prevent loss by vaporization. In the electric furnace 107 grams of gold lost 52 grams in a very short time !

The advance which this discovery represented may be gauged from the fact that the hour required to melt one pound of iron in Siemens' furnace thirteen years before was now reduced to a few minutes. Even 4 lb. of the far more difficultly fusible chromium were melted in an hour, and on one occasion no less than 22 lb. of molten metal were obtained. Many other substances, such as manganese, tungsten, molybdenum, titanium, vanadium, and silicon, were chemical curiosities, and had only been obtained previously with great difficulty in small quantities. They all have a profound influence on steel, and are essential constituents, with chromium and nickel, of most of the special steels which are now made in such quantities, and which have had such an important influence on modern manufacture.

Thus guns, projectiles, armour plate, tools, tyres, axles and other parts of machinery owe to a large extent their progress to the quantity and cheapness of substances which improve the quality of the steel used in their construction. Chromium confers toughness, tungsten and molybdenum hardness, titanium soundness, and vanadium strength to the material when added in appropriate amount. Manganese in quantity greater than 8 per cent gives exceptional hardness combined with ductility, and destroys the magnetic properties of the steel. Silicon in small percentages produces a suitable steel for springs. A new industry has thus been created to supply rich alloys of iron with chromium, tungsten, molybdenum, titanium, and vanadium, which are added to steel to render it more suitable for some specific purpose.

But to limit the world's debt to Moissan to the creation of but one industry would be to understate the case. His work forms the starting-point for a dozen. In the course of the investigations which culminated in the preparation of artificial diamonds, he repeated and extended Berthelot's experiments on varieties of carbon. After proving that there were only three forms of pure carbon—amorphous carbon, graphite, and the diamond—he showed that the first and the third are converted into graphite at the temperature of the electric furnace. Natural graphite is somewhat scarce ; it occurs in inaccessible districts, and there has been an increasing demand for it in recent years. It is now made in quantity at Niagara merely by passing a powerful electric current through anthracite, and is one of numerous substances now manufactured by electric processes where cheap power is available. The pencil with which you write, the " blacklead " used to polish the grate, the material that reduces the friction of the machinery in the neighbouring factory, the cores of the carbons in the arc lamps which illuminate the streets of the town in which you live, may all have had their origin in the dark recesses of an American mine. Torn from its hiding-place by giant powder, packed in trucks and hauled up the shaft at 600 feet a minute, it is whirled half-way across the continent to the foot of the Falls. There it is charged into a brick chamber, and subjected to the glowing energy of an electric furnace, which converts the hard black lumps into a fine, impalpable material, soft and greasy to the touch, and capable of innumerable uses for which it was originally unsuited.

Forty years ago, when the present writer was serving his time in the works, one of the most useful tools was an emery wheel. It was made of a hard natural oxide of aluminium mixed with some binding material and compressed into discs. Rotating at a high speed, it was capable of rubbing off rough edges of metal with the production of showers of sparks, quickly causing the metal to become red-hot. Since that time grinding has become one of the most accurate and useful workshop processes, capable of the delicacy required for scientific instruments, and the energy necessary for truing up an armour plate. But modern wheels are mostly made of a new substance, called carborundum, composed of carbon and silicon, and having the formula CSi. In 1893 Mr. Acheson of Niagara produced $6\frac{3}{4}$ tons, and nine years later 2,700 tons. His furnace was extremely simple. It was not a permanent structure, but was built up for each operation. A brick pit or box 15 feet long, 7 feet wide, and 7 feet deep had fixed in each end sixty carbon rods each 3 inches in diameter and 2 feet long, mounted in bronze sockets. The furnace was filled with about 10 tons of a mixture containing 34 per cent coke, 54 per cent sand, 10 per cent sawdust, and 2 per cent salt. The sawdust rendered the mass porous. Between the carbon poles was placed a core of finely broken coke along which the current passed. On dismantling the furnace the carborundum was found in a zone round the carbon core. Outside this zone was another substance, called siloxicon, having the composition C_2Si_2O. It is highly refractory and is used for furnace linings.

Probably the most important product of the electric furnace is calcium carbide, CaC_2. The value of this substance lies in the fact that on the addition of water it yields acetylene, C_2H_2, a gas of high calorific and illuminating value, the use of which for welding and cutting has been described in Chapter VII. A very common method of obtaining the gas is to use a holder similar in principle to the arrangement found in chemical laboratories for producing sulphuretted hydrogen. Most boys are probably familiar with Kipps' apparatus so frequently employed for this purpose. An acetylene generator consists of a gas-holder inverted over water, and a perforated box containing the carbide. As the latter is decomposed slowly by moist air, gas is being formed even when the apparatus is not in use. It is necessary therefore to limit their charge of carbide to the full capacity of the holder. Acetylene can also be obtained compressed in steel cylinders, but before this could be accomplished some difficulties had to be overcome. The formation of the gas is accompanied by absorption of heat, and it is therefore somewhat unstable. When it decomposes this heat is evolved. A chemical change that is accompanied by an evolution of heat is invariably more easily effected than one in which heat is absorbed. In the early attempts to compress acetylene in the same way as oxygen, hydrogen, and other gases are compressed, some explosions occurred. The heat engendered by the compression was so liable to cause an explosion that the method had to be abandoned. It was found, however, that the gas was very soluble in acetone, so the practice now followed is to compress it into a steel bottle partly filled with this liquid, which yields up its excess when the valve is opened. Acetylene is used for lighting country houses, and in this case the low-pressure system first described is used. For welding and cutting metals either a low-pressure generator, or *acetylene-dissous*, as the compressed gas is called, is employed, but for large work of this kind the compressed gas in necessary.

Calcium carbide is produced by heating lime and carbon in an electric furnace, and as such furnaces have been in operation since 1885 it is curious that its value was not recognized before. A considerable quantity of it must have been formed incidentally and regarded as waste material. In fact, Professor Vivian Lewis states that the boys at the Cowles Aluminium Works were playing with it in 1887, at least five years before it was known to be of commercial importance. Since 1903 it has acquired a new interest. As explained in another chapter it absorbs nitrogen at a temperature of about 1,000° C. forming calcium cyanamide, a valuable manure sold under the name of " nitrolim ". The nitrogen for this purpose is obtained by the liquefaction of air and subsequent distillation.

The number of different furnaces which have been patented can be counted by the dozen. The production is rising by leaps and bounds, and well over 250,000 horse-power is utilized in the industry.

THE ELECTRICAL MANUFACTURE OF STEEL

In addition to the application of the electric furnace to the production of alloys of chromium, tungsten, molybdenum, and other metals with iron, to which reference has already been made, the last ten years has seen a more ambitious development in connection with the manufacture and refining of steel. The Hérault furnace, shown diagramatically in Fig. 141, has been established at Froges, in France. Another type, using three-phase current, and therefore requiring three electrodes, is shown in Fig. 142, with the molten metal pouring from a vent near the bottom. It is really a refining furnace, and is based on the principle involved in the original furnace of Sir William Siemens.

FIG. 142.—ELECTRIC FURNACE POURING.

To face page 170.

FIG. 145.—BIRKELAND-EYDE FURNACES AT NOTODDEN.

To face page 171.

The material—in this case molten steel—forms one electrode, and the other is a pair of large rectangular blocks of graphite, dipping into it. By raising the electrodes until they just fail to touch the surface of the metal a pair of arcs is obtained, but by lowering them into the liquid metal heat is produced by the resistance. The lower part of the container is curved and provided on the outside with teeth which engage with those of a straight rack, so that the furnace may be tipped for pouring. Several similar furnaces have been designed.

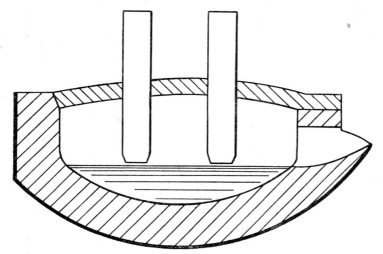

Fig. 141. HERAULT REFINING FURNACE FOR STEEL.

One disadvantage of this type, however, is the expense of renewing the carbon electrodes. A furnace which not only avoids this difficulty, but possesses the merit of constituting a very remarkable scientific achievement, is that invented by the Swedish engineer, Kjellin, though Mr. Ferranti was the first to suggest the method. The reader will probably be aware of the principle upon which an ordinary Rhumkorf or sparking coil works, but if not the following explanation will make the mode of operation of the furnace clear.

If two wire coils of any shape, but preferably round, are placed in the same plane or parallel with one another, then stopping, starting, or varying the strength of an electric current in one coil will produce currents of electricity in the other. An alternating current sent through one coil "induces" an alter-

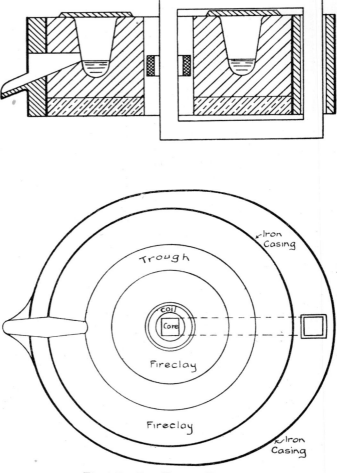

Fig. 143. The Kjellin Furnace.

nating current in the other. The total electrical energy which passes through a wire in a given time is equal to the strength (measured in amperes) multiplied by the pressure (measured in volts). In the two coils considered the quantity "induced" is very nearly equal to the quantity inducing it, but the number of volts in each is proportional to the number of turns of wire, and the number of amperes is inversely proportional to the number of turns. That is to say, if the second coil has half as many turns as the first the pressure will be half and the number of amperes will be doubled. The heating effect of a current flowing through a conductor is proportional to the square of its strength, so that if the strength of current is doubled the heating effect is four times as great, and if the strength is trebled the heating effect is ninefold, and so on. It is therefore possible by means of a current of small strength and high voltage to produce a current of low voltage but very great strength. Moreover, if the second coil is composed of a substance which conducts electricity less readily than the first, the heating effect will be greater—varying directly as the resistance. Thus, if the first coil is of copper and the second coil of iron, the heat produced would be six times as great as if both coils were of copper.

In the Kjellin furnace, see Fig. 143, a coil of wire with a core of soft iron is fixed at the centre of a ring-shaped trough of refractory material containing the constituents in the form of iron, scrap steel, etc., in the proportion necessary for the grade and class of steel required. When an alternating current is sent through the coil, very strong currents are induced in the ring of steel in the trough, and in a short time this is reduced to the molten condition.

Apart from the fact that there is no expense in carbon electrode, there is the accompanying advantage that no carbon at all need come into contact with the metal. The ring is covered with fireclay blocks, practically no air enters, and the resulting steel has the composition which was intended with a very high degree of accuracy. These furnaces were first used soon after 1900, and since then much larger ones have been erected at Gesinge, in Sweden, at Krupp's Works, in Germany, and in many other places, Though they are not very economical, this is of small consequence when water power can be obtained cheaply, and there are several firms in Sheffield which employ such a furnace for making special steels. According to the *Scientific American* the quantity of steel produced in electric furnaces rose (in round numbers) from 48,000 tons in 1909 to 129,000 tons in 1911.

This is, of course, very small in comparison with the world's production, but it consisted almost entirely of high-class tool and other special steels, and on that account is of great importance. The greatest increase, moreover, occurred in Sweden, where the Kjellin furnace had a monopoly.

Though the steel becomes quite fluid in the Kjellin furnace, it is hardly hot enough for some purposes, and has been improved by Roechling and Rodenhauser, Frick, and others. The Roechling-Rodenhauser furnace has two coils and two circular troughs which run into one another, forming a figure 8. The heat from the induced currents is supplemented by a current flowing through the bridge or central trough of the 8 from carbon electrodes fixed in the opposite walls of the trough.

Whatever the future may show, the invention must be regarded as one of the greatest achievements of electrical science. Everyone has become familiar with the fact that a crackling spark, or even an electrical tremor in a long wire will flash signals across oceans and continents, but probably few realize that a coil which can be handled with impunity may be radiating energy that will reduce to the molten condition a mass of steel placed a foot or so away.

THE ELECTRICAL MANUFACTURE OF NITROGEN COMPOUNDS

The process next to be described has for its object the capture of the nitrogen in the atmosphere, and in view of its connection with the food supply of the world (see Chapter XI) must be regarded as one of the most important inventions which have ever claimed the protection of the Patent Office. It differs from those which have hitherto been described in that it does not arise directly out of Moissan's work. For many years it has been suspected that the small quantities of nitrous and nitric acids contained in rainwater were the result of electrical discharge in the upper regions of the atmosphere, and though there is some doubt whether lightning is really the cause, Sir William Crookes showed in 1892 that an electric arc fed with an alternating current produced a flame, in which nitrogen and oxygen did actually combine. Unless these gases are removed as quickly as formed the heat decomposes them again. In 1895 Lord Rayleigh used the method to obtain argon. His apparatus consisted of a large glass globe into which the rods for forming the arc were passed. Air, together with additional oxygen required to combine with the nitrogen, was passed in by one tube and the oxides of nitrogen were absorbed in a solution of caustic soda, which, entering by another tube and impinging on the top of the

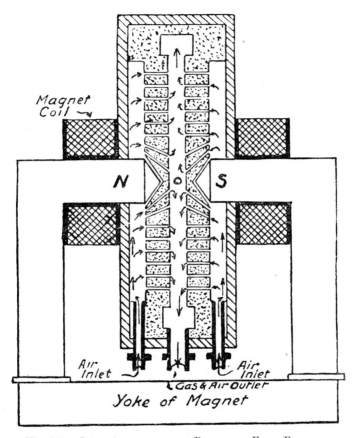

Fig. 144. SECTION THROUGH THE BIRKELAND-EYDE FURNACE.

globe, spread out in a thin film over the sides, and thus offered a large surface.

Sir William Crookes, in his Presidential Address to the British Association at Bristol in 1898, drew attention to the coming scarcity of nitrogenous manures, and emphasized the importance of rendering available the huge store of nitrogen in the air. In 1902 Messrs. Siemens and Halske, the famous Berlin firm of electrical engineers, patented a method by which air was passed through a flame arc which are spread out by a magnet so as to offer a large surface. The stream of glowing gas between the poles of an arc behaves as a flexible conductor, and a suitably arranged magnet will blow it out into a flare. The air passes through this and the oxides of nitrogen formed can be absorbed in water or alkalies. Similar processes are in operation in the United States and Italy.

Another method, differing somewhat in detail, was devised by Professor Birkeland and Dr. Eyde in 1905, and has been developed to a considerable extent in Norway. The furnace is shown in Figs. 144 and 145. It consists of a narrow circular brick chamber in an iron casing. The distance apart of the walls is only an inch or so, but the diameter of the latest type of furnace is 15 feet. The arc is formed between the closed ends of two copper tubes which enter radially, and are cooled by water circulating through them. Air enters by a number of narrow passages in the brick-work on both sides of the disc-like space, and leaves by a passage at the circumference. The arc is produced by an alternating current in which the direction is reversed fifty times a second between the poles of a powerful electro-magnet, and is blown out into a flaring half-disc, now above and now below the level of the copper tubes. The air therefore passes through a thin, almost continuous disc of flame, and on emerging contains from 1 per cent to $1\frac{1}{2}$ per cent of nitric oxide, NO. This gas combines spontaneously with a further proportion of oxygen to form the well-known red peroxide of nitrogen, NO_2, which if dissolved in water forms a mixture of nitrous and nitric acids. Nitrous acid is very unstable and absorbs oxygen to form nitric acid.

The temperature of the furnace is about 3,000° C. and the gases leave it at a temperature of from 800° to 1,000° C. They are led under steam boilers, and then passed through four towers containing quartz over which water trickles. The nitric acid formed here is converted into calcium nitrate by adding it to limestone. The gas which escapes absorption by water passes through two towers in which it is exposed successively to the action of milk of lime (lime suspended in water) and sodium hydrate, and at the end of the process not more than 3 per cent of the oxide of nitrogen escapes. The general arrangement of the power plant and the extent of the industry have been described in Chapter I. Another type of furnace has been invented by Herr Schoenherr, and is used at Notodden alongside the Birke-land-Eyre furnace, of which there were no fewer than eighty-three in 1912. The Schoenherr furnace consists of concentric tubes between which the discharge passes and the air circulates. This process has now been largely replaced by the catalytic process for making ammonia described on p. 251.

In order to appreciate at its true value this new industry of the twentieth century let us glance for a moment at its ramifications. Fig. 146 with its key shows diagrammatically the materials produced by the Norwegian Company and the uses to which they are directly or indirectly applied. The nitrates of lime and soda, the phosphates of lime and ammonia are important and, indeed, necessary fertilizers, while the nitrates of potash and ammonia which can be used for this purpose are generally reserved for the preparation of materials which command a higher market price. Nitrate of ammonia, for example, is an invariable constituent of so-called " safe " explosives, and yields " laughing gas ", which is used by the dentist to produce temporary insensibility to pain. Nitrate of potash is one constituent of ordinary gunpowder, the others being sulphur and charcoal. Nitrate of silver is used in photography, in water analysis, for silver-plating, and for cauterizing wounds, e.g. after the bite of a dog.

Aluminium nitrate has a narrower range of utility, being chiefly employed in dyeing and calico-printing. The fibres used for spinning and weaving contain fine pores. When such material is dipped into a dye it takes up the colouring matter, but does not fix it, so that the latter is removed on washing. Some hydrates, however, have the property of combining with the dye to form a coloured compound, and as the hydrate of aluminium not only possesses this property, but is itself colourless, it is particularly suitable. The hydrate is called a mordant, and the process is carried out by soaking the fabric in an aluminium salt and then forming the hydrate, in the presence of the dye, within the pores, by adding an alkali. The colour is then fast.

The nitrates of barium and strontium are used in the manufacture of fireworks, the former giving a crimson, and the latter

N

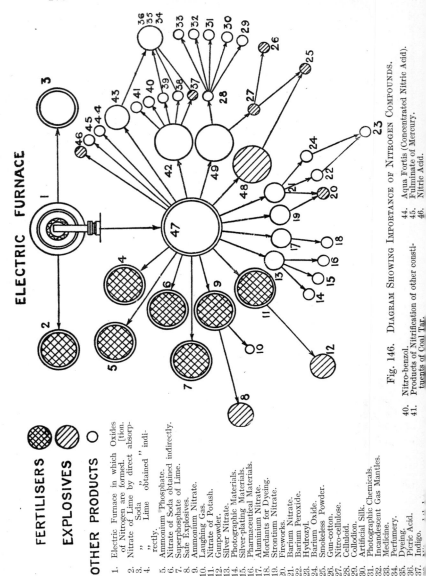

Fig. 146. Diagram Showing Importance of Nitrogen Compounds.

FERTILISERS

EXPLOSIVES

OTHER PRODUCTS

1. Electric Furnace in which Oxides of Nitrogen are formed.
2. Nitrate of Lime by direct absorption.
3. „ Soda „ „
4. „ Lime obtained indirectly.
5. Ammonium Phosphate.
6. Nitrate of Soda obtained indirectly.
7. Superphosphate of Lime.
8. Safe Explosives.
9. Ammonium Nitrate.
10. Laughing Gas.
11. Nitrate of Potash.
12. Gunpowder.
13. Silver Nitrate.
14. Photographic Materials.
15. Silver-plating Materials.
16. Pharmaceutical Materials.
17. Aluminium Nitrate.
18. Mordants for Dyeing.
19. Strontium Nitrate.
20. Fireworks.
21. Barium Nitrate.
22. Barium Peroxide.
23. Hydrozyl.
24. Barium Oxide.
25. Smokeless Powder.
26. Gun-cotton.
27. Nitro-cellulose.
28. Celluloid.
29. Collodion.
30. Artificial Silk.
31. Photographic Chemicals.
32. Incandescent Gas Mantles.
33. Medicine.
34. Perfumery.
35. Dyeing.
36. Picric Acid.
37. Indigo.

40. Nitro-benzol.
41. Products of Nitrification of other constituents of Coal Tar.
44. Aqua Fortis (Concentrated Nitric Acid).
45. Fulminate of Mercury.
46. Nitric Acid.

a green, colour to the flame. An essential constituent of all fireworks is, in fact, a nitrate or chlorate, more frequently the former, which provides the oxygen necessary for the rapid combustion of charcoal, sulphur, and other combustible material in the mixture. The sparkling effects are produced by coarse filings of magnesium and iron, which burn readily under these conditions.

From the comparatively harmless firework to the dangerous explosive, again, is but a short step, and the diagram indicates that when glycerine or cotton-wool is treated with nitric acid, nitroglycerine or gun cotton is produced. If these two are mixed together the well-known dynamite, or " giant powder " of the American miner, is obtained. This is a yellowish, waxy substance which has to be " thawed " in very cold weather before it is used When gun cotton is dissolved in a mixture of alcohol and ether the solution is known as collodion, and is used in photography, in medicine, and in the manufacture of incandescent gas-mantles. Its use depends upon the fact that when exposed to air, the alcohol and ether evaporate rapidly, leaving a thin film or skin of nitrocellulose of great strength. Treated with camphor, nitrocellulose gives celluloid, which is now used for such a vast number of articles—for combs and collars, for paper-knives and electric storage batteries. The ready inflammability of celluloid was humorously emphasized in *Punch* a few years ago. A little boy wearing a celluloid collar was standing aloof from the merry-makers round the Christmas-tree, in obedience to his mother's warning to keep away from the fire and light !

For some years collodion was (and still is to some extent) employed in the manufacture of artificial silk. The silkworm exudes a gummy substance from a fine orifice and this, on drying, forms the silk fibre. As cellulose comprises the woody tissue of all plants, it is easily obtained, but cotton-wool is perhaps more usually employed for the manufacture of artificial silk. The method originally devised was to dissolve nitrocellulose in ether and then to expel it through a fine hole. The solvent evaporated almost instantaneously and a fine thread of nitrocellulose remained which, on conversion into cellulose, can be spun and woven, though it is a little more brittle than the natural variety. This process has now given place largely to one in which the xanthate of cellulose is used instead of the nitrate. There is also a third process. Incidentally the curious fact may be noted that a large quantity of artificial silk is exported to China, the home of the natural material.

Another group of important substances is obtained by treating the products of the distillation of coal-tar with nitric acid. These form the starting-point in the manufacture of an almost uncountable number of colouring matters—the aniline dyes— and depend absolutely upon nitric acid for their preparation. From this source also comes most of the perfumes which are used in the manufacture of toilet soaps ; many flavouring essences, and a score of useful drugs. Then, again, nitric acid is employed in the manufacture of sulphuric acid, which has perhaps a wider application in industrial chemistry than nitric acid itself. Here, however, another process is available.

It should be observed here that another body, almost as useful as nitric acid, is obtained through the agency of the electric furnace. Calcium cyanamide, to which reference was made on page 170, yields ammonia when heated in a current of steam, and even a brief statement of the numerous uses to which ammonia is applied in arts and manufactures would take up almost as much space as has already been given.

Perhaps sufficient will have been said to indicate the profound influence upon modern industry of compounds of nitrogen. A perusal of Chapter XI will show also to what extent these compounds are necessary for maintaining the supply of food, and the dire consequences of a shortage in the supply of natural nitrates. It is really only during the last twenty years that man has adopted the plan of taking stock of natural resources. True there have been individual cases of scarcity in the world before, but on these occasions a nation or a tribe shifted its ground or took what it required from its neighbour. But these primitive forms of acquisition are less possible now ; war is prompted by some meaner or certainly less fundamentally human motive ; and the settled habits of the twentieth century prevent migration on a large scale. Moreover, in cases like the supply of nitrogen compounds migration would not have met the case. And all the tricks and trickery of politics and statecraft pale into insignificance beside the importance of the operations by which man is fed, the means by which he maintains his existence and works out his destiny in the great and intricate scheme of Nature.

FIG. 147.—A LARGE ALUMINIUM TANK.

FIG. 148.—WELDED COILS OF ALUMINIUM.

To face page 180.

FIG. 151.—ENGINE ROOM OF LARGE COLD STORE.

THE ARROW POINTS TO THE SUCTION VALVE OF THE PUMP, WHICH IS ALWAYS THICKLY ENCRUSTED WITH HOAR FROST FROM THE MOISTURE IN THE ATMOSPHERE.

To face page 181.

ALUMINIUM

The manufacture of aluminium, though the first electric smelting process to be developed commercially, is still conducted by the aid of electricity, but on a different principle. The first commercial electrical furnace was established by Cowles in America in 1885. It was based on the principle of Siemens' furnace of 1879, and was charged with a mixture of aluminium oxide and carbon. At the high temperature produced by the passage of the electricity through the badly conducting material the aluminium oxide was reduced to the metal. In 1886 C. M. Hall discovered that aluminium oxide, or alumina, as it is called, would dissolve in fused cryolite, a double fluoride of sodium and aluminium, 6 NaF, Al_2F_2, and that when an electric current was passed through the liquid, metallic aluminium separated at one pole. The process is therefore similar to that in which copper and other metals can be deposited by the aid of electricity from aqueous solutions. In all furnaces other than those for preparing oxides of nitrogen the electricity is first converted into heat, and the high temperature is the cause of the reactions which occur. In the aluminium furnace the electricity acts mainly *as* electricity, and the process is said to be *electrolytic*.

It is curious that though aluminium is one of the most widely distributed constituents of the earth's crust it should be so difficult to obtain. The oxide forms from 10 per cent to 20 per cent of all clays, but the difficulties of separating it from other materials is so great that one mineral forms the source of all the aluminium made in the world. This is bauxite, which occurs in Ireland, France, and N. America. Before it can be introduced into the furnace it has to be purified, and this is an expensive process which absorbs 40 per cent of the cost of manufacture. The purified alumina and fused cryolite are fed in at the top of the furnace and aluminium is drawn off at the bottom. The process is continuous, more alumina being fed in as required. The workman is informed when this is necessary by a simple device ; as the metal is drawn off the resistance of the furnace alters and so causes an electric lamp to light up. The readiness with which aluminium burns in air renders it extremely undesirable that any of it should rise to the surface, but this is difficult to prevent owing to the lightness of the metal. Its density in the liquid condition is only 2·54 and the average density of the other materials differs very little from this.   If the furnace becomes too full there is considerable loss.

The story of the rise and progress of the aluminium industry reads like a romance.   It is difficult to believe that the clay which we trample underfoot should contain rich stores of the beautiful white metal which combines in remarkable degree lightness and malleability.   And it is galling to think that a material which is so widely diffused and so useful should offer such obstacles to its recovery.   Still, the genius of Hall, Hérault, and others and the special properties of bauxite have enabled great progress to be made.   In 1855 aluminium was £28 a pound, and now even the best and purest samples are very little more than 1s. a pound.   In 1833 only 83 lb. were produced, and in 1885 only 283 lb.   After that the process described came into use, and in 1902 the amount obtained was 8,000 tons.   A year after that there were nine factories, three in America, two in France, and one each in Scotland, Germany, Switzerland, and Austria.

The early uses of aluminium were limited because the metal corroded rather easily and could not be welded ;   and these disadvantages detracted from the value arising out of its lightness and the non-poisonous character of its compounds.   Increasing purity of the metal, however, and the discovery of a process twenty years ago by which it can be welded, have led to an extraordinary increase in the range of usefulness.   For cooking utensils it is unsurpassed, provided the housewife or the cook avoids cleaning it with soda.   The replacement of rivets or lapped joints by the smooth weld enables it to be cleansed easily with a brush and hot water, and other materials are quite unnecessary. It has an advantage over enamelled ware in that it does not chip. For military and traveller's outfits its lightness renders it peculiarly suitable.

Its resistance to attack from the acids contained in foodstuffs has enabled it to be used for the manufacture of foods on a large scale, and it is now employed in the preparation and storage of Meat Extracts, Beer, Mineral Waters, Edible Oils, Margarine, Milk Preparations, Jams, Preserves, Chocolate and General Confectionery, Yeast, Sugar, Patent Foods and Emulsions. The illustrations, Figs. 147 and 148, show a large tank capable of holding 150 barrels of beer and two large welded coils.   Comparison with the men in both figures gives some idea of the size of the aluminium vessels, which can be made without difficulty. It is being used successfully in Soap Works, and in the manufacture of Candles and Waxes, Pharmaceutical Preparations, Organic Oils, Essences, Ethers, Perfumes, Colouring Matters, Varnishes, Rubber Preparations, Camphor, Acetone, Artificial Silk, Collodion, Celluloid, and Explosives.

Again, both the pure metal and its alloys are used in the manufacture of optical and scientific instruments, for the gear cases of motor-cars, the crank cases of aeroplane motors, the framework of airships, and many other examples of engineering work in which lightness is essential. Judging from experiments which have recently been conducted, it may come into extensive use for electrical transmission, for though it is not so good a conductor as copper it is much lighter and a thicker wire can be used. So long as copper and tin remain at their present high prices there will be no limit to the variety of purposes which the new metal can serve.

In addition to the foregoing substances there are many others prepared in smaller quantity which now rank among the productions of the electric furnace. Thus carbon bisulphide, formed by running melted sulphur over red-hot carbon, is wholly prepared in this way. Of the thousand tons of phosphorus which are required by the world's chemists every year, half is obtained by reduction in a resistance furnace, and a large quantity of glass is also made in arc furnaces. Wherever water-power is available for the production of electricity at low cost, there electric smelting furnaces are springing up. The relative scarcity of such power prevents any great development in this country, but Switzerland, Norway and Sweden, Canada and the United States are reaping a rich harvest.

CHAPTER X

THE ARTIFICIAL PRODUCTION OF COLD AND ITS APPLICATIONS

ON 7th May, 1913, the Cold Storage and Ice Association held their fourteenth annual dinner in London, and among other items the menu contained turtle soup from Queensland, salmon from Canada, lamb from New Zealand, beef from the River Plate, quails from Egypt, potatoes from the Canary Islands, pineapples from Jamaica, and apples from Australia. This list contains only a few of the articles of food that reach Great Britain from distant places. In days when the country was less thickly populated than it is now, and people lived farther part, the land within a radius of a few miles from each homestead produced all that the family had to eat. The industrial revolution of the eighteenth century led to the rapid growth of towns, and each town was supplied with food mainly by farmers in the immediate vicinity. Before the advent of railways food was mostly eaten within twenty miles of the land which produced it. Moreover, it had to be eaten quickly. Many articles of food will keep fresh and sweet only for a limited time, and in hot weather this is very short indeed. Until the middle of last century the only way to keep meat sweet was by salting it. But too much salt food is not wholesome ; mediaeval armies, deprived of unsalted food for some time, and the sailors who went on long voyages to remote parts of the earth, suffered terribly from scurvy. The value of ice for preserving food and in relieving fever was doubtless well known, for attached to many old mansions in England is an ice store. This consists of a thick-walled underground or semi-underground building generally located in a wood, and thickly thatched to keep out the warmth of the sun. Here were stored blocks of ice cut during winter frosts, to be used when the summer came round again. Then when the importance of preserving food became greater—when the population of the towns grew so that food had to be stored—ice was brought from Northern Europe in ships. And even to-day the south of France derives much of its ice from the glacier quarries of the Alps.

184

In spite of the fact that we eat more meat than any other nation it is improbable that even a simpler diet would have sufficed for the rapidly increasing population of the last century. The great industries of the country could not have been developed by a half-starved race of workmen. A growing anxiety made itself felt in the 'fifties, and about that time live cattle were first brought to England. It was estimated that the home production amounted to 910,000 tons per annum, or 72 lb. per head of the population. An importation of 44,000 tons of live cattle from America brought this up to 75 lb. per head. But America was not the only country which produced more than its people could eat. Australia and New Zealand had rich pastures, and were raising mutton faster than they could eat it, and faster than they needed to produce all the wool they could sell. The trouble was—how to get it to the old country ? Live cattle could be brought from South America—and as the population of the States grew, less and less beef was exported, and more and more tended to come from the rich grazing lands of the southern continent. Huge structures called " lairages " where the cattle were slaughtered on arrival, were erected at Birkenhead and elsewhere, and until the prevalence of foot-and-mouth disease in the Argentine led to an embargo on the importation of live cattle a few years ago, the lairages at Birkenhead found employment for 2,000 men. But Australia and New Zealand were a long way off.

Many attempts were made at this date to preserve meat otherwise than by salting it, but most of them were doomed to failure. A more successful plan was to cook and can it, and to-day an enormous amount of tinned meat comes into this country. The home population, however, wanted uncooked fresh meat, and the colonists were willing to supply it ; so as it was known that meat kept better in frosty weather, the new and comparatively little-known process of freezing was tried, and the first cargo of frozen beef was brought from America in 1877. Three years later the first shipment of mutton from Australia reached these shores, and since then the trade has gone up by leaps and bounds. In 1910 we imported 13,000,000 carcases of lamb and mutton from Australia and New Zealand, and 250,000 tons of beef from the Argentine.

Nor does the story end there. Rabbits come from Australia, apples from Tasmania, fish, fruit, and dairy produce from

Canada, and fruit from South Africa. How people would live without these vast supplies of food is an interesting problem. But the fact that they have not to make the attempt is due to the inventions and discoveries which have made cold storage possible, so we proceed, without more ado, to inquire how this is achieved.

THE ARTIFICIAL PRODUCTION OF COLD ; THE MANUFACTURE OF ICE

A moderately low temperature may be produced by mixing salt and snow or pounded ice. The mass speedily liquefies and the temperature falls. This will be familiar to most people who have seen salt used to clear the streets of snow. The liquid is easily washed down the drains, but so long as it remains it causes great discomfort to those who have to walk in it. The liquid soaks into the leather of the boots and makes the feet very cold. The lowest temperature is obtained when the mixture contains 23·5 per cent of salt, and this is about 8° F. below freezing-point. Calcium or magnesium chloride gives a still lower temperature.

Every reader will have observed how rapidly the puddles formed by rain disappear in warm or windy weather. What happens to pools of water would happen to pools of any other liquid to a greater or less extent. Some liquids evaporate and pass into invisible vapour more readily than others, and the process is in all cases hastened by heat. Under ordinary circumstances heat converts a solid into a liquid and a liquid into a vapour, though there are a number of solids which on being heated pass directly into the gaseous condition without first becoming liquid. The fact that puddles evaporate on cold as well as on warm days shows that water is slowly converted into vapour at ordinary temperatures. For each liquid there is a definite pressure exerted by its vapour at a given temperature ; when the vapour of a liquid in an open vessel becomes equal to the pressure of the atmosphere the liquid boils, and no matter how much more heat we apply, the temperature remains constant until the whole of the liquid has boiled away.

The effect of wind in drying up puddles is due to the fact that the vapour is swept away from the surface as rapidly as it is formed. As the air can only take up a definite quantity of moisture at any given temperature the fresh air blowing over the surface encourages the formation of more vapour. The drier the air the more rapidly will the vapour be absorbed and removed, and the more quickly will the puddle dry up. There is a well-known method, based on this fact, for ascertaining the direction of the wind when it is so light as to be hardly noticeable. If the finger is wetted and held up it feels cold on the side on which the wind blows. The air takes up moisture more rapidly on the side upon which it impinges than upon the other side. But why does this produce a feeling of cold ?

It has been remarked that the passage from the liquid to the gaseous condition ordinarily requires heat—that the gaseous condition is associated with a greater amount of heat than the liquid. The probable explanation therefore appears to be that if a liquid is caused to evaporate by mechanical means, without an artificial supply of heat, then the heat corresponding to the gaseous condition must be absorbed from somewhere. There are several simple and beautiful experiments which illustrate this theory. A little water is poured upon a wooden block so as to form a pool ; in this pool is placed a glass vessel or beaker, containing ether (which, by the way, is a dangerously inflammable liquid) ; air is forced through the ether by means of a pair of bellows. This causes the ether to evaporate rapidly, the heat ordinarily necessary for the formation of vapour is absorbed from the glass, water, and wood, and in a minute or two the beaker is found to be frozen hard to the block.

Another very beautiful experiment requires a piece of apparatus devised by Wollaston, called a cryophorus, which can be purchased from any instrument-maker cheaply. It consists of a bent glass tube with a bulb at each end, Fig. 149. A small quantity of water is introduced and boiled until all the air has been expelled. It is then sealed up ; and the whole of the apparatus is thus occupied by water and water-vapour. The lower bulb is immersed in a freezing mixture of ice and salt. The vapour in this bulb is condensed, and as the pressure in the apparatus is thereby reduced, the water in the upper bulb becomes cooled. This goes on until in a short time a layer of ice forms in the upper bulb. The water has thus been frozen by its own evaporation.

Two further scientific facts must now be noted. If air or any other gas or vapour is warmed it expands and occupies a greater volume. This expansion is a measure of the heat which has been supplied to it. But if the gas or vapour is caused to expand without heat being artificially supplied, it abstracts heat from its surroundings, or itself becomes cooled. The conversion of a liquid into a vapour and the increase in volume of a gas vapour, are both therefore processes which naturally take place when heat is supplied, so that if the processes are carried on without the artificial supply of heat the liquid or vapour or gas becomes cooled. In other words, if a liquid or gas is stretched by mechanical means, it is cooled. Similarly, if a gas is compressed, or a vapour is converted into a liquid by pressure, some heat is, as it were, squeezed out of it, and bodies in the neighbourhood become hotter.

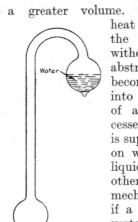

Fig. 149. WOLLAS-TON'S CRYOPHORUS.

There is a universal tendency for all bodies to acquire a uniform temperature, and no body can remain for very long hotter or colder than its surroundings without heat being continually supplied or taken away. A red-hot poker, taken from the fire, gradually cools, and a lump of ice gradually melts, at the ordinary temperature. The poker gives up its heat, and the ice receives heat. Bodies in the neighbourhood of the poker become warmer, and bodies in the neighbourhood of the ice become cooler, than they were before. These bodies in turn affect bodies more remotely situated, and the whole tendency in the universe is towards a dead level of temperature.

If, therefore, it is possible by means of a machine to convert a liquid into a vapour, or to cause a vapour, or gas to increase in volume, then that machine will produce cold. The substance usually employed is one of three gases: ammonia, sulphur dioxide, and carbon dioxide. The first two, as most schoolboys are aware, possess extremely pungent smells and are highly injurious if breathed even in small quantities; the third, if breathed in quantity, would cause suffocation. Great care has therefore to be exercised in the construction of the apparatus to avoid the possibility of leakage. Ammonia gas can be liquefied at ordinary temperatures by moderate pressures. Thus at 10° C.

a pressure of about 100 lb. on the square inch is sufficient for the purpose, while at the ordinary pressure of the atmosphere the boiling point of the liquid is –33·5 C. At –74° C. the liquid freezes to a white mass. Sulphur dioxide can be liquefied by passing it into a vessel immersed in a freezing mixture of ice and salt, when it condenses to a clear, mobile liquid which boils vigorously if the tube containing it is warmed by the hand. The boiling-point under ordinary atmospheric pressure is –10° C. In both cases the boiling-point is, of course, much lower if the pressure is reduced, that is, if the vapour is pumped away as fast as it is formed.

Provided carbon dioxide is not hotter than 31° C. it can be liquefied by pressure. At that temperature a pressure of nearly 1,120 lb. per square inch is necessary. If the gas is no hotter than 13° C. a pressure of about 750 lb. per square inch will suffice. When a stream of the liquid (contained in a steel bottle) is allowed to escape through a canvas bag part of it evaporates, and this causes the remainder to freeze to a solid white mass which remains in the bag. This " carbonic acid snow " does not evaporate very rapidly and can be used for maintaining very low temperatures. Mixed with ether it forms a pasty mass having a temperature of –80° C., while if the vapours are pumped away the temperature falls to –100 °C.

Sulphur dioxide is largely used in creameries and bacon factories because the pressure is low, there is a good deal of condensation, and the pump lubricates itself and requires very little attention.

It is clear that carbon dioxide involves much lower temperatures and higher pressures than either ammonia or sulphur dioxide. For these gases the apparatus may be of cast iron, but for carbon dioxide every part which has to bear the pressure must be wrought out of forged steel.

It will be sufficient to describe two commercial methods that are used for the production of cold. In the first the ammonia gas is compressed into a spiral tube or worm by a pump, and the heat resulting from this compression is removed by circulating cold water through a vessel containing it, Figs. 150 and 151. The same pump then reduces the pressure in another worm into which the liquid flows and then evaporates. Air or brine circulating round the second worm is cooled and is then sent to the chamber or tanks, the cooling of which is required.

The pump is called the compressor. In the diagram it is shown as double-acting, taking in the ammonia through the inlet valves, and compressing it through the outlet valves at both ends. The gas passes at every stroke of the piston into the condensing worm, which is cooled by water. Thence it passes

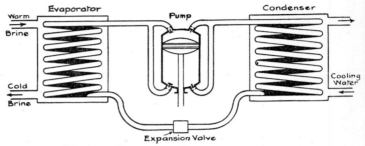

Fig. 150. DIAGRAM OF REFRIGERATING APPARATUS.

through a regulating valve, which reduces the pressure somewhat, into the evaporator, where it vaporizes. The heat required for evaporation is abstracted from the air or brine which surrounds this worm. There is thus no loss of material, the gas or air and brine being kept continually circulating through the apparatus.

A plant of this kind has been used in the United States for an interesting purpose, which was originally suggested by Lord Kelvin in 1852. If, instead of water, air is used to cool the gas after compression the air becomes warmed. This warm air is then circulated through a building in winter, while air from the refrigerator is circulated in a similar way in summer.

Another method dispenses with the pump. It depends upon the fact that ammonia is very soluble in water, which dissolves, at the freezing-point, 1,160 times its volume of the gas. If this solution is heated the gas is given off, and if the apparatus is closed the pressure rises. The compressed gas is cooled in a worm or coil of piping as before and then allowed to evaporate in another coil. The evaporation is hastened by the gas being placed in communication with the weak liquor from which the gas has previously been expelled. Except for small details —chiefly for separating moisture from the gas in the first half of the process—the plan is very similar to the one already described. Instead of the backward and forward strokes of a pump, the ammonia solution is alternately heated by passing steam round the vessel containing it, and cooled by substituting water for steam. The method can only be used for ammonia ; if sulphur dioxide or carbon dioxide is to be used, a pump is necessary.

If for any reason it is inconvenient to use air to convey the cold, a liquid which does not freeze at the temperature it is required to produce must be used, and for this purpose brine is employed. This brine is not always the solution of common salt to which the name is usually applied. Such a solution cannot be reduced in temperature below –8° F., and then only with 23·5 per cent of salt.   With a higher percentage salt separates out before this reduction of temperature is reached and with a lower percentage ice separates out from the solution. A solution of calcium chloride containing 25 per cent of the salt is more generally suitable.   This can be reduced to 18° F. below freezing-point without any separation of ice or salt occurring. In recent years a solution of magnesium chloride has found increasing favour. A 25 per cent solution remains liquid down to 22° F. below freezing-point and the solubility of the salt does not vary much with the temperature. In this respect it differs from calcium chloride and is similar to sodium chloride, but it can be used for lower temperatures than either.

The use of a refrigerating plant may now be described under the following heads :—

A. The manufacture of ice and the supply of refrigerating materials.

B. The maintenance of cold stores on land and sea.

C. Civil engineering and mining.

D. The liquefaction of the permanent gases.

THE MANUFACTURE OF ICE

Formerly nearly all the ice used for the preservation of food and other perishable articles, for various processes of manufacture requiring a moderately low temperature, and for the alleviation of fever in hospitals, was saved from the previous winter or brought by ships from Northern Europe. About the middle of last century small machines capable of producing a limited quantity of ice were in use ; but it is within the last thirty-five years that large plant for the purpose of producing many tons per day has been installed.   While the process

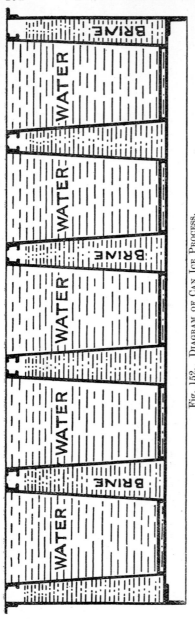

Fig. 152. DIAGRAM OF CAN ICE PROCESS.

consists essentially in freezing water in metal vessels by immersing them in brine at a low temperature, there are many details to which attention must be given if the ice is to find a sale at a remunerative price. If it is to come into contact with food or to be used for the table it must be prepared from water which is itself free from objectionable impurities. For although in the process of freezing water throws out dissolved substances, these are liable to be caught and encased in the interior of the block. Then, again, clear transparent blocks look better, keep better, and sell better than those which are opaque, Opacity is generally due to the enclosure of small bubbles of air, which are separated at the moment of freezing, and are entrapped by fresh ice before they have time to escape. The reader will probably have observed that ice formed over still water is almost invariably opaque, while that formed on running water is clear and transparent. The enclosure of air bubbles in artificial ice can be avoided by using distilled water, or by gentle agitation up to the moment of freezing.

In one method of manufacture which is adopted, the water is contained in thin metal boxes or cans (Fig. 152), which are slung in rows on iron rods, and lowered into a tank through which brine at a temperature of 12° F. to 25° F. is circulating. The

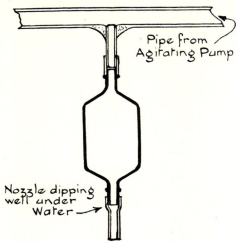

Fig. 153.   AGITATING APPARATUS.

water in the cans is agitated by thin wooden paddles which move backwards and forwards automatically, and are removed before the water freezes, or by the ingenious method shown in Fig. 153. The vessel is made like a double-necked bottle, and one is used for each can.   The lower ends dip well below the surface of the water, and the upper ends in each row are all connected with a pipe leading to the agitating pump.   As the pump works, the water in the cans is alternately sucked into and forced out of the bottle, which is removed in time to prevent it being frozen in.   The ice forms quickly at first, about an inch being produced in the first hour. After that the process is slower, 10 hours being required for 4 inches, 36 hours for 8 inches, and nearly 80 hours for 12 inches of ice.   The cans are made in various sizes, but in the factory described each holds 1¼ cwt. There are twenty in a row, and forty-eight rows in all, so that the total capacity is 60 tons.   As 48 hours are required to complete the process from start to finish, 30 tons of ice are produced per day.

o

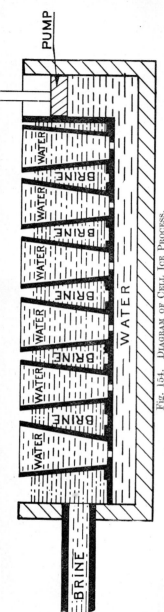

Fig. 154. Diagram of Cell Ice Process.

Another method is illustrated in Fig. 154. There the water is contained in cells which are fixed to the false bottom of a tank through which cold brine can circulate. The whole of the cells together with the false bottom of the tank are filled with water, cold brine is circulated, and the agitating pump is set working. With each stroke of the pump water is alternately forced into and out of each cell through the hole in the bottom, and this keeps the water in constant motion until it becomes solid. Before this point is reached, a piece of rope with an eyelet is suspended in each cell, and is frozen into the ice block. The ice is loosened from the cells by switching off the cold brine and circulating warm. The blocks can then be lifted out by the rope eyelets.

A most interesting method has been adopted in America. A number of jets of water are allowed to escape into a large cylinder from which the air has been extracted. The nozzles rotate and move up and down along the axis so that the water is sprayed evenly over the inner surface. As the cylinder is immersed in cold brine the water freezes into a tube about 6 feet long and 4 feet outside diameter, with walls a foot thick, which is sawn into blocks of saleable size. As this ice is formed in a vacuum there are no air bubbles, but the ice is opaque on account of its highly crystalline structure.

An ice factory may produce and distribute other cooling agents besides ice. Small refrigerating plant is sometimes made with liquefied gases in strong steel bottles, which are charged at the ice factory. These are used very largely for experimental research at low temperatures. Again, a portable freezing agent used for minor operations in surgery is ethyl chloride, which is sent out in strong glass tubes closed with a metal cap. When the cap is unscrewed the ethyl chloride volatilizes, and the " spray " of vapour is directed on to the part to be operated upon. The flesh is thus frozen. At that temperature the pain of the surgeon's knife is not felt, and an inestimable boon is conferred on suffering humanity.

It is obvious that the demand for ice is greater during the summer than in the winter, and the factory is liable to be relatively idle for a considerable part of the year. For several years now there have been skating rinks with real ice in London and Manchester. The latter is near an ice factory, from which it takes the surplus power during the winter. It consists of a large hall covering a shallow tank filled with water through which cooled brine passes. Twice a day the ice is swept, flooded, and re-frozen, to keep the surface in good condition. Here for a small sum a most delightful winter pastime and health-giving exercise can be enjoyed, and practice in a graceful art can be obtained by many to whom an annual visit to Switzerland or Northern Europe would be an impossibility. So Discovery and Invention defy the English climate and contribute to the health and pleasure of mankind.

COLD STORES ON LAND AND SEA

When heat is required to pass from one fluid to another, a thin-walled metal plate is the best form of partition, and this has been used in the production of ice. But when a cold store is to be erected, in which perishable articles are to be kept at a low temperature, the walls must be so constructed as to permit as little heat as possible to pass through from the outside. The building has, in fact, to be insulated. The walls are therefore made of considerable thickness (see Fig. 155) and are lined with 7 inches or more of flaked charcoal, silicate cotton, granulated cork, or some other material which offers a high resistance to the flow of heat. This insulating material is held up by an inner skin of boards. All steel girders, pillars, and other masses of metal are similarly covered with non-conducting material.

These large rectangular buildings with five or six floors (see Fig. 156) are usually cooled by a current of air which passes over the brine pipes ; but sometimes the pipes themselves

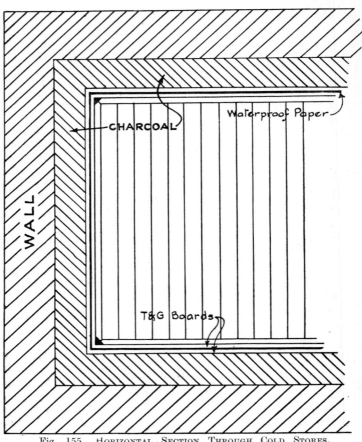

Fig. 155. Horizontal Section Through Cold Stores.
Thickness of wall exaggerated to show construction.

enter the building with or without cold air. The circulation of air ensures better ventilation, but it is liable to be very dry, and this causes shrinkage of the food. The air is cooled by being blown through a network of brine pipes over which brine is trickling. In most cases a freezing temperature is not required. In the case of food the temperature need only be such that it will arrest the processes of decay ; in the case of furs to prevent the hatching of moth's eggs ; and in the case of wines to maintain that temperature at which the flavour develops most perfectly. The table [1] on page 198 will convey some idea of the enormous variety of goods with which the cold storage manager has to deal.

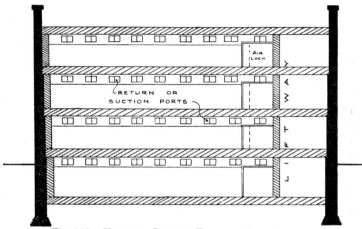

Fig. 156. VERTICAL SECTION THROUGH COLD STORES.

But the table indicates something more than this. It conveys some idea of the great range of experiments which must have been undertaken in order to determine the most suitable limits of temperature for each article stored. Few outside the actual business of food supply and distribution can have any idea of the influence of cold storage on their habits. For example, probably most people imagine that the fowls, geese, and turkeys which appear in magic multitude in the poulterer's shop window at Christmas time were gaily picking up their food in the farm-yard a week before. But the writer was in a cold stores in the middle of last November and saw hundreds of Christmas birds plucked and frozen as hard as blocks of wood, waiting for the

[1] From J. Wemyss Anderson's *Refrigeration.* (Longmans.) Much work on the most suitable temperature has been done during the war, and Professor Anderson kindly revised the table for the second edition.

STORAGE TEMPERATURES
IN DEGREES FARENHEIT

10°	15°	20°	24°	26°	28°	30°	32°	34°	36°	38°	40°	45°	50°
Frozen Mutton		Ham and But-ter (frozen)	Poultry Game for storing		Chilled Meat		Celery, Oysters	Vegetables, Fresh Fruit, Berries, Canned Goods, Furs (undressed), Syrup			Dates, Figs, Dried Fruits, Sugar, Wines, Flour		Clarets
Fish to freeze 0° to 5°	Furs (long stor-age)	Frozen Meat (gene-ral), Rabbits		Fresh Fish, Furs (dressed)			Apples, Eggs, Cheese, Milk		Bananas, Tomatoes, Peaches			Oranges	
Game to freeze 5°				Woollen Goods, Carpets				Cigars, Tobacco, Cider, Grapes, Potatoes, Lemons, Onions				Ale, Beer, Porter, Wines, etc. (in bottle)	
Frozen Eggs		Margarine			Hops	Butter (short period), Lard, Pork, Ham			Porter, Beer, Ale in (casks), Nuts				

demand which experience has shown would come.   Moreover, there were thousands of rabbits from Australia, packed in the same boxes in which they had travelled in the icy hold of a refrigerating steamer, to supplement the supply of English rabbits.   Here also was a piece of venison waiting for some aldermanic feast, and capercailzies, pigeons, and other delicacies, ready at short notice to be exposed for sale in the shops.   It has been said that cold storage has had something to do with keeping up the price of food, and certainly it prevents material being sold at any price to clear.   The dealer in fresh meat, fish, and other perishable articles can keep a smaller stock in his shop, because he knows where to get more with very little delay.

In addition to these large stores, which are like furniture repositories in that they accept suitable goods at a rental until they are required by their owners, there are numerous specialized installations attached to particular industries, in which the manufacturing processes can be carried on most effectively at low temperatures.   Thus refrigerating plants are often attached to dairies and butter and cheese factories, to bacon factories, to breweries, to dyeworks, and to many chemical works.   It has been found that while the temperature of boiling water is sufficient to destroy all forms of animal and vegetable life, the roots and bulbs of many plants are not adversely affected by many degrees of frost.   So long as they are exposed to a low temperature they remain dormant, but immediately they are removed to more genial conditions, they start into growth and come into flower in shorter time than under natural conditions. In this way plants can be retarded and flowers can be obtained over nearly the whole year.   The trade in retarded plants has risen to enormous proportions.   Every year thousands of bulbs of Japanese lilies are placed in cold store, while the number of crowns of lily of the valley so treated is measured in millions.

Turning now from provision on land to that on sea, every passenger vessel undertaking long journeys carries large quantities of fresh food in cold chambers.   This not only relieves the tedium of a long voyage, but contributes materially to the health of the passengers and crew.   Scurvy is now almost unknown ; salt junk and hard ship's biscuits are now memories of a former age ; and many disadvantages of ocean travel described by Clark Russell and other writers have been removed by the artificial production of cold.

But cold storage at sea has a wider significance than that of comfort to travellers.   There are more than 250 vessels engaged in conveying beef and mutton to the teeming populations of our manufacturing centres.   From Canada, South America, South Africa, New Zealand, and Australia comes wholesome food to nourish and strengthen those who toil in our workshops, factories, and mines.   Without this regular supply of the products of far-off lands our trade could not have developed, our wealth could not have grown, and our sons and daughters must either have starved or emigrated to the spacious and fruitful lands across the sea.

The earliest methods of refrigeration in use on the Atlantic were the circulation of water cooled by ice and salt, and the circulation of air which had been compressed and afterwards cooled by expansion.   In the 'eighties ammonia compression machines came into use, and now either ammonia or carbon dioxide is regularly employed.   The first source of our foreign meat supply was the United States.   The cattle were killed in Kansas City or Chicago and conveyed in refrigerator railway cars to Boston or New York for shipment.   By 1901 the amount obtained in this way reached 160,000 tons.   But the United States was growing rapidly, and from this time forward the trade was transferred gradually from North to South America.

In meat-carrying vessels the main body of the ship's hold is divided up into a number of chambers, along the roof and sides of which pass the pipes carrying the cold brine, Fig. 157.   While mutton, rabbits, and pork may be frozen hard without ill-effects, beef deteriorates considerably owing to the bursting of small blood-vessels.   Practically all the beef, therefore, which has come to this country since 1899 has been submitted to a temperature of 28° F. to 30° F., and is known as *chilled* beef.   Under these conditions its quality is unimpaired, and it will last three weeks in good condition.   The fact that this is just long enough to enable the meat to come from South America and be consumed in this country is not without significance.

Extraordinary precautions are taken to secure cleanliness and to maintain a steady temperature.   Each kind of meat carried must have a separate compartment, and while frozen carcases can be packed like ordinary cargo, chilled meat must be hung.   The chambers are sealed and the seal must be unbroken at the end of the voyage.   The temperature must be constant

FIG 157.—A "CHILLED BEEF" CHAMBER ON A MEAT-CARRYING VESSEL.

To face page 200

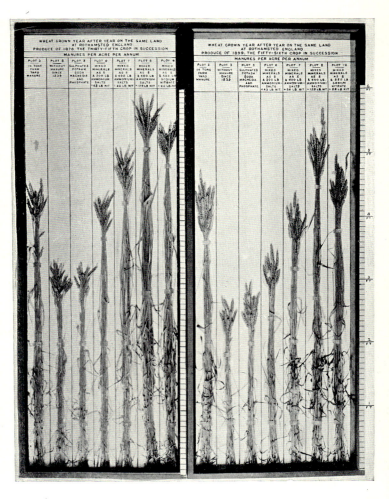

FIG. 162.—RESULT OF CONTINUOUS WHEAT-GROWING EXPERIMENTS
AT ROTHAMPSTEAD, ENGLAND, FOR 56 YEARS.

To face page 201.

within a degree, and the difficulty of securing this in a vessel which crosses the equator can be imagined. The machinery is duplicated or even triplicated to provide for breakdown, and the complete equipment for a large vessel may cost £50,000. In some cases self-registering thermometers have been used, but they are not always dependable. Another plan is for the engineer to take the temperature at stated times by lowering a thermometer down a tube leading from the deck into a hold ; but the warm, moist sea air meeting the dry, cold air from below causes deposits of snow in the tube. Some years ago a very ingenious device was patented. An ordinary thermometer illuminated by an electric lamp is hung near the lower end of the tube, and the scale is reflected up so that the temperature can be read by an observer on deck. The human element is avoided by closing the upper end with a camera. At stated periods the *temperature* and the *time* at which the observation is made are photographed, and the film, when developed, is a faithful record of the conditions during the voyage.

Another industry which would be quite impossible on a modern commercial scale, without the processes which have been described, is sea fishing. There are some 1,500 trawlers engaged round the British coast, and many of these are equipped with refrigerating apparatus. A voyage extends from two to four or five weeks. If you buy a *fresh* plaice from the fishmonger's it is not likely to have been caught less than a week ago, and it might be a month old.

Refrigeration serves a special purpose in the Royal Navy. Cordite, the powerful explosive used in modern guns, deteriorates if kept in a temperature higher than 70° F. In a warship required to do duty in any part of the world, it would be impossible to keep the magazine below this temperature unless artificial means were available. Consequently, among the enormous amount of machinery with which the vessels of the Navy are filled, are to be found refrigerating machines which not only preserve the food, but also maintain ammunition in good condition.

CIVIL ENGINEERING AND MINING

The engineer has frequently to excavate in boggy ground, or in ground so wet and soft that the sides fall in before he can complete his task. He overcomes the difficulty by sinking brine pipes in the soil and freezing it solid, so that he can complete his excavations, lay his foundations, and build his walls. Once he has reached solid ground he can fill up the cavity with masonry and defy the bog or quicksands which formerly barred his way.

Again, the miner desires sometimes to reach coal or other minerals which lie beneath a bed of water-bearing strata. When he attempts to sink the shaft the water rushes in faster than it can be pumped out. In these circumstances a ring of vertical brine pipes is buried round the spot where the shaft is to be sunk, and the ground frozen solid in the form of a cylinder which holds back the water. As the sinking is carried on, the sides are bricked or " tubbed " with an iron casing. The completion of the ice-wall takes from four to ten months, and it has to be maintained solid for from six to fifteen months to allow of the shaft being completed. Magnesium chloride is used in preference to calcium chloride or common salt because it can be relied on to a greater extent not to form any deposit in the pipes, which cannot be taken up for inspection and cleaning when once the process has been started.

LIQUID AIR

In none of the cases so far considered has the temperature been very low—certainly not lower than the natural cold of the Arctic and Antarctic regions. Far lower temperatures, however, can be obtained, and the practical results have been of first-rate industrial importance. From what was said in the earlier part of this chapter it will be obvious that the simplest and most effective method of producing cold is by the rapid evaporation of a liquid. The colder this liquid is and the more rapidly it evaporates the lower will be the temperature produced. Now the only substances which evaporate rapidly at low temperatures are those which at ordinary temperatures exist in the gaseous condition. Consequently, the problems to be studied are associated with those which occur in the liquefaction of gases.

In the first quarter of the last century Faraday liquefied chlorine by a very simple method. This gas was led into water contained in a vessel immersed in a freezing mixture such as is described on p. 186. It formed a crystalline compound with the water. The crystals were placed at one end of a bent tube and the other end was sealed up. On warming the end containing the crystals the chlorine gas was evolved and condensed to a yellow, oily liquid at the other end. Using the cold produced by a mixture of ice and salt, he succeeded in liquefying sulphur dioxide, ammonia, and other gases.

Returning to the subject again in 1845, and using a mixture of solid carbon dioxide and ether, he liquefied sulphuretted hydrogen, nitrous oxide, and hydriodic and hydrochloric acids. The high pressures used involved no little danger. Faraday himself had some narrow escapes, but there seemed to be a fascination about the experiments, which attracted many scientific men. Caignard de la Tour and Colladon worked at the problem, and Thilorier had an assistant killed by the bursting of a cast-iron vessel containing liquid carbon dioxide which had been prepared for a lecture in Paris. The great problem was to liquefy the so-called permanent gases, oxygen, hydrogen, and nitrogen, which had so far resisted all attempts, though Natterer, for example, employed a pressure of nearly 60,000 lb. per square inch. The most important step was taken by Andrews in 1868. He showed that in the case of carbon dioxide no amount of pressure would cause it to liquefy unless the temperature were below 31° C. This temperature is called the critical temperature, and the pressure required to liquefy the gas at that temperature is called the critical pressure. All failures to liquefy the permanent gases had failed because the temperature had not been sufficiently low.

The first actual liquefaction of oxygen was accomplished independently and nearly simultaneously by Pictet and Cailletet. Both subjected the gas to enormous pressure and then allowed it to expand. Pictet merely opened a cock and permitted the gas to escape. This expansion caused intense cooling, and a stream of liquid was obtained. It could not, however, be kept. Cailletet compressed the gas in a glass tube and then suddenly increased the volume by rapidly unscrewing a plunger. A mist, and then a meniscus separating gas and liquid, formed in the tube.

Further progress was made by Wroblewski and Olszewski working together, and subsequently by the latter alone. To high pressures they added intense cooling, by surrounding the vessel containing the gas by another liquefied gas which was kept boiling by a pump. In this way oxygen and nitrogen were

liquefied and some of the properties of the liquids were determined. Soon afterwards Linde in Germany, Dewar and Hampson
in England, and Tripler in America succeeded in obtaining liquid
air in quantities, and the problem which had baffled scientific
workers for a century was solved.

In order to understand how this result has been achieved
it is necessary to recall an experiment by Lord Kelvin and Dr.
Joule. If a gas expands without performing work, no appreciable
cooling takes place. The cooling is a measure of the external
work performed. For if there is an attractive force between
the particles internal work must be done, and an equivalent
amount of heat must be absorbed. Kelvin and Joule passed the

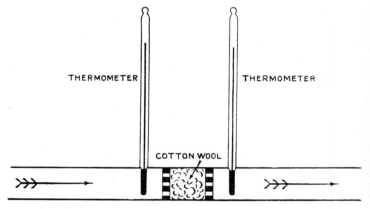

Fig. 158. KELVIN'S POROUS PLUG EXPERIMENT.

gas through a porous plug in a tube fitted with delicate thermometers as in Fig. 158. The gas passed in the direction shown
by the arrows. The slow diffusion of the gas involved no external
work, and any difference between the readings of the two thermometers would be due to internal work. All gases except hydrogen
showed a lower temperature (about $-0.25°$ C.), indicating that
there was attraction between the molecules which had to be
overcome during expansion. Hydrogen gave an increase
of temperature, and this could only be explained on the assump-
tion that the molecules of hydrogen repelled one another. At
a lower temperature it has since been found that hydrogen
behaves in the same way as other gases.

The important and far-reaching principle just described
enabled Linde, in 1895, to liquefy air by means of the apparatus
shown in Fig. 159. Highly com-
pressed air is passed through the
inner of two concentric tubes,
whence it issues from a fine
orifice. This orifice serves the
same purpose as the porous plug in
Kelvin and Joule's experiment, and the
issuing gas is cooled. The air passes
back through the outer tube, thus
lowering still further the temperature
of the air before it leaves the orifice. As
the air leaves the inner tube, therefore,
it is continuously lowered in temperature until at last liquid air drips from
the end of the inner tube into the vessel
below.

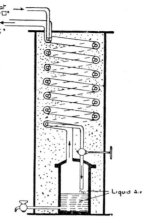

Fig. 159. LINDE'S APPARA-
TUS FOR LIQUEFYING AIR.
(After Mellor.)

That this liquid is air is hardly conceivable ; the air that passes freely in
and out of our lungs ; the air that
rustles through equatorial forests and
moans through northern pines ; the air that devastates the
southern states of America and hurls giant steel ships on the
rock-bound coast ; and yet lying here tamed and with all the
fire taken out of it. Now and then a turbulent bubble breaks from
bondage, or an angry tremor ripples across its surface. Still,
so long as it is kept in an open vessel it will pass away quietly
into gas ; but if an attempt is made to confine it in closed vessels
it will burst its bonds and scatter its prison in a thousand
fragments.

Liquid air boils at –181° C. It is clear and colourless. A
test-tube full of mercury plunged into the liquid is frozen into
a solid rod which can be hammered into various shapes like
wrought iron. Inelastic bodies become elastic, and indiarubber
becomes so brittle that it can be broken with a blow of a hammer.
A small quantity of the liquid poured into the boiler of a model
steam-engine evaporates rapidly and causes the fly-wheel
to spin round as though under a high pressure of steam, while
the boiler becomes crusted with ice from the moisture in the
atmosphere. At this temperature the electrical resistance of

metals decreases to such an extent as to suggest that at a still lower temperature it would become a vanishing quantity. A soap bubble blown on the end of a thistle funnel and held in the vapour over the liquid becomes frozen, and when struck upon a hard substance such as table-top it is shivered into invisible particles.

In the course of his researches Professor Dewar devised a flask, Fig. 160, which enables liquids to be retained for a considerable time at a low temperature. The Dewar flask has double walls, the space between which is exhausted of air. The object is of course to prevent the heat entering the flask through the walls, and if the air were left in, warmth would be conveyed

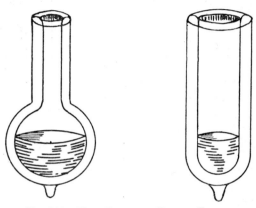

Fig. 160.   Two Forms of Dewar Flask.

to the inner vessel by conduction, and by convection currents, as well as by direct radiation. The first two processes are prevented by pumping out the air, and the effect of the latter is largely reduced by silvering the inner walls. Professor Dewar found that if a small quantity of mercury was introduced into the space, its vapour condensed in a brilliant mirror which acted in a similar way.

Any arrangement which will prevent heat passing in one way will be effective in preventing its flow in the opposite direction, so that a body can be kept hot just as well as one can be kept cold. The well-known Thermos flask is, in fact, simply a Dewar flask enclosed in a metal and leather case, and provided with a cover.

The liquefaction of air has had at least two interesting industrial applications. Oxygen gas has for many years been a regular article of commerce. It is required for blowpipes for brazing and welding, for the limelight, for scientific investigation, and for use in hospitals. The method of manufacture until recently was by Brin's process, which depended upon the fact that barium monoxide absorbs oxygen at a high temperature, forming barium dioxide, and gives it up at a higher temperature, re-forming the monoxide. There are practical difficulties connected with alternations of temperature, so that alternations of pressure were employed instead. Air was pumped into a series of cylinders contained in a brick furnace maintained at a temperature of 700° C. The oxygen was absorbed and most of the nitrogen escaped through a valve at the end of the series. After ten or fifteen minutes valves were opened and others closed so that the effect of the pump was reversed. The first portion of the gas which came off was nitrogen, and this was allowed to escape into the air through a valve. This valve was then closed and the oxygen passed into a gasholder. Oxygen prepared in this way contained from 4 to 10 per cent of nitrogen, but was quite pure enough for all ordinary purposes. The modern process was devised by Linde in 1895. It depends upon the fact that nitrogen boils at a lower temperature than oxygen, so that when air is liquefied and allowed to boil, the nitrogen passes off more rapidly than the oxygen. The apparatus is shown in Fig. 161. A continuous stream of liquid air

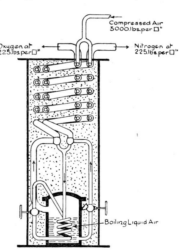

Fig. 161. APPARATUS FOR SEPARATING NITROGEN FROM LIQUID AIR. (After Mellor.)

flows into the vessel and the constituent gases are separated. The oxygen is obtained practically pure, but the escaping nitrogen contains over 7 per cent of oxygen. By a process patented by Claude in 1903 this nitrogen is freed from all but a trace of oxygen, and then passed over calcium carbide heated to a temperature of 1,000° to 1,100° C. so as to form calcium cyanamide. The importance of this substance as a fertilizer will be discussed in Chapter XI.

Before Moissan in 1892 and Linde in 1895 began their experiments, the highest temperature which had been obtained was about 2,000° C., and the lowest that could be maintained for any length of time was about 100° C. below zero. The limits are to-day about 4,000° C. and –270° C., giving a range of 4,270° C. It is now possible to examine the properties of matter and to carry on experiments from the point at which only one substance remains liquid, to the point at which metals boil and every known substance passes into vapour. It has been estimated that the lowest temperature of inter-planetary space is –260° C., and the temperature of the sun about 6,000° C. This gives a range of 6,260° C. Man is thus able to produce a temperature 10° C. lower than the minimum in the Solar System, and to climb two-thirds of the way to the temperature of that luminous body whose energy, transmitted across 92½ millions of miles of space, preserves the earth from the relentless grip of eternal ice.

CHAPTER XI

SOIL AND CROPS

IF man had been born into a world in which the climate was always genial, the soil for ever fruitful, and the vegetation plentiful and varied, or if he had been permitted to stay in the Garden of Eden, he might have avoided labour. The original man on the earth worked because he was hungry, and also, perhaps, because he was cold. He slew such beasts as he could with primitive weapons, clothed himself in their skins, ate the flesh, and varied his diet with fish or fowl, and fruit and roots from the primeval forest. So long as he was a more or less solitary wanderer, these supplies did not fail him. But as the family or small group grew into the tribe, and more mouths were to be fed, food had to be collected from a wider area and stored in the form of flocks, and herds, and granaries. Instead of hunting and killing his meat when he wanted it, he kept it alive in captivity ; and instead of searching for and seizing fruits and roots as hunger assailed him, he conceived the idea of growing them near his abode. In this way arose the Agricultural Arts, upon the practice of which every man, peer or peasant, depends for his food and clothing.

In order to satisfy these needs, flint instruments were replaced by those of copper and iron. Iron, in turn, helped man to obtain more fuel and more iron ; the latter gave him the steam engine ; the steam-engine developed manufacture, and manufacture made greater demands upon agriculture to supply the raw material by which so many of the later wants of civilization are satisfied. The concentration of people in towns and their employment in the manufacture of goods threw the burden of providing food upon the shoulders of fewer men, without whose labours no manufacture could be carried on. Agriculture is therefore called the mother of industries, and still claims the larger share of human energy, human knowledge, and human skill. Even in England—the Workshop of the World—it is the largest occupation, and in every other country except Belgium it is more important than all other industries together.

If the term Agriculture is used in its broadest sense it includes the tilling of the soil and the cultivation of all the plants which yield material for food or manufacture. The cotton fields of the United States and Egypt, the rubber, coffee, cocoa, tea, and banana plantations of the tropics, the fruits of temperate and sub-tropical climates, timber, flax, hemp, jute, and the numerous plants that are grown for their fibre, and the cereals or flowering grasses which in so many instances form the staple food of man are all included. Flocks and herds are usually omitted because except in densely populated countries where every inch of ground has to be utilized, and where the demands of large towns render it profitable, pastoral and agricultural pursuits are each confined to more or less separate areas.

The history of England shows in a striking way the effect of manufacture upon agricultural practice. From the time of the Norman Conquest or earlier the land was cultivated in open fields, in which each man had a share, and to the labour on which all contributed, except the lord of the manor. In the sixteenth century much of the land was enclosed by the land-owner and used for wool production, and the remainder became less productive and less capable of producing the food that the nation required. From the middle of the eighteenth century further enclosures took place, so that within another hundred years the open field system had entirely disappeared from the English landscape. But this time the movement was accompanied by improved methods of cultivation. Large farms arose, the small occupier was driven to the wall, and new systems of farming, more expensive, but capable of yielding a higher return from the soil, came into being. Still, no possible methods could produce the food necessary to maintain the growing population, and instead of remaining an independent self-feeding community, Great Britain is to-day more dependent upon imported foodstuffs than any other country in the world.

The problem of economical and successful home farming, then, is one of profound importance, and the soil which has yielded so generously of its fruits during the centuries in which the nation was being created, has now to be cultivated with that skill and foresight which scientific knowledge alone can supply. In a country so thickly populated, and with so many of its inhabitants engaged in mining, manufacture, transport, and their attendant services, every acre of land is a precious possession, and must of necessity bear a heavier burden than the virgin soils of the vast plains of North and South America that have more recently been brought under the subjection of man. During the last thirty years the knowledge obtained as to the relation between plants and the soils in which they grow has shed a new light on the operations of the oldest of industries, and we shall now proceed to examine in brief outline some of the more striking discoveries on which modern practice is based.

THE FOOD OF PLANTS

The improvements in English farming to which reference has been made, were based on a recognition of the facts that plants require food like human beings, but not upon any exact knowledge of the constitution of this food nor of the mechanism by which it was obtained. Observation had shown that the soil on which a crop had been grown for a number of years became less fruitful, and that its productiveness could be regained by allowing it to lie fallow or idle for a year, or by growing different crops in rotation. The order of the rotation was the result of experience, and so also was the use of farmyard manure to restore the impoverished soil to its former condition, or to increase the yield of the crop on fertile soils.

It was Boussingault who found that plants absorb carbon dioxide from the air and not from the soil, and the great German chemist, Justus von Liebig, who emphasized the discovery in a report to the British Association for the Advancement of Science in 1840. The gas enters the plants through minute openings, called stomata, in the under surface of the leaves and, under the influence of sunlight, the carbon is appropriated and the oxygen is evolved. During the night this process ceases. It was known, too, that the roots absorb water from the soil, and with it any substance that the water holds in solution. In this way they obtain their mineral constituents, consisting chiefly of lime, potash, and phosphorus compounds together with more complex bodies containing nitrogen.

The interior of a plant is a miniature chemical factory, taking in material from the air and soil, and building up with marvellous regularity and precision the complicated substances which determine its constitution. Among its products are timber, fibres, sugar, aromatic essences, perfumes, deadly poisons, healing drugs, and dyes that rival the rainbow in hue. The vegetable world, using throughout substantially the same raw materials, but in different proportions, specializes its manufactures, each workshop, from the oak to the microscopic fungus, concentrating its energy upon a limited but characteristic series of finished goods. The only condition imposed is that the raw material shall be supplied in a soluble form so that the root hairs can convey it into the system. No matter how rich the soil may be in the elementary constituents necessary for growth, if these are not in a palatable and digestible form they are useless for the end in view.

A chemical analysis of the plant will reveal the relative quantities of the various materials that are required for its growth, and manuring consists in supplying to the soil just those materials that are necessary to supplement the amount which is available, and which have been found to be necessary for the maximum yield. From the time of Liebig it has been recognized that potash and phosphorus compounds are necessary, and their effects have been fully well understood, but the theory of action of nitrogen compounds was for long a subject of acute controversy. This, however, and many other problems of cultivation have been solved by the long series of experiments which have been carried on in the Lawes Agricultural Experiment Station since 1903.

It has been established in the case of wheat that potash gives increased vigour and power to resist drought, damp, and rust, while phosphatic compounds promote root development in the early stages, and hasten ripening at a late period in the life of the plant. Nitrogenous manures encourage leaf growth and the attainment of vegetable maturity; without nitrogen there is no progress beyond the seedling stage. A certain plot of ground on the station at Rothamsted has grown wheat continuously for seventy years, and still produces 13 bushels per acre. If a manure containing all the necessary mineral constituents but without nitrogen is added the yield is only increased to 15 bushels. The use of a nitrogenous manure alone raised the yield to 21 bushels, and the addition of both the mineral and nitrogenous manures to 35 bushels. These facts are illustrated in Fig. 162.

Of all the manures used by the farmer those containing nitrogen are the most costly and remain for a shorter time in the soil.

Some of the mineral substances such as superphosphate, basic slag, etc., serve for more than one season, but the nitrogenous manures are decomposed or washed out by the winter rains.

Let us then review the various ways in which nitrogen required for growing crops is supplied to the soil. The ammonia, nitrous and nitric acids which are found in rain water, and which have been supposed to be formed by electric discharges in the upper regions of the atmosphere, provide an infinitesimal fraction of the amount required by the vegetable world. The vast ocean of nitrogen in the air, which amounts to 33,000 tons per acre, is not, except to a small extent in a way to be described later, capable of being assimilated directly by plants. The amount of animal and vegetable refuse containing the nitrogen which has been abstracted from the soil is limited. The ammonium sulphate obtained as a by-product in gas and coke manufacture might be very largely increased in amount, and commanded before the war from £12 to £15 a ton. Here the nitrogen from the plants of the carboniferous age is being used to nourish and sustain their descendants.

While the decay of animal and vegetable matter does yield some return to the soil, this return is neither immediate nor complete. Some artificial manuring is necessary sooner or later even with virgin soils, and nitrogenous manures are the most expensive which the farmer has to buy. For many years now the chief sources have been the guano beds in certain islands of the Pacific and in Peru, and the beds of sodium nitrate in Chile. Guano is the excrement of countless generations of sea birds. It is only found in limited areas, and cannot be relied upon for many years longer—at any rate, so far as the more extensive agricultural operations are concerned. The Chile " saltpetre " beds were in all probability formed by the drying up of a large lake which, occurring in a rainless district, left behind great deposits of salts which would ordinarily have been washed away. In several other parts of the world there are similar tracts of desert in which the soil is impregnated with salts of sodium and potassium, and where the absence of water has proved a great obstacle to exploration. The present production of the Chile beds is about 2,000,000 tons per annum, of which four-fifths is used as a manure. (It should be observed that these beds, until atmospheric nitrogen became available, were the chief source of supply of nitric acid, which is prepared by acting on sodium nitrate with sulphuric acid, and of iodine.)

The possibility of this raw material being exhausted within thirty or forty years may well give pause to those who study the conditions under which food is produced and the rate at which the demand for it is increasing.

This was the position when in 1898 Sir William Crookes in his Presidential Address to the British Association sounded a note of warning. The virgin soils of Canada, the Western States, and Argentina, rich as they are in nitrogen, cannot go on producing wheat indefinitely without overdrawing the nitrogen account in the Bank of Nature. And even if they could, the older countries could not continue to grow even a reasonable amount of home produce without supplying to an impoverished soil the material which plants need for their growth and sustenance. The only soil in Great Britain which is capable of producing a good yield of wheat without assistance is the black soil round the Wash, which contains ten times as much available nitrogen as the soil in any other part of the country.

It is characteristic of our age that the solution proposed by Sir William Crookes should be an accomplished fact within five years or so of its announcement. In Chapter IX, descriptions are given of methods devised by Siemens and Halske, and Birkeland and Eyde, by which a small proportion of the nitrogen in the air is caused to combine with a little of the oxygen, and the compound used to form calcium nitrate. Again, in Chapter X it is shown how the nitrogen, which is an otherwise useless by-product in the manufacture of oxygen by Linde's process, is passed over calcium carbide and converted into an available plant food called nitrolim. There are yet other processes to which reference will be made in the next chapter, and it is clear, now that the main and well-nigh inexhaustible reserve store of nitrogen has been tapped, there will be no lack of material to recoup the earth for the depredations of former years.

THE BIOLOGY OF THE SOIL

The application of artificial fertilizers is, of course, only a means of increasing productiveness, and though the general principles upon which successful cultivation depends have been established by centuries of practice in the agricultural arts, it is only within recent years that a real insight has been obtained in the secrets of fertility. Beyond the breaking up of the surface to enable air to obtain access, and to provide a medium into which the roots can penetrate freely, and draining to prevent accumulation of stagnant water, there is now a vast amount of information about the changes which the materials undergo, and the causes to which they are due. A soil may contain all the necessary elementary constituents of plant food and yet be unfertile ; it may lose its fertility temporarily by over-cropping and regain it by lying fallow, by bearing another crop, or by the addition to it of materials in which it is deficient. What, then, are the processes by which plant food is manufactured in and below the surface of the ground ?

Generally speaking, a soil may be regarded as a mass of inert mineral matter containing about 15 per cent of water. The water contains certain mineral substances which it has dissolved, and the particles of soil in varying size form a framework over which the solution spreads as a thin film. Apart from the water there may be 80 per cent of mineral matter and 5 per cent of organic material—the decaying remains of vegetable and animal life. We have already seen that except for the infinitesimal amount of nitrogen compounds which falls with rain, and that which is added by the cultivator, the main source of nitrogenous food is the organic matter in the soil. The fertility depends largely, therefore, upon the decayed animal and vegetable material which accumulates in or is added to the land.

Many years ago Pasteur showed that the soil contains bacteria —minute forms of vegetable life—which may exist in such numbers that they produce profound and far-reaching chemical changes in the material in which they live. Their size is about 1,000 of a millimetre or 25,000 of an inch. They multiply under suitable conditions with extraordinary rapidity, one dividing into two every 35 minutes, so that at the end of 12 hours one bacterium may have 12,000,000 descendants. A cubic inch of soil may contain several hundred millions of them. And in addition to these there are other lowly microscopic forms of vegetable life, such as fungi, and protozoa or similarly small members of the animal kingdom. We have to deal, therefore, not with a mixture of substances such as is ordinarily examined in the chemical laboratory, but with a teeming population living, working, dying, competing for nutriment, breaking down the material in and on which they dwell, and effecting changes which are so numerous and complex that it almost passes the wit of man to unravel them.

The first suspicion that certain changes in the soil were biological rather than chemical arose in 1878. It was known that as a rule the form in which nitrogen is most easily assimilated is that of a nitrate, and that if ammonium salts are added they are rapidly concerted into nitrates. When, therefore, it was found that nitrification, as the process is called, did not start immediately in an artificial soil, but required 20 days for commencement, it was suggested that the change depended upon the growth and multiplication of some form of life. And in 1887 Warrington in England, and Winogradsky in Russia, working independently, isolated the microbes responsible for the work.

The change takes place in two stages, each due to the action of a particular bacterium. One converts the ammonia into nitrous acid and the other converts the nitrous into nitric acid. Neither of them is able to effect the complete conversion ; they must work in co-operation.

Ten years later another variety of the microscopic flora of the soil was discovered to serve a further purpose. The roots of plants belonging to the order leguminosæ, comprising, among others, peas, beans, clover, and vetches, possess nodules (Fig. 163) on their roots, and these nodules were found to contain colonies of bacteria (Fig. 164) capable of absorbing nitrogen from the air and converting it into protein for the use of the plant. They feed their host in return for a habitation and a home.

In contradistinction to the *nitrifying* organisms first described, these are called *nitrogen-fixing* bacteria. The same or a similar variety has been found on the roots of forest trees, and some are also found free in the soil. There is, in addition, a third form which in some way decomposes nitrogen compounds, producing free nitrogen which escapes into the air, and helping to maintain that uniformity of composition which is one of the most important and striking properties of the atmosphere.

During the last twenty years this underground society has yielded up still another of its secrets. Instead of working in apparent harmony and co-operation it would appear that some of them prey upon the others. In 1888 Frank showed that if soil was heated to 130° C. its productiveness was decreased, but that if the temperature was not more than 100° C. its productiveness was more than doubled, and the soluble constituents were increased. Five years later Hiltner and Sturmer showed that treating the soil with carbon bisulphide altered the microscopic

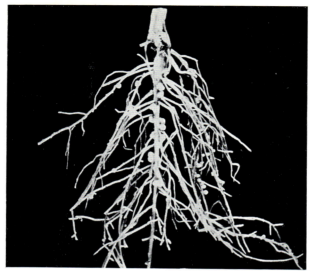

Photo by Dr. H. B. Hutchinson.

FIG. 163.——PHOTOGRAPH OF ROOT NODULES OF BEAN.

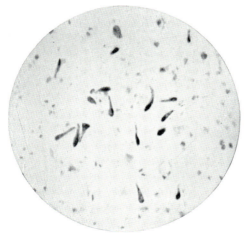

Photo by Dr. H. B. Hutchinson.

FIG 164.——MICRO-PHOTOGRAPH OF BACTERIA FROM
ROOT NODULES OF CLOVER.

To face page 216.

FIG. 169.—MODEL OF FROTH-FLOTATION APPARATUS FOR
SEPARATING DIRT FROM FINE COAL.

To face page 217.

flora. The number of bacteria which could be counted decreased by 75 per cent, but when the carbon bisulphide had evaporated their number increased until they became more numerous than before. At a later date toluene and other substances were found to have a similar effect.

The whole question has been minutely investigated by Drs. Russell and Hutchinson of the Rothamsted Experiment Station. A microscopic examination of the soil before and after heating or other treatment showed the presence in the former case of protozoa, algæ, fungi, and other low forms of life, which were absent after heating or other treatment. The protozoa are extremely minute members of the animal kingdom, and two varieties which are recognized—*colpoda cucullus* and *amoeba nitrophila*—are known to devour bacteria. The algæ and fungi may also operate in other ways which are unfavourable to the growth of more useful forms of vegetation. But if the protozoa are killed then the bacteria can increase, and so far as these are concerned in the manufacture of plant food, the soil will gain in fertility.

The introduction of animal and vegetable refuse into the soil, therefore, benefits it in two ways. Part of it is converted into carbon dioxide, ammonia, water, and nitrogen, and part tends to accumulate, increasing by its texture the power of the soil to retain moisture. Some of the ammonia is absorbed by the clay constituents of the soil, forming a curious compound the nature of which is not yet known, and some of it is converted by the nitrifying bacteria into nitric acid. Part of the carbon dioxide is assimilated by the bacteria and other forms of plant life in the soil, and part escapes into the atmosphere to suffer a similar fate. The nitrogen is attacked by the nitrogen-fixing bacteria or escapes.

Dr. Russell divides the microscopic life into three groups :—

(*a*) Saprophytes, which live on and decompose organic matter.

(*b*) Phagocytes, which devour living bacteria.

(*c*) Larger organisms, which in other ways than (*b*) are inimical to plant growth.

Raising the temperature of the soil to 98° C. or treating it with carbon bisulphide, toluene, or other substances kills the members of groups (b) and (c) and allows the members of the group (a) to increase and do their beneficial work more vigorously.[1]

The tendency in uncultivated lands is for nitrogen compounds to accumulate. Clearing and ploughing let in light, air, and rain. Some of the carbon dioxide and ammonia escape, and much of the soluble nitrogenous material is washed out of the soil. Deterioration goes on in new countries in all cases until wheat is displaced by rotation of crops. The exhaustion of the soil is not produced merely by the wheat crop, but also by the method of cultivation altering the microscopic flora of the soil and destroying the natural balance of food supply and food demand.

On so-called sour land there are doubtless other influences than those we have outlined at work. In the absence of calcium carbonate the decomposition of organic matter may produce poisonous substances which hinder plant growth, and lack of fertility may be due to this cause rather than to lack of available food. But in the main the explanation which has been given is in accord with the greatest number of facts. It will be clear, however, that we are only on the threshold of a vast field of knowledge, the existence of which has been revealed by a glimpse into the underground world of the animal and vegetable kingdom. When the old agriculturalist spread manure over the land and grew his crops in rotation, he knew by long experience that the results would be good ; but he was profoundly ignorant of the fact that he was altering the balance of microscopic existence down in the corridors and caverns of the soil. He did not realize that just as the highly organized plant he tended with such care built up from the materials in the air, and about its roots the food, medicine, or fibre it required, so also the more lowly forms of life were busy preparing for consumption the food to be enjoyed by the aristocratic giants of their race.

THE PHYSICS OF THE SOIL

From 1830 to 1880 the problem of fertility was regarded mainly as one of Chemistry. It was thought that if the mineral matters which plants abstract from the soil were provided in sufficient quantity and in a more or less soluble form, crops would flourish year after year in the same ground. The next stage was recognition of the influence which low forms of life such as bacteria,

---

[1] The bacteria are killed, but not their spores.

fungi and protozoa have upon the growth of larger plants, and
consequent development of the biological theory of fertility
which has been described. At the same time it was understood
that these were only two aspects, and that a number of causes
are necessary to explain the facts of experience in different
climates and on a variety of soils. For example, growth is most
vigorous between certain narrow limits of temperature, and
conditional entirely upon a certain minimum supply of water ;
while the production of available food material by bacterial
agency is again dependent upon temperature and does not occur
in a water-logged soil. Consequently much attention has been
given in recent years to investigation of the conditions which
favour the retention of heat in the soil and which control the
water content and drainage.

The whole problem has become more complex because many
of the phenomena belong partly to physics, and partly to that
borderland between physics and chemistry which are described
in the next chapter. Soils vary in texture from loose sands to
compact clays and in all except the former a proportion of colloid
material, both inorganic and organic, is present. We shall not
attempt, therefore, to do more than indicate a few of the more
important physical properties in their relation to cultivation.

If a narrow glass tube is dipped into water or any liquid that
wets the surface, it rises to a higher level inside the tube than
outside, and the narrower the tube the higher the water will rise.
In the case of mercury, which does not wet glass, the level inside
the tube is depressed. This capillary elevation—so-called from
capillus, a hair—is due to surface-tension. The free surface of
every liquid is stretched. Consequently a drop, falling freely,
always assumes the shape of a sphere, because a sphere has a
smaller surface than any other shape for a given volume. This
surface behaves as though it were an elastic skin, and it is not
difficult to see why this should be so. There must be some force
which holds the particles together or the liquid would fall to
pieces. And it is easy to understand that while the particles
inside a liquid are pulled equally in all directions, those at the
surface are attracted inwards. They are held more firmly than
the internal particles, they resist separation, and a needle will
rest upon the surface of the water, though once it breaks through
it sinks.

Again if a marble or other object is suspended by a thread,

dipped into water and then lifted out, it will be seen to be enclosed within a film, and all that can be removed by shaking is the surplus drop at the bottom.

Now the soil must be regarded as a number of particles of varying size with tiny capillary spaces between them. Movement of water within the soil takes place along these passages, and water is retained in thin films on the surface of the particles even when the passages are empty. If the soil is sandy it drains well and water rises freely in it provided the particles and therefore the pores are small. But for that reason it also loses water readily by evaporation, and crops upon it will suffer in a dry season. Further, since dry soil requires much less heat to raise its temperature than wet soil, such a soil is warm and will raise an earlier crop than a wet clay soil. On the other hand, a clay soil, being more retentive of moisture for reasons which will appear later, will support growth to a later period in the autumn.

The ancient practice of breaking up the surface-soil during the growing season has a scientific explanation. The liquid film which, in a firm soil, may be regarded as continuous, is broken and evaporation is checked. Moreover, the dry particles separated by air spaces form an excellent non-conductor and keep the soil about the roots at a uniform temperature. If any rainfall is to be beneficial during a severe drought, it must penetrate this dry surface layer to the soil below. Failure to do this is the cause of crops wilting even after a shower ; and it is obvious that if a watering-can has to be used in the garden it should be used thoroughly.

Part of the water which falls as rain sinks into the ground and runs into wells or drains away ; part is evaporated directly from the surface ; part goes to form root and stem, leaves and fruit ; and part is transpired. The amount transpired is on the average 300 lb. for each pound of dry vegetable matter produced. It follows, therefore, that a vast quantity must be required per acre—more in fact than is available from the annual rainfall in some districts. Wheat, for example, transpires a 6-inch, and mangolds a 10-inch rainfall. Some land will not support continuous cropping. But in the semi-arid regions of the United States and Australia a plan of dry farming has been developed, based on the practice in the relatively dry counties of the south and south-east of England. When the land is bearing a crop the surface is frequently stirred in order to conserve the moisture, and the land is allowed to lie fallow for one year in two or three in order to accumulate a year's rainfall in the soil.

Unfortunately the whole of the moisture in a soil is not available for the use of the plant. A crop begins to wilt while a certain percentage of water remains. This quantity may be only 1·5 per cent in sand, 10 per cent in clay, and as high as 40 per cent in peat. The " free " water available for plant growth is that contained in the pores or capillary spaces ; the " hygroscopic " water exists as a film which surrounds the particles, and in all soils equally dry this film is about 0·00003 inch in thickness. But the problem of water-retention, of texture, and of other factors is complicated by the fact that all soils except the loose sands contain colloidal substances [1] which " adsorb " or attract moisture (and solutions) to their surfaces. In fact, the film which surrounds the particles is generally a colloid of the " gel " class, the members of which are capable of taking up and holding strongly considerable quantities of water. The readiness with which some of these bodies pass from the colloidal to the flocculent condition and *vice versâ* in the presence of other substances which do not, in the ordinary sense, exercise a chemical action upon them, explains two very important facts which have long been known. These are the power of a soil to retain certain soluble salts used as fertilizers, and the bad effect which certain salts exercise upon the texture of the soil to which they are supplied. Thus all ammonium salts and organic compounds of nitrogen are retained—i.e. they do not pass out with the drainage water. In the case of the ammonium salts it is the ammonia which is adsorbed, and the liberated sulphuric acid attacks the calcium carbonate in the soil, and converts it into soluble sulphate which is washed away. The result of continued applications of ammonium sulphate to a clay soil is to remove the calcium, which as bicarbonate is effective in keeping the clay particles in a flocculent condition.

Again, other salts, such as potassium sulphate and sodium chloride, either have an alkaline reaction or undergo changes in the soil which result in the formation of alkalis. The effect is to deflocculate the clay, and to create a colloidal condition with a close texture which is unfavourable to plant growth. For that reason a mixture of potassium sulphate and ammonium

[1] See Chapter XII.

sulphate is better for clay land, because the influence of the one upon texture neutralizes that of the other.

The removal of calcium carbonate from the soil which is inevitable with the potash and ammonium fertilizers is due also to a very interesting reaction which occurs in the soil owing to the presence of zeolites. A zeolite would be described by a chemist as a hydrated double silicate of aluminium and an alkali or alkaline earth metal ; but it is no worse for that. The zeolites are the result of weathering upon the mineral felspar in igneous rocks. They have the general formula

$$RO, Al_2O_3, 3SiO_2, 2H_2O,$$

where R stands for Ca, Mg, K_2 or Na_2. If the zeolite contains calcium it has only to be placed in a solution of sodium chloride in order that sodium shall replace calcium in the mineral. Similarly, if the sodium compound is exposed to calcium chloride solution the calcium goes back home and the sodium comes out. It may be noted in passing that the permutite processes for softening water consists in removing calcium compounds by means of a sodium zeolite. There seems to be no doubt that calcium is removed from soils in this way.

But these are only a few amongst the hundreds of facts which have accumulated from the patient labours of agricultural chemists in many parts of the world.

PEDIGREE WHEAT

When man is faced with a big problem like that of the world's supply of wheat, he is not satisfied with a successful attack in one direction, but must needs seek other ways of extending his power and dominion over Nature. He knows that there is not one variety, but many varieties of wheat, and that a fertile soil, cultivated by the most enlightened methods, cannot produce either the greatest quantity or the highest quality from an inferior strain of seed.

For the sake of simplicity the matter may be considered from two points of view—quantity and quality. The amount of wheat produced in the world was investigated nearly twenty years ago by Professor G. F. Unstead, and Professor H. N. Dickson, in his presidential address to the Geographical Section of the British Association in 1913, drew attention to the fact that supply was not increasing so rapidly as the demand. The opinion was expressed that with existing varieties and methods of farming the wheat-growing areas of the world would, sooner or later, be taxed beyond their capacity.

Such a result could only be deferred for a time by the use of artificial fertilizers. For, in addition to suitable food, the wheat plant requires a stiff soil to support its long stem, a wet season of growth, and a warm dry period in which to ripen—conditions that are only found in certain regions of the globe. In many other districts the soil might be suitable, but the summer is too short or too wet, so that the grain would not ripen, or the disease called *rust* would make its appearance. The fact, however, that existing varieties have very definite requirements does not mean that other varieties, less fastidious in their needs, are unobtainable. All those that are grown now—and their name is legion—have developed from four which flourished in olden times, and where there has been so much change there is possibility of more.

Consider now the question of quality. Everyone knows that some varieties of flour are better for baking than others, because they make a larger and better shaped loaf, and it has long been the practice to ascertain the quality of flour by an actual baking test before purchase. The theory of baking itself is not without interest. Flour consists essentially of three constituents—*starch*, a gummy substance known as *gluten*, and about 1 per cent of *sugar*. When it is mixed with water, and *yeast* is added, the latter feeds on the sugar, producing carbon dioxide, which fills the mass of dough with small bubbles. The heat of the oven causes these bubbles to expand, and the final result is a light spongy framework of hardened gluten impregnated with grains of starch.

The sugar is formed from the starch by the action of a non-living ferment called diastase, which appears to exhibit varying degrees of activity in different varieties of flour. The amount of sugar is fairly uniform, and when this has been used up by the yeast, the production of a further quantity of gas is dependent upon the action of the diastase in manufacturing more sugar. If the ferment is active, the yeast grows quickly, produces a large volume of gas, and makes a big loaf. But if the ferment is sluggish, the production of gas is slow, and a small loaf results. Professor T. B. Wood, of Cambridge, who has been working on the problem of the strength of wheat for many years has shown that the amount of gas evolved in a given time furnishes a very good guide to the size of loaf that will be produced, and he has devised a method of testing the baking strength of flour which can be performed on the grain from a single ear.[1]

But the size of the loaf is only one aspect of strength, for shape and texture are of considerable importance, and these depend, not on the diastatic fermentation of the starch, but upon the character of the gluten. All attempts to trace the result back to the chemical properties of gluten have failed. It is one of those curious substances known as colloids, the physical properties —appearance, texture, etc.—of which are profoundly modified by the presence of small quantities of acids or salts. It occurred to Professor Wood, therefore, that the variable character of the gluten in different kinds of flour might be due to the presence of a particular acid or salt in the grain. The final result of his work is to indicate that the shape and texture of the loaf are closely connected with the presence of phosphates, which are found in larger quantities in strong than in weak wheats.

The result of unravelling the meaning of the term " strength " has been that millers and bakers are taking steps to confer upon weaker wheats those properties of strength which they so much desire. Malt extract, for example, contains an energetic diastatic ferment and, by spraying it over flour, the rate of formation of sugar, and therefore of gas, is increased. And in the other direction, certain phosphates are being mixed with flour in order to secure the shape and texture that brings the largest trade.

But to return to the main problem. Hardier, earlier ripening, disease-resisting forms of wheat, producing a strong flour, are clearly desirable, and to these ends many minds are being directed. The problem is of peculiar importance to this country because English wheats lack strength, and have to be mixed with a hard Canadian or other variety for which a higher price is paid. It is of importance to the world at large because it is estimated that the loss from rust alone is equal to one-third of the world's harvest. So the farmer, who has long recognized the importance of pedigree in cattle and sheep, has now realized the importance

---

[1] The whole question is admirably treated in Professor Wood's little book, *The Story of a Loaf of Bread.* (Cambridge Press.)

of plant breeding in enabling him to supply the workers in mines, in factories, and in transport with food.

It has already been remarked that the breed of wheat is generally mixed, and it follows that valuable land, manure, and labour are being expended upon varieties which yield an inadequate return. Out of this fact two separate problems arise —one is to replace the inferior by superior varieties, and the other is to improve even the superior varieties themselves. The first problem is mainly one of selection. A variety that possesses the requisite qualities is singled out and seed is sown. As the plants arrive at maturity seed is gathered, sown in the same way as before, and again used to increase the stock ; and when sufficient has been accumulated, it is distributed to farmers, who discard the varieties upon which they have hitherto depended.

Until the last twenty years the practice of plant breeding was carried on by the method of trial and error, and occasional success was a small oasis of comfort in a vast desert of failure. Yet the key of the hidden chamber was found nearly sixty years ago by Gregor Mendel, and was only rediscovered in the last year of the old century. Mendel's Law of Heredity is rather complicated, and space will not allow of its full statement or explanation here. It must suffice to say that the various characters possessed by plants and animals are of two kinds— dominant and recessive, and that the former are capable of being handed down from generation to generation, while the latter are liable to appear and disappear. When two plants are crossed the first generation is, generally, intermediate in character between the parents. If the seed from these plants is sown a second generation is obtained in which the characteristics of the parents are arranged in many possible combinations, but in certain definite proportions. Some of these types are fixed, and will continue to breed true. Others are not fixed, and produce a variety of progeny in each generation. But the essential fact is that Mendel's Law enables the fixed types to be picked out in the second generation, so that the labour of selection, trial, and error is largely avoided. The second generation, in fact, reveals and separates the dominant and recessive characters of the parents.

Though the characters which confer quality, immunity from rust, and other valuable properties are not so simple as those

Q

which determine the size, colour, shape, etc., of the grain and
ear, Professor R. H. Biffen, of Cambridge, has applied the
method with great success to the hybridization of wheat. He
has collected varieties of wheat from all parts of the world,
grown them, and crossed those which seemed to possess features
which would be desirable in combination. In this way he
has succeeded in producing, amongst others, two varieties
" Yeomen " and " Little Joss ", the one having the same valuable
baking quality as the hard wheats of North America, and the
other unaffected by yellow rust. " Yeoman " holds the record
in this country for a yield of 96 bushels per acre—a yield which
may be compared with a world's average of thirteen. For his
work on plant breeding, Professor Biffen received in 1920 the
Darwin Medal of the Royal Society. Again in December, 1921,
the Essex farmers presented silver bowls to Professor Biffen for
his wheats and to Mr. E. S. Beaven, of Warminster, for his barley,
" Plumage Archer ". At that meeting it was stated that half
the wheat in this country was grown from the two varieties which
have been named.

But finality has not been reached. Quality, for example,
seems to depend in some way upon soil, and the hard wheat of
Manitoba does not reproduce its characteristic property when
grown on the manured land of England. Many workers are
engaged upon this and similar problems. It is a race between
demand for a higher standard of living and the supply of the
nutritious and palatable foods for which man craves. And the
forces which create demand have a start over those which
control supply.

Let us emphasize this point again. In this chapter and the
last we have learnt something of the attempt which has been
made to solve what is, to the white races of mankind, the great
problem of the future. Western nations have learnt how to
produce power, they have devised thousands of ingenious manufacturing processes ; they have covered a large area of the earth's
surface with a network of railways, of steamship lines, of telegraphs, and of the means of wireless communication ; they have
increased enormously in population ; and they have developed
a desire for a much more varied dietary than that which satisfied
them in years gone by. But, by imprisoning their people in
workshops, factories, and mines, and in tying them down to lines
of communication, they have limited the proportion available

for producing the basic essential of material food. As sources of power become exhausted in one place, people may move to another. But they cannot much longer continue to increase in numbers without either a larger section being engaged in food production or improved methods of winning from the soil that which is essential to their existence.

CHAPTER XII

The Borderland of Modern Chemistry

COLLOIDS

It is a good plan never to be afraid of a word, for it may be, and often is, not half so bad as it looks. One of the characters in a French play expressed astonishment on hearing that he had been talking prose all his life, and doubtless many people would be equally credulous if they were told that they had been familiar with colloids all their days. Yet such is the fact. People eat colloids, drink colloids, wear colloidal clothes, and walk on collodial earth. In every nerve and muscle, in every drop of blood, in every single microscopic cell in their bodies they have colloidal matter. So that if you are very angry with a man, and wish to express yourself forcibly about him, it is quite truthful to describe him as a colloidal aggregate. For reasons which will appear later, it is not so safe to say that he has colloidal notions.

The word comes from *colla*, meaning glue, and was given by Graham in 1861 to substances like glue, gelatine, and gum-arabic, which diffused with exceptional slowness through water. On the other hand, substances like common salt, which spread through water with great rapidity, he called crystalloids, because they were invariably bodies which could easily be obtained in a crystalline form. His apparatus, shown in Fig. 165, consisted of a sheet of parchment stretched over a ring of wood, or other material to form a shallow dish or basin. This was floated on pure water contained in a larger vessel, and the solution under examination was poured into the dish with a parchment bottom. While crystalloids readily passed through the membrane into the outer vessel, colloids refused to do so, and Graham concluded that either the molecules of colloids were larger than those of crystalloids—which is certainly true in some cases, or the " molecule " of a colloid consisted really of a group of molecules.

In the course of his work he succeeded in preparing what seemed to be solutions of such bodies as silicic acid and ferric hydroxide, which in the dry state are insoluble in water. These preparations, however, were perfectly clear and transparent, and they passed through filter paper just as common salt does when in solution ; yet the apparently dissolved substance was retained by the parchment. He called them colloidal solutions, or " sols ". Examples of such solutions had been known before his time, but he was the first to make a full and exhaustive inquiry into the phenomena. Thus, Faraday in 1857 had prepared a clear red liquid by adding a few drops of a solution of phosphorus in ether to a dilute solution of gold chloride. Ordinarily any reducing-agent added to a solution of gold chloride would have pre-cipitated the gold as a brown powder, soluble only in " aqua regia ", a mixture of nitric and hydrochloric acids. But in this liquid no particles could be seen, nor were any deposited on standing. Nevertheless Faraday believed that the solid particles of gold were there, though so small as to be invisible even under the microscope.

Fig. 165.   GRAHAM'S DIALYSER.

A few years later, to be precise, in 1869, Tyndall was studying the effect of fine particles of dust, etc., in air and other gases ; and, for the purpose of detecting particles which were beyond the range of microscopic vision, he passed a powerful beam of light through the containing vessel. When this was viewed at right angles to the general direction of the light, which was screened from the eye, a visible beam revealed the presence of solid or liquid particles by the scattering of light from their surfaces. While this method is still used, it has been replaced by a much more effective instrument, which is based on the same principle. The ultra-microscope, as it is called, was invented by Zsigmondy and Siedentopf in 1905. It consists (Fig. 166) of a microscope mounted vertically and having on the stage a small glass or quartz trough in which the liquid under examina-tion can be placed. The liquid is illuminated by a narrow slit in a small screen upon which are concentrated rays from the sun

or a powerful electric arc.   With this arrangement no light can
enter the objective of the microscope unless it is reflected from
solid or liquid particles in the trough.   If these are present,
the observer sees, not the soft glow that characterizes the Tyndall
beam, but flashing points of light like moving stars in a velvet
sky.   About this movement something will be said presently.

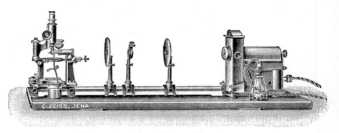

Fig. 166.   The Ultra-Microscope.

A simpler, though, in some respects, a less useful form of
apparatus, is one in which the microscope and the beam of light
are in the same straight line.   For this purpose a paraboloid
condenser is employed.   This condenser, shown in Fig. 167,
is a lens, paraboloid in form, fixed in the stage of the microscope.

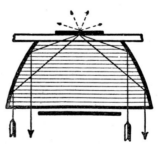

Fig. 167.   Diagram showing Passage of Light through Paraboloid
Condenser.

The middle of the lower face is covered by an opaque circular
disc, called a " stop " so that parallel rays entering from below
impinge on the sides and are reflected to the focus.   The focus
lies in the upper face, and the rays strike this face at such an
angle that they are reflected to the other side of the lens and out

again at the lower face.   But if the condenser has a layer of water or cedar oil instead of air above it the rays are not reflected, because the critical angle is not the same for glass and water or cedar oil as it is for glass and air.   Consequently, when a slide containing a shallow cell is laid upon the water or oil the light enters the cell obliquely and illuminates anything which it contains from the sides.   These oblique rays meet the cover glass at such an angle that they are reflected downwards, and the only light which can enter the objective is, again, that which is reflected from the surfaces of small particles within the cell.   The plan is similar to that devised by early British microscopists for examining minute objects such as diatoms, and is commonly known as dark-ground illumination.

The invention of the ultra-microscope gave a great impetus to the study of colloids, and by this and other methods it is clear that their properties arise from the fact that they are composed of particles varying from microscopic dimensions down to single molecules.   The diameter of the smallest object visible in an ordinary microscope is $1/100,000$ of an inch ; that of the smallest particle in the ultra-microscope is about $1/25,000,000$ of an inch ; the diameter of a molecule of hydrogen is $1/250,000,000$ of an inch.   Yet a molecule of hydrogen would probably be seen if only a sufficient amount of light could be concentrated upon it.   It follows, however, that a cube composed of 27 molecules of hydrogen or a square slab of 9 molecules could be seen if it were suspended in a medium from which it differed sufficiently in refractive index, so that the light which fell upon it were reflected at its surface.

A colloid is characterized by the smallness of its particles, and the word refers rather to a *form*, or *state*, or *phase* of matter than to a particular class of substance.   The ultimate structure of matter, at least so far as the chemist is concerned, is molecular.   All the " stuff " of which the world is composed consists of molecules, and these molecules are groups of atoms which, in elements, are of the same kind and are usually two in number.   Only a few elements with monatomic molecules are known.   In compounds the atoms are of two or more kinds.   Both atoms and molecules are too small ever to be observed singly, and their existence can only be inferred from the properties of solids, liquids, and gases in bulk.   The distinction between these is merely one of temperature.   Every substance which does not decompose on heating has been liquefied by cold. A colloid is matter in a fourth state—a state of small molecular aggregates, and in a sol these aggregates are " dispersed " with relatively wide spaces between them.

Colloids, then, owe their special properties—properties which distinguish them from matter in the solid, liquid, and gaseous conditions, to the extremely small size of their particles. Suppose a cube of one inch edge be divided into eight equal cubes. The area of a surface of the inch cube will be six square inches, of one of the small cubes $1\frac{1}{2}$ square inches, and of all the eight small cubes 12 square inches. Next, suppose that each of the smaller cubes is further subdivided into eight equal cubes. The surface of one of the smaller cubes will be $\frac{3}{8}$ square inches, and as there will be 64 of them, the total surfaces will be 24 square inches. Finally, suppose that the process be repeated with each of the small cubes. The edges now will be $\frac{1}{8}$th inch, the surface of one of them $\frac{3}{32}$ inch, and the total surface 48 square inches. It will be obvious that by subdivision the surface of each particle decreases less rapidly than the volume or weight, and that, if the inch cube weighed one ounce, the surface of the same weight of material will have increased from 6 square inches to 48 square inches. Meantime each particle will have only $\frac{1}{512}$th volume and therefore will weigh only $\frac{1}{512}$ of an ounce.

Two consequences follow from this enormous extension of surface for the same weight of material. A particle of such a size that the ratio of surface to weight is large, sinks very slowly in a liquid or gas because the friction is relatively high. Even visible particles of small size remain suspended for a long time, as anyone will have observed who has travelled along dusty roads in summer, watched the smoke from a pipe or a chimney, or shaken up very fine mud in a cylinder of water. The effect of friction, therefore, upon colloidal particles must be very much more pronounced. But it is not sufficient to account for the stability of sols, and there must be some additional reason why the particles do not settle in accordance with the law of gravitation.

The explanation has been found in a half-forgotten discovery of Robert Brown, a botanist, in 1827. He observed that fine particles of non-living matter of the most varied description suspended in water were in a state of rapid motion. The experiments were repeated by other workers at intervals and

several theories were proposed.   Fig. 168 shows the movements
of a particle as mapped by Victor Henry.   As early as 1863
Wiener, and later others, suggested that it was due to the bombardment by the molecules of the liquid which, according to the
kinetic theory of matter, are themselves in rapid motion.   This
theory asserts that the particles of all bodies are in motion—
most violently in a gas, less violently in liquids, still less vigorously

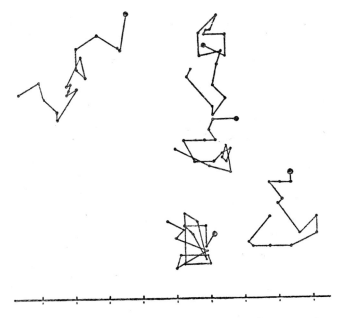

Fig. 168.   Diagram showing Successive Positions of a Single Particle
at Intervals of $\frac{1}{20}$ of a Second.

in solids, but at rest only at the absolute zero of temperature,
to which reference has been made in Chapter X.   But while
this view accounts for many of the properties of matter, and is as
firmly established as any conception in physical science, no one
had dared to hope for such visible evidence of molecular motion
as the Brownian movement supplies.   In fact, the idea was so
astounding that it has only been accepted after every other
possible explanation has been shown to be untenable.

The second consequence of extended surfaces is that a number of chemical reactions may occur which do not obey the ordinary laws which govern chemical changes. Thus freshly ignited charcoal, which is a porous material and thus has a large surface, will take up 170 times its own (apparent) volume of ammonia at 0° C. and under normal pressure. The condensation of this volume of gas within the pores and on the outer surface of the charcoal under these conditions is so great that some of the gas at least is supposed to be liquefied. At any rate, 170 cubic inches of ammonia gas at normal pressure and temperature cannot be reduced to the space in the pores of one cubic inch of charcoal by pressure without liquefaction. Other gases, such as ethylene, carbon dioxide, carbon monoxide, oxygen, and nitrogen, are also taken up, but in smaller quantities. In all cases the quantities are greater at low than at high temperatures.

Charcoal—more especially animal charcoal—also takes up colouring matter, quinine sulphate and lead salts from solutions, and is used to remove the brown colouring matter from syrup in the manufacture of sugar, and fusel oil in the manufacture of whisky. Filter paper, which is also a porous substance, abstracts lead salts from solution, and china clay and fuller's earth are used for purposes similar to that in which charcoal is employed. The hydroxide of aluminium and similar substances take up the colouring matter from the solution of a dye and form substances called " lakes ".

In all these cases a preference is shown. It is as though there were chemical affinity between the bodies and yet the combination—if it can be called combination—is loose and does not take place according to the laws of chemical change. The carbon is still carbon, and the ammonia or the oxygen is still ammonia or oxygen. But a substance condensed upon a surface in this way is more active chemically than it is in the free state. Thus, hydrogen and chlorine combine very slowly under the influence of light ; but if charcoal which has been " soaked " in chlorine gas is brought into contact with hydrogen combination takes place at once. Sulphuretted hydrogen and oxygen can be mixed together at ordinary temperature and pressure without any change occurring ; yet if a piece of charcoal which has been " soaked " in sulphuretted hydrogen is introduced into a jar of oxygen the combination is so vigorous that the charcoal bursts into flame. It would appear therefore that condensation on a surface brings substances into closer contact and facilitates chemical change.

The condensation of one substance upon the surface of another without, in the chemical sense, combining with it, is called *adsorption*. It may be due simply to molecular attraction, but it is selective and yet not in accordance with the laws of chemical combination. It may be, in some cases, electrical in origin. When an electric current is passed through some sols the particles move towards the positive electrode, through others they move to the negative electrode, through still others there is no movement or " cataphoresis " at all. In the first case the particles must carry a charge of negative electricity, in the second a charge of positive electricity, in the third, no charge at all—or rather, the charges may be equal. It has been found that a positive colloid adsorbs a negative colloid, and *vice versâ*. Further, a positive colloid adsorbs the negative part of a salt in solution, e.g. chlorine from sodium chloride, so that the solution becomes alkaline, and a negative colloid adsorbs the positive or basic part of a salt, so that the solution becomes acid.

Again, it must be remembered that while in ordinary chemical reactions we are dealing with the reaction between individual molecules, in colloidal phenomena we are dealing with molecular, and in some cases electrical, forces between lumps of matter. Thus it was pointed out on page 219 that in the free surface of a liquid the molecules are attracted to the interior and thus form a sort of elastic skin. This free surface is really a boundary between liquid and air. There must be similar boundaries between liquid and solid, liquid and liquid. The tendency of a stretched surface is always towards a reduction of area, or rather —and this is the one respect in which the elastic skin analogy is incorrect—a reduction of length. Any change in the surface tension will have an important effect in the stability of an emulsion, i.e. of a liquid dispersed in a liquid. This aspect of the question is too difficult to be dealt with here, but one important application will be described at a later stage.

There are obviously two methods of preparing small aggregates. One is to disintegrate a mass, and particles of colloidal dimensions are actually produced by grinding. Another is to build up the aggregate from individual molecules. As an example of the first method sols of metals can be most easily obtained by forming an electric arc between two rods of the metal immersed in the liquid in which the colloid is to be suspended. A simple method of preparing hydrosols (colloidal solutions in water) of gums and similar substances is to add a few drops of a solution of the gum in alcohol to about 100 cc. of water.   A milky fluid is obtained from which the particles do not settle and from which they cannot be removed by filtration through ordinary filter paper.   And as an example of the opposite process, any chemical reaction which produces an insoluble substance may, if performed with dilute solutions, produce a sol. The molecules of the insoluble substance are formed singly and are so few and so distributed that the aggregates formed are ultra-microscopic in size.

Many sols are only stable in the presence of an electrolyte. To explain fully what an electrolyte is would involve an excursion into the theory of solution for which space cannot be spared ; and it must suffice here to say that pure water is a non-conductor of electricity, that all soluble acids, bases, and salts render pure water a conductor, and that they are called electrolytes.

A simple illustration of the effect of electrolytes upon a certain class of colloids is afforded by the conditions under which ferric hydroxide sol is destroyed and formed.   If ammonium hydroxide is added to a solution of ferric chloride, a reddish brown precipitate of ferric hydroxide is obtained according to the equation

$$6NH_4OH + Fe_2Cl_6 = Fe_2(OH)_6 + 6\ NH_4Cl.$$

If the precipitate is filtered off and thoroughly washed with water a point is reached at which a red filtrate is obtained instead of a clear liquid.   This red liquid is ferric hydroxide sol.   The explanation is as follows : the precipitate was formed in the presence of an electrolyte (ammonium chloride) and retained some of this after filtration ; but as the electrolyte was washed away the precipitate began to break up into smaller aggregates until these were able to pass through the pores in the filter-paper. The addition of a little ammonium chloride causes the precipitate to reform.

All the metallic hydrosols and many others, are similarly destroyed by the addition of an electrolyte and the metal is deposited as a powder which will not pass again into the colloidal condition.

Many other substances like gelatine, glue, etc., set to a jelly on standing, while silicic acid also forms a jelly by loss of water or the addition of an electrolyte.   In this condition they are called gels to distinguish them from sols.   As a rule they can be redissolved to form the sol.   The gels are an extraordinarily interesting class of substances.   Gelatine, and some of the gums for example, absorb water and swell, and as the volume of swollen substance is less than that of the original gel plus the water, contraction must occur, and enormous forces brought into play.   They appear to possess a definite structure like a sponge and to differ from sols in which a solid is dispersed in a liquid in that the liquid phase is distributed throughout a solid or semi-solid framework.   In gelatin gel the framework is elastic, in silicic acid gel the framework is rigid.

Sols in which a solid is dispersed in a liquid are called suspensoids, but there are a large number of cases in which both the dispersed substance and the medium are liquids.   These are called emulsoids.   Oil, soap, and many other liquids may be dispersed in water and, *vice versâ*, water may be dispersed in many other liquids.   Perhaps the commonest example of an emulsoid is milk, which contains globules of fat dispersed throughout the butter-milk—the butter-milk also containing two colloids called casein and albumin.

Primarily milk is a colloidal solution or an emulsoid of fat in water, the fat amounting to 3·6 per cent.   By churning milk or cream, the fat globules are collected, and the resulting butter is a solid emulsion of water in fat, containing 84 per cent fat and 13 per cent water.

Milk also serves to illustrate another remarkable fact.   Certain substances are only stable in the colloid condition when another colloid is present.   The latter is called a " protecting " colloid. In milk the albumin acts as a protecting colloid to the casein, and when the albumin is destroyed by rennet or by acids (sour milk contains lactic acid produced by the action of bacteria) the casein is precipitated and the milk is said to curdle.   This has an important bearing on digestion.   Asses' milk for example contains about three times the amount of albumin and only about one-fifth of the amount of casein in cows' milk, and can be digested by infants far more easily.   The " protection " of the casein particles in asses' milk is more complete and curd is not only less readily formed, but when formed is in smaller flakes.

One of the most interesting cases of protective colloid is afforded by ice-cream.   If this delectable substance is made without eggs, gelatin, gum tragacanth or similar colloid (generally called a "filler" in the trade) it is, or soon becomes, gritty or sandy in texture ; but if one of these "protectors" is present a smooth, velvety texture is secured and retained. It is interesting to note that the added colloid serves also to prevent aggregation of the small crystals of ice which are present and which in time would tend to coalesce. Of the substances mentioned, gelatin is far the most effective, and half of one per cent is sufficient to achieve the purpose. The ice-cream maker is probably more familiar with the facts than the explanation.

The separation of the fat globules from milk to form cream can be prevented by "homogenizing" the milk. This process consists in forcing milk at a temperature of 50°–60° C. through very fine orifices under considerable pressure, which causes a reduction of the fat globules to one-hundredth of their original size. Such milk cannot be churned for reasons which would take us too far afield. It is also possible to homogenize cream which cannnot then be churned, nor, again for reasons which must again be omitted, can it even be "whipped" unless a colloid such as gum-tragacanth is added.

Perhaps the most familiar of all colloidal substances are soaps, which are the sodium or potassium salts of oleic, palmitic, or stearic acids. The sodium salts are hard soaps, the potassium salts are soft soaps. If calcium or magnesium be substituted for sodium or potassium the resulting soap is insoluble. That explains the curd produced when ordinary soap is used in hard water. The soluble soaps form colloidal solutions with water, and owe their detergent properties to this fact. The mechanism by which dirt is removed is too complex to be described in a few words, and while there is probably more than one cause there is no doubt that the dirt is absorbed by the colloidal particles of soap.

One of the most important industries which finds an explanation of its processes in the properties of colloids is Tanning. Skin, after the hair and epidermis have been removed, is composed of *collagen*, a near relative of gelatin, and its texture is a net-work of fibres which are really colloidal jelly. When the hide is dried the fibres cling together, forming a horny mass. The process of tanning consists in separating the fibres so that the leather may be flexible and porous, and so treating them as to render this condition permanent. The former is effected by soaking in dilute acids or alkalis and the latter by adding substances which alter the chemical character of the fibres or cover them with a protecting coating. The latter is a mineral substance such as chromium hydroxide, a vegetable substance such as tannin (both of these are colloids) or an oil (which may be a colloid).

The artificial silk industry, which has had such a meteoric development in recent years, again uses colloidal material. Real silk is an exudation of the silkworm, and consists of the cellulose originally contained in the leaves upon which the worm feeds. Woody fibre of all plants is composed mainly of this substance. Mashed up and mixed with a little size, spread out in a thin layer and dried between hot rollers it forms paper. In the stem of the flax plant and the boll of the cotton plant it is found already in the form of fibres which can be spun and woven into linen or cotton cloth. The difference between woody fibre, linen, cotton, and silk is almost entirely one of structure. All of them are composed of cellulose, and cellulose is, or rather behaves like, a colloidal substance.

Reáumur, in 1734, was the first to suggest that silk might be formed artificially by solution of a gum. At that time the chemical composition of substances was unknown. Acids, alkalis, and salts were recognized, but the exact difference between them was a matter of speculation. In 1855 a Frenchman named Andomars made cellulose nitrate by treating it with nitric and sulphuric acids, and dissolved this in a mixture of alcohol and ether. If this was forced through a fine orifice the alcohol and ether evaporated, leaving a fine thread of cellulose nitrate which could be reconverted into cellulose without appreciable alteration of form. The technical experience necessary to develop this process was not available at the time, and it was left to Count Hilaire de Chardonnet to make artificial silk on a commercial scale in this way. Before the war Belgian factories were producing nearly 6,000 lb. a day. This kind of artificial silk is very resistant to water, but the nitric thread is highly inflammable. Its chief disadvantage, however, was the high cost of the alcohol and ether, but effective methods for recovering these solvents are now employed.

A cheaper process is that in which the cellulose is dissolved in an ammoniacal solution of cupric hydroxide. Cotton fibre is, however, used in this process in place of wood pulp. Another method is to use cellulose acetate, and the silk so produced is called celanese.

The annual production of artificial silk to-day is about 1,000,000 tons, and 80 per cent of this is the variety known as viscose. This is made from cellulose zanthate, which is formed by treating cellulose with carbon disulphide. Cellulose zanthate was first made by Cross, Bevan, and Beadle in 1892, and it is the application of this discovery more than anything else which has led to the enormous expansion of the industry, to which there is hardly a parallel in the history of applied science.

It will be observed that in all these processes the cellulose, in the form of wood pulp, or cotton, is first converted into a compound soluble in a particular solvent. It forms a more or less viscous or gummy solution which then is forced through a fine orifice, Fig. 168a, or a number of fine orifices in a nozzle, Fig. 168b. If the solvent is volatile like the ether-alcohol mixture or acetone, the filament may be dried in air. With non-volatile solvents the filaments pass into a bath filled with a substance that abstracts the solvent. For cellulose nitrate both dry and wet methods are used. The filaments are so fine that a number of them have to be wound together to form a yarn sufficiently strong for wearing.

The discovery of the Chinese perhaps 5,000 years ago that the cocoon of the silk worm could be used for making beautifully light, soft, and warm garments for human wear is a wonderful example of observation, patience, and ingenuity. But the material was always so rare that only the wealthy could afford it. Indeed, silken garments were among the privileges of Paradise that Mahomet promised to his followers. Yet the development of chemical science during the last hundred and fifty years has enabled man to produce something which, while more brittle and less durable than the natural product, is so nearly like it as to render it almost as desirable. And the price brings it within the reach of all except the very poorest people.

Many colloidal substances are now used in medicine. Among these are colloidal silver, iron, copper, arsenic, manganese, sulphur, iodine, and quinine. Their action differs in many respects from that of the element, or its salts administered in the ordinary condition. Colloidal iodine, for example, does not stain the skin ; it can be obtained as a suspensoid in oil which softens the cuticle while the alcohol in " tincture of iodine " hardens it. Again while ordinary silver nitrate—the lunar-caustic of the druggist—stains the skin black, colloidal silver does not stain the skin and is effective in allaying inflammation and promoting the healing of wounds. Colloidal iron has been used effectively for catarrh of the nose and throat, and an injection of colloidal manganese is one of the best remedies for boils.

Since most animal and vegetable substances occur normally in a colloid condition it will be obvious that the conceptions

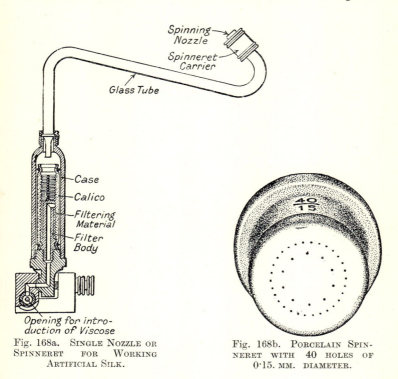

Fig. 168a. SINGLE NOZZLE OR SPINNERET FOR WORKING ARTIFICIAL SILK.

Fig. 168b. PORCELAIN SPINNERET WITH 40 HOLES OF 0·15. MM. DIAMETER.

of colloid chemistry have given a great impetus to the study of Physiology and Biochemistry, and a new vista has been opened in the study of vital processes during the last twenty years. Moreover, when it is realized that most substances which ordinarily occur as crystalloids—even so common a one as sodium chloride—can be obtained in colloidal form, it will be equally clear that there is no region of chemical industry and hardly

R

any sphere of chemical activity which is not penetrated by the new ideas. There are now three great branches of Science where formerly there was one. Firstly, there is the Science of Atoms and Molecules built up on Dalton's Atomic Hypothesis of 1808 ; secondly, there is the Science of Small Aggregates which has grown out of the observations of Faraday in 1823 and Graham in 1867 ; thirdly, there is the Science of Atomic Structure arising from the discovery of radium by Madame Curie, which is described in Chapter XX.

We may look at the phenomena in another way. If two individuals meet, their behaviour will be determined by their personalities, by the degree of intimacy between them, and by the affinity they possess for one another. If small groups of people meet, a member of each group will not act as though he or she were alone, but will be influenced by the character of and relation with his, or her, associates. If the groups are relatively large individuals will again tend to act in a more independent and personal way. Though analogy with the world of men cannot be pushed very far, it may help to create a picture of what goes on in the world of inanimate matter. An atom is an individual ; a molecule is a family ; a colloidal aggregate is a group of families or a family group which for the most part seems bent upon energetic reform of environment and sometimes becomes dissolved in the process ; matter in the mass is a crowd.

THE UTILITY OF FROTH

We have referred several times to the peculiar properties of liquid surfaces, and to the tension which is manifested at the junction between liquid-gas and liquid-solid. Further, we have explained this as arising from the mutual attraction of the molecules, by which those at the surface are under a resultant force that draws them inwards. In order that each molecule shall approach as near as possible to the centre a portion of liquid suspended freely in another liquid or a gas becomes a small sphere, because of all shapes the sphere has the smallest surface for a given volume. Again, when a liquid is broken up by blowing air or some other gas through it spherical bubbles are formed, but only if they are formed singly. If they are formed in close contact they suffer more or less deformation owing to their own weight, though if they are small the spherical form is maintained. A mass of air bubbles forms a froth.

The froth formed by stirring up or blowing air through water lasts only for a short time, but if certain substances are present, such as soap, oil, etc., the froth is much more permanent. Colloids are especially effective both in promoting the formation of froth and in conferring lasting properties upon it.

The phenomena are not noticeable in gases because the molecules, by reason of their kinetic energy or energy of motion repel one another. In solids, on the other hand, they have so little freedom that the ordinary effects of surface tension are masked. Nevertheless, the surface tension of a liquid is profoundly modified in contact with a solid, and the influence is manifested by the different degrees to which different liquids will " wet " the surface of a particular solid. From these two facts—the formation of froth and the " wetting " of solid surfaces have arisen one of the most important groups of industrial operations which have seen the light within the last twenty years.

Many ores of metals occur in narrow veins from which they cannot be removed free from rock and they require to be " concentrated " before smelting is possible. The usual method of carrying this out has depended upon the fact that the ore has a higher specific gravity than the gangue with which it is associated. The material from the mine, therefore, was crushed and subjected to the action of running water, which washed the lighter particles of rock away while the particles of ore sank to the bottom. But the separation was never very complete. The particles varied in size and the lighter ore particles were washed away with most of the gangue while some of the latter remained behind. It was rarely possible to recover more than about 80 per cent of the ore contained originally in the material.

The new process takes advantage of the facts (a) that a froth can be produced in water containing the crushed ore, (b) that this froth can be rendered permanent by the addition of a very small quantity of an oil or mixture of oils, (c) that the oil wets the particles of ore more readily than it wets the particles of gangue, and (d) that the oil all goes into the froth. Consequently when a mixture of crushed ore is stirred up with, say, ten times its weight of water containing a small quantity of oil as stabilizer, the ore becomes attached to the air bubbles which are formed and rises up into the froth, while the lighter gangue sinks to the bottom of the water. The separation is more complete if a small quantity of acid, say, sulphuric acid, is added, for this wets the surface of

the gangue more effectively than water, so both oil and acid are
used. The process takes less room than the older one, is quicker
in action, and far more efficient. It has come into widespread use
for the treatment of copper, zinc, and lead ores. At Anaconda,
the percentage of ore extracted is 95 compared with 76 under the
old system, and in other places ores which formerly would hardly
pay to concentrate are rendered profitable.

The process was first worked out for sulphide ores. It could
not be applied without modification to other compounds in which
metals occur because the relation between the oil-water surface,
the ore and the gangue are not necessarily the same. But
materials have been found which enable the cassiterite of the
Cornish Tin Mines to be separated from its gangue and
experiments on a large scale were begun some years ago.

A very important extension of the process has been made to the
separation of fine coal from the dirt with which it is invariably
accompanied. Fig. 169 shows a model in the Manchester College
of Technology. The coal is crushed and sifted through a mesh
with not fewer than ten holes to the inch, and delivered into about
four times its weight of water in the vertical chamber. To the
water is added cresol and paraffin oil in the proportion of one
pound of the mixture to a ton of coal. The whole mass is then
violently stirred, a black froth containing practically the whole of
the coal is produced and can be run off from the outer vessel
into the upper dish and the dirt is run off from the bottom into the
lower dish. By this means the amount of dirt in the coal can be
reduced to one-quarter of that contained in the original coal.
Four plants are in operation in this country, and one of them is
capable of dealing with 600 tons of coal slack per day.

Of all the phenomena of everyday life surely froth is one of the
most familiar—and one about which the least is known. Baby
hands play with the fragile foam which forms on the surface of
the water in the bath, and the glorious colours in a soap bubble
are a source of never-failing delight to young and old. Whipped
cream over a dish of fruit confer upon it an additional attraction,
and the " head " on a glass of beer is accepted as evidence of
condition.   At the seaside the white-capped waves convey to
our minds something of the endless succession of events, of the
unchangeable onward movement of time from far back in the
beginning to the distant future into which no one can see. But
who, dwelling upon these phenomena twenty years ago, could

have imagined the unseen forces of the bubble being utilized to control the separation of hundreds of thousands of tons of coal and ore a year ?

CATALYSIS

Many chemical changes take place at a lower temperature or more readily in the presence of a substance which apparently undergoes no alteration in composition itself. Thus potassium chlorate requires to be melted in order to yield oxygen readily ; but if it is mixed with a little manganese dioxide the mass does not melt and the manganese dioxide is found to be unchanged. Again, oxygen and hydrogen do not combine at ordinary temperatures unless they are ignited ; yet combinations occur in the presence of a strip of platinum at 50° C. A hundred years ago Dobereiner constructed a self-lighting lamp in which a jet of hydrogen impinging on spongy platinum immediately took fire.

The essential feature of catalytic action is that a very small quantity of the catalyst is capable of facilitating the change in an almost unlimited quantity of the reacting substances. Thus starch is converted into sugar by boiling with a dilute acid, alcohol is converted into ether by boiling with dilute sulphuric acid, a small quantity of the oxides of nitrogen promotes the formation of an enormous quantity of sulphuric acid from sulphur dioxide, oxygen (in air) and steam in the " chamber " process of manufacturing that substance. Scores of examples could be given. In every one the catalyst has the same composition at the end as it had at the beginning, and, excluding losses which can be explained, the quantity is unaltered.

Berzelius, in 1835, was the first to draw attention to these phenomena and to apply the term catalysis to them, and Faraday was the first to draw attention to the importance of surface. The more spongy and porous the material the more effective it was in promoting the change. Thus, if a piece of asbestos is soaked in a solution of platinic chloride and then heated the platinic salt decomposes and platinum in a finely divided condition is left on the fibres. The " spongy platinum " thus formed is more effective than a sheet of the metal. This suggests that " adsorption " which was described in the last section but one is the primary cause, though until we know more about adsorption it does not carry us very far. Still, wherever a solid catalyser is used its value depends largely upon its extended surface, and ingenious ways of " mounting " the fine particles have been devised. In some cases brick is soaked with a solution and dried ; in other cases asbestos is employed. But perhaps the most ingenious method is to soak hydrated magnesium sulphate (ordinary Epsom salts, $MgSO_4, 7H_2O$) in a solution of a platinum salt, and then to heat strongly in a current of SO_2. The water of crystallization in the magnesium salt is driven off and the salt swells up, the platinum salt is decomposed and the fine particles of the metal are distributed over a porous mass many times the volume of the original magnesium sulphate.

Another view of catalytic action is that the catalyst enters into an unstable union which immediately breaks down or interacts with another substance to produce the final results. Thus, Williamson showed in 1854 that when ether is prepared from alcohol in the presence of sulphuric acid the first stage is the formation of a compound of alcohol and sulphuric acid, called ethyl sulphonic acid or sulphovinic acid, and that this substance reacted with more alcohol to form ether and sulphuric acid. A similar explanation involving the formation of an unstable intermediate compound has been suggested to account for the action of the oxides of nitrogen in the chamber process for the manufacture of sulphuric-acid.

There is no doubt that this is what occurs in many other cases. Nor is the theory ruled out in the case of solid catalysts. Thus if crystallized manganese dioxide is used to facilitate the preparation of oxygen from potassium chlorate it is found at the end of the reaction that the crystals have become reduced to powder ; and the fine platinum gauze which is used in some industrial processes is found after some time to be deeply pitted or to present a rough blistered surface.

It will be obvious that the possibility of speeding-up a process or of carrying it out at a lower temperature is of enormous importance from an industrial point of view. But the transfer of a process from the laboratory to the works is attended with two difficulties. In the first place the rare metals, platinum and palladium which were found to be so effective in the earlier experiments are too expensive to be used in large quantities, and extensive researches have had to be undertaken to discover cheaper materials. Nickel, copper, iron, and many metallic oxides have been used, and in some cases a second substance has been found to increase the activity of the catalyst itself—such a substance is called a " promoter. " In fact, so many substances have been found to possess some activity and so many reactions are effected by their presence that Ostwald has expressed the view that " there is probably no kind of chemical action which cannot be influenced catalytically, and there is no substance, element or compound, which cannot act as a catalyser ". Even if this is an over-statement it indicates the importance of these phenomena in both chemical science and chemical industry.

The second difficulty in commercial applications arises from the fact that an impurity in the reacting substances rapidly reduces the effect of the catalyst or " poisons " it. Though this fact had been discovered in the laboratory it became of far greater importance when working with large quantities on a commercial scale because of the difficulty and expense of securing pure materials ; and applications to industry have been delayed until this part of the problem could be solved.

It would not be possible within the limits of this chapter to mention the numerous examples in which catalysis is employed in the laboratory or in the works to-day. Only a few cases can be considered and they will be confined to industrial processes which have become of enormous importance during the last twenty years—processes which yield at a reasonable price products that are essential to modern industrial development.

Thus sulphuric acid has been manufactured by the chamber process since about 1750 and every text-book of chemistry contains an account of it. The space occupied by the plant is very large ; the capacity of a single set of three chambers may be 150,000 cubic feet. The direct union of sulphur dioxide SO_2, and oxygen under the influence of platinum to form sulphur trioxide SO_3, was suggested by Davy 1812 and attempted by Phillip in 1831 ; but success was not attained until 1901 and development has been so rapid since then that the older process is threatened.

In the contact process, as it is called, the sulphur dioxide and air from the pyrites burners have to be thoroughly freed from dust, finely divided sulphur, arsenic, antimony, and lead by washing and cooling. They are then led in through a " converter " containing the catalyst, care being taken to secure a temperature between 400° C. and 450°. As catalyst burnt pyrites (Cu O, Fe_2O_3) has been used for the preliminary conversion but the final product is always obtained by the aid of platinum. This metal has become so expensive in recent years—five times the price of gold—that the largest possible surface with a given weight of metal is necessary. Some manufacturers deposit the finely divided metal upon coarse nets made of asbestos string, twelve to twenty or more being stretched across the converter. This involves from $1\frac{1}{2}$ lb. to $3\frac{1}{4}$ lb. of platinum which would cost well over £300. Another method is the Schroder-Grille plan described on page 246. The heated magnesium sulphate may contain only 0·2 per cent to 0·3 per cent of platinum, but is so active that 5 grains of platinum enable a ton of acid to be produced a day. The gases are absorbed in sulphuric acid, which is more effective than water.

The acid obtained by the contact process is called oleum on account of its appearance. It is stronger than the ordinary acid, having the composition $H_2S_2O_7$—like the old Nordhausen sulphuric which used to be prepared by heating crystals of green vitriol $FeSO_4\,7H_2O$. It is essential for certain chemical processes for which the ordinary acid is less suitable. The ordinary acid can readily be obtained from it by the addition of water. By the contact or chamber process the production of sulphuric acid before the war reached 4,000,000 tons per annum.

Hydrogen gas is usually prepared in the laboratory by dissolving zinc or iron in an acid, but this method is too expensive when large quantities are required for industrial purposes. Another plan which is employed is to pass steam over red-hot iron. A still more recent plan involves the aid of a catalytic agent. Thus, as explained in Chapter II, if steam is passed over red-hot coke in a converter (Fig. 170) a mixture of carbon monoxide and hydrogen is produced. If this mixture, together with steam, is passed over finely divided nickel, cobalt, tin, or oxides, or a mixture at a temperature of 400° C. to 600° C. under a pressure of 4 to 40 atmospheres (60 lb. to 600 lb. per square inch) the following reaction takes place :—

$$CO + H_2O + H_2 = CO_2 + 2H_2.$$

The CO_2 is absorbed by passing the gases, still under pressure through water or milk of lime. It is interesting to note that a third of the energy used for driving the pump is recovered by causing the liquid in the absorption vessel to drive a Pelton Wheel. The process is continuous and once it has started the temperature is maintained by the heat produced by the reaction. Moreover, the hydrogen is of a high degree of purity and extremely suitable for use in the next process to be described.

Animal and vegetable oils have been used for centuries for making soap, and especially within the last thirty years for

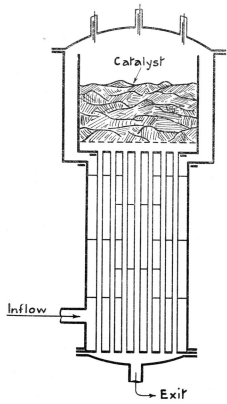

Fig. 170.   DIAGRAM OF CATALYTIC APPARATUS FOR THE MANUFACTURE OF HYDROGEN.

making margarine. They consist of an organic acid combined with glycerine, and displacement of the glycerine by soda or potash results in a soap.   But the growth of the margarine industry made great inroads upon the raw material of the soap manufacturer, who began to look round for other materials. One well-known soap manufacturer complained that he no sooner obtained a new fat for soap making than the manufacturer of margarine also began to use it. Finding olein and stearin from beef fat becoming expensive he began to use cocoanut oil or palmitin and other natural oils or fats.

Now these natural products occur in two parallel series. In each series the molecules of the successive members differ by one atom of carbon and two atoms of hydrogen—in other words by the group—CH_2. The series differ from one another by two atoms of hydrogen. That is to say, in one series each member has two atoms of hydrogen less in the molecule than the corresponding member in the other series.

Comparing the lower members of each series, the substances which are short of hydrogen are liquids, and those which have their full complement of hydrogen are solids, or semi-solids. The latter are called saturated compounds and are most useful ; the former are called unsaturated and are less useful. The solid substances are used to make candles. The possibility of hardening fats, or converting an unsaturated substance into a saturated one is therefore of great industrial importance.

The foundation was laid when Sabatier and his colleagues, particularly Senderens, showed from 1897 onwards that many unsaturated compounds would take up hydrogen in the presence of nickel and many of its compounds and become saturated. Though the early conversions were accomplished by passing hydrogen and the vapour of the unsaturated body over the catalyst in a heated tube, it was known by 1903 that the same result followed if hydrogen were passed into an unsaturated liquid in which the catalyst was suspended, and the process, with many modifications, is now in extensive use. The hydrogen must be cheap and pure—hence the importance of the process just described ; it is forced through the liquid under high pressure— 25 atmospheres (360 lb. on the square inch) or more, and at a temperature which is not usually higher than 200° C. A large amount of useless material has by this process been rendered useful, and an increased consumption of necessary materials rendered possible without a corresponding increase in price. People to-day are eating with relish fats which twenty years ago the candle-maker could not, and the soap-maker would not, use .

To the demand of the world's crops for nitrogenous materials was added, during the war, the demand of the world's armies for explosives, and the processes for manufacturing nitric acid from atmospheric nitrogen which have been described in Chapter IX were developed at a far greater rate than would have been possible in times of peace. Particularly was this the case in Germany, cut off as she was from the Chilean nitrate fields, and but for the energetic efforts of her industrial chemists, she could not have maintained her position half so effectively nor for half the time. It was, indeed, fortunate for her that, in addition to the Birkeland-Eyde and similar processes, a German named Haber had invented another one, the details of which had not leaked out. This process was the direct combination of hydrogen and nitrogen under the influence of a catalyst to form ammonia. The difficulties of the process are exceptional and success has been described as " one of the greatest triumphs of modern physical and engineering chemistry ". A committee of British chemists was established to investigate the problem, though they had little but the fact that the process *could* be carried out successfully to guide them ; in all the allied countries a small experimental plant was established ; and in more than one case the problem was solved—at least on a small scale.

The problem was to prepare both hydrogen and nitrogen of a sufficiently high degree of purity ; to ascertain the most effective catalytic agent ; and to determine the temperature and velocity of gases over the catalytic agent which give the highest yield, besides many other details. In patent specifications many metals have been proposed as catalyst, though iron in some form or other is probably employed. Generally speaking metals of high molecular weight which are known to combine with hydrogen or nitrogen are preferred. The temperature is 500° C. or 600° C., and the gases are circulated round and round, through the converter (Fig. 171), the small quantity of ammonia formed each time being removed by liquefaction or solution in water, and the pressure is between 100 and 200 atmospheres—1,500 lb. to 3,000 lb. per square inch.

The Haber process was most fully developed by the Badische Anilin und Soda Fabrik which had the original patents and enjoyed a long start. But an American Company have made considerable progress, and when the war ended the British Government were ready to begin. In December, 1919, however, a French improvement was announced. M. Georges Claude claimed to have succeeded in securing a yield four times as great by increasing the pressure to 1,000 atmospheres, or 14,000 lb. on the square inch.

For many years chemists have been performing experiments over a wide range of temperature, but these have for the most

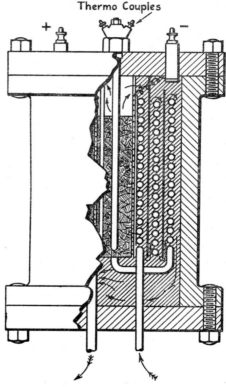

Fig. 171. Diagram of Catalytic Apparatus for the Manufacture of Ammonia.

part been conducted in open vessels or, at any rate, under a small range of pressure. The steel maker has carried out what are practically chemical operations in glowing furnaces, the temperature of which may reach 2,000° C. or more. In his internal combustion engines the engineer employs a chemical action, at, say, 1,000° C. and 1,000 lb. on the square inch. But the process to which reference has been made requires not merely a temperature of 500° C. to 600° C., but a pressure of 14,000 lb. per square inch—surely one of the most marvellous results of discovery and invention that have ever been recorded.

The importance of the Haber or Claude process for producing synthetic ammonia lies in the fact that it is the most effective method of obtaining nitrogen compounds from the air. The ammonia can be oxidized to nitric oxide by mixing it with air or oxygen and passing it over a catalyst—usually platinum in the form of a fine gauze at a temperature of 600° C. to 700° C. The nitric oxide combines with oxygen in the air and water to form nitric acid. The nitrogen in the air can therefore be converted into nitric acid by two different processes—firstly, by direct union under the influence of the electric arc, and secondly by (a) liquefying air and separating from the oxygen, (b) combining with hydrogen in the presence of a catalyst to form ammonia, (c) oxidizing the ammonia with air, in the presence of a catalyst. Both of these have been rendered possible by the scientific knowledge of the Twentieth Century. They were not possible before.

FERMENTATION

The term fermentation was first given to a number of chemical changes which were produced by the action of living things. Thus *yeast*, which is a low form of plant life, akin to the moulds, converted starch into sugar, carbon dioxide being evolved and causing the liquid to froth. Again the mould *mycoderma aceti* converts alcohol into acetic acid. These processes have been employed for many years in the manufacture of beer, stout, wines and spirits, vinegar ; and many other changes occur in the presence of bacteria which are still lower forms of the vegetable world.

Changes of this kind, however, were known to occur in the presence of minute quantities of non-living substances, and under conditions in which the possibility of life were ruled out. These substances are called unorganized ferments or *enzymes*. The identity of the two processes was first indicated by Buchner, who, in 1896, squeezed out from ruptured yeast cells an enzyme called zymase, which acted in precisely the same way as the original yeast. The view to-day, therefore, is that the organism, animal, or vegetable, mould, yeast, or bacterium produces the enzyme, and the enzyme causes the chemical change in much the same way as the catalysts which have already been described.

Enzymes are colloids, and are extremely complex bodies which have probably never been prepared in a pure condition. Being products of a living organism they are always associated with life, and are intimately concerned in vital processes—the changes which occur in living things. They are the active directive agents by which plants build up from earth and air their structure, colour, and perfume ; and in man and other animals they cause the changes which convert the food he eats into substances which can be assimilated by bone, and muscle, and nerve. Those which are obtained from animal sources are most active at a temperature of 40° C. ; those from the vegetable kingdom are most effective at a temperature of 25° C. ; those from cold-blooded animals exert their influence most powerfully at 15° C. At 0° C. most of them are practically inert ; so that putrefaction of meat, the decay of vegetable matter and the souring of milk can be prevented or delayed by cold storage.

Their close association with vital processes is also emphasized by the fact that most enzymes are destroyed by a temperature of 60° C. and all are instantaneously " killed " at 100° C. Hence meat will keep longer when it is cooked because the agents of putrefaction are destroyed and decay can only begin by the entry of fresh bacteria or other forms of life which will form a fresh source of the ferments. Perhaps one reason for attributing the action to bacteria, or moulds, rather than to their product, for so long, is the fact that the action can be prevented or stopped by " poisons ". Substances like hydrocyanic acid, sulphuretted hydrogen, mercuric chloride, iodine and many others are effective. It is upon this principle that the antiseptic treatment of wounds is based, and substances less harmful to man than those which have been mentioned are used for preserving food, e.g. borax and salicylic acid. It is possible to destroy the enzyme without killing the organism and *vice versâ*.

Perhaps the commonest example of an everyday operation involving the use of an enzyme is the manufacture of cheese. A substance called " rennet ", which is produced by the glands in the mouth and stomach of animals, when added to milk causes clotting, and for this purpose one part of rennet will curdle 400,000 parts of milk. The character and flavour of the cheese depends upon the enzymes which are present. These matters are not now left to chance. Each form of lowly life produces its own particular enzyme, which attacks the material at hand in a particular way. Thus for Cheddar Cheese the Bacillus *acidi lactici* discovered by Pasteur is alone necessary, while the green mould *penicillium glaucum* is required for Gorgonzola. In order to produce the correct flavour both in butter and cheese-making, pure " cultures " are employed, and a bitter or sour flavour or special colour are produced by organisms whose presence is not welcome.

Other applications are to be found in the curing of tobacco, and in the separation of the fibres of flax and hemp. A process that may attain very great importance is the fermentation of starch (from maize, rice, acorns, chestnuts, etc.) by enzymes from certain organisms with the production of acetone and fusel oil, the latter being a mixture of alcohols with a larger number of carbon and hydrogen atoms than ordinary alcohol. Acetone has become important for the manufacture of explosives, chloroform and dopes for aeroplanes, and fusel oil is employed for varnishes. Both of these substances are necessary in the process for the production of artificial rubber—a process which has been carried out in the laboratory but not yet on a commercial scale on account of the cost of the raw materials.

Perhaps no more wonderful effect of the influence of enzymes can be imagined than the explanation which has been given during the last dozen years by Dubois, McDermott, and others of the origin of light in the firefly. The body of the insect contains a substance which has been called *luciferine*, and this is oxidized under the influence of an enzyme, *luciferase*, which is secreted by the fly.

But, after all, this can hardly be more wonderful than the changes which go on in our own bodies, by which food is built up into the structures and tissues of which we are composed. This is a problem which is rapidly changing, appearing in new aspects as it is viewed in the light of modern knowledge. But it will be well to describe some of these aspects in a separate section.

VITAMINES

From a chemical point of view, the foods of man may be grouped into four divisions :—

Carbohydrates, such as starch (e.g. in bread, potatoes, etc.) and
sugar (of which there are several varieties) containing only
carbon, hydrogen, and oxygen.

Fats, such as butter, lard, the fat of meat and fish, all animal in
origin, and vegetable fats contained in seeds of plants,
again containing only carbon, hydrogen, and oxygen, and
consisting of an organic acid combined with glycerine.

Proteins contained mainly in meat, eggs, milk, and cheese. They
are complex bodies containing nitrogen.

Mineral matter.

The purpose of the carbohydrates is to produce energy. They
are converted, ultimately, into carbon dioxide and water—
burnt up, in fact, by the oxygen inhaled from the air. The
purpose of the fats is also to produce energy, but in this matter
the carbohydrates enjoy preferential treatment—they are used
first. But fat can be, and is, stored up to be used when carbohydrates are deficient. The purpose of the proteins is to build
up the tissues and to repair waste, though when the carbohydrates
and fats are exhausted, the proteins may also break down and
produce energy. The purpose of the mineral matter is more
varied and its consideration may, for the purpose of this chapter,
be omitted.

Now let us examine more closely the transformation and
destination of each of the first three types of materials. The
proteins are broken up in the stomach and small intestine into
bodies called amino-acids, of which eighteen are known though
only a few have as yet been fully investigated in relation to the
maintenance of life. These amino-acids find their way through the
walls of the small intestine into the blood, by which they are
carried to all parts of the body and to every one of the million
or more cells distributed throughout bone, and muscle, and nerve.
In the cells they are reconverted into proteins, but not necessarily
the original ones. They are recombined, but not in the same way.
All of them may not be used. The cells are fastidious as to both
character and quantity. They accept what they need, and reject
what they do not require. What is rejected is waste and passes
out of the body without performing any service.[1]

If the various proteins are labelled A, B, C, D, etc., and the

---

[1] A small quantity of protein reaches the liver and is "burnt up" like the
carbohydrates.

amino-acids are designated by a, b, c, etc., to r, then A may yield
a, b, and k ; B may form a, d, f, n, and o ; from C we may
obtain e, l, m, and r.   In other words, different proteins give
rise to different amino-acids, though a particular amino-acid
may be contained in different proteins.   But since the food value
lies in the amino-acid all proteins are not equally valuable.
Some, indeed, may be useless.   In 1814, the French Government
tried to feed the patients in hospitals on the well-known cheap
and easily obtained protein called gelatin, with disastrous results.
It is now known that the amino-acids *tryptophane* and *tyrosine*,
which are yielded by many proteins and are essential to life,
are absent from gelatin.   Without these there can be no growth
and no repair of wasted tissue.

One important difference between proteins and amino-acids
is that the former are colloids and the latter crystalloids.   Colloids
will not pass through a membrane ; crystalloids will.   The wall
of the small intestine is a membrane, and the result of the action
of the digestive juices upon proteins is to convert colloidal
substances into crystalline substances which can pass through the
membrane into the blood.

The carbohydrates consist of several complex, and several
relative simple, substances.   The most complex, cellulose, which
exists in such varied forms as woody fibre and cotton wool, is not
digestible at all.   Starch, cane-sugar and lactose, or milk-sugar
are converted by the digestive juices into the simpler bodies
glucose, fructose or fruit-sugar, and galactose, from milk sugar.
These bodies pass into the blood with the amino-acids and proceed
to the liver, where they are converted into glycogen, or animal
starch, which is stored up and produces energy as required by
undergoing conversion into carbon dioxide and water.   Though
these sugars present a simpler problem than the eighteen amino-
acids their relative value has not yet been worked out with
precision.

Fats from animal and vegetable sources contain, as we have
said, an organic acid united with glycerine.   When such a
substance is treated with an alkali, glycerine is liberated and the
alkali combines with the acid to form a soap.   A similar change
is brought about by the digestive juices, and glycerine, fatty
acid and soap are formed.   These are carried by the blood to
the cells, where they undergo reconversion into fat, which is
stored up as adipose tissue.   Should a man use up energy at a

s

greater rate than it is provided by the carbohydrates in his food, he draws upon his store of fat and becomes thinner, while in a case of starvation he draws upon the protein of his tissues and becomes emaciated.

From experiments upon men working in closed chambers, and measuring the heat produced, the amount of energy required from food can be calculated. This varies from 2,200 calories for a man at rest to 5,000 calories or more for a man engaged in the heaviest labour. For a woman the figures vary from 1,800 to 3,200 calories. By labour, physical labour only is meant, because the experiments fail altogether to measure the energy consumed in brain work. The bulk of this energy is supplied by the carbohydrates, and a little from the fats, though to what extent fats are necessary to life, and whether there is any difference between them, is still largely a matter of conjecture. It was in the course of an investigation into this matter that Professor Hopkins of Cambridge lighted upon a great discovery.

At the end of last century and the beginning of this one, a number of investigators were busy inquiring into the relative value of different proteins. For this purpose rats are generally employed. They only live about three years, and arrive at maturity in 280 days. Consequently, it is possible to find out a great deal about the effect of food materials upon growth by experiments which do not occupy an abnormally long time. To obtain results over the same range with human beings, experiments on a man would have to extend over a period of sixty years. But with rats it is easy to find out, by varying one constituent of the food at a time, the effect of each constituent upon the growth of young, or maintenance of weight in old, rats. The results of such experiments show that for a protein to have any value as food it must be capable of yielding one of the amino-acids, trytophane, tyrosine, cystine, and lysine, all of these being essential to life.

In 1906 Professor Hopkins fed two sets of rats, one with an artificially prepared food mixture—protein, fat, sugar, and mineral salts—which had the same chemical composition as milk, and the other with the same mixture together with a small quantity of fresh milk. Those without the milk lost weight, and those with the milk gained weight. On the eighteenth day, the diets were changed over, with the result that the rats which were getting thinner began to grow, while those which had formerly grown began to lose weight (Fig. 172). The amount of milk was only two cubic centimetres a day, yet that made all the difference between growth and starvation. The fresh milk contained something without which the most carefully selected artificial diet of (apparently) the same composition as a natural food was incapable of supporting life.

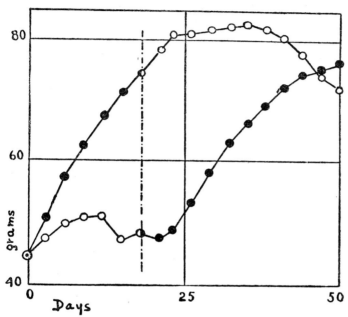

Fig. 172.   DIAGRAM SHOWING GROWING OF RATS.

Three years later Stepp discovered that rats fed on bread made with milk would grow ; but that if the bread were first extracted with a mixture of alcohol and ether, growth stopped and weight was lost. If the alcohol-ether extract was added, growth recommenced.

Experiments by Dr. Osborne of the Connecticut Experiment Station, Dr. Mendel of Yale, and Dr. McCollum, showed that while a diet containing lard might be unsatisfactory, it was at once effective in promoting growth if the lard was replaced by butter-fat. It is inferred, therefore, that butter-fat and other

fats which can replace it, contains a substance called a vitamine
—distinguished in this case by the special name fat-soluble A.

Further investigation by Funk and McCollum has revealed the
presence of a second vitamine which is ordinarily present in
milk-sugar, but which is absent when the sugar has been highly
purified. It is present to a greater extent in yeast. From the
latter it can be abstracted by water and is called water-soluble B.
A third vitamine called water-soluble C has been shown to be
present in most fresh fruit and vegetables. No vitamine has
yet been isolated ; no one knows what they are or what they
look like. Their presence can only be inferred by the difference
between the effects of fresh food and food which has been heated
or extracted by solvents upon the body in health and disease.

Thus, a disease called beri-beri, which is a sort of general
paralysis, was very common in the East and not unknown in other
parts of the world. It had long been recognized that it was in some
way connected with the diet. Forty years ago the men in the
Japanese Navy suffered severely from it. During a cruise of nine
months there were 169 cases and 25 deaths on a Training Ship.
When the next ship went out the diet was changed at the
instigation of Takaki, who had been medical inspector-general of
the Navy. The quantity of rice was reduced, and milk and meat
were added, so that the food approached more nearly that
supplied in European navies. The result was that the only cases
of beri-beri which occurred were among the 14 men who had
refused to eat the new food.

Again, in 1897, a Dutch medical man in Java had some fowls
which became sick from a disease called polyneuritis, which is
very similar to beri-beri in man. On inquiry he found that they
had been fed with some cooked rice left over from the kitchen.
He fed them with ordinary rice in the husk and they speedily
recovered. Subsequently he found that polished rice from which
the outer skin has been removed would always produce the
disease in birds. By investigation in prisons he found that
whenever polished rice was given beri-beri appeared, but when
unpolished rice was used the disease was rare. These results were
confirmed by Dr. Frazer and Dr. Stanton, in the Malay States
in 1908–9, and again by Major Chamberlain in the Philippines
between 1910 and 1913. Before the former year any number from
100 to 600 of the 5,000 men were always suffering from beri-beri.
By giving unpolished rice and beans for polished rice the cases

FIG. 173.—A BIG C.P.R. LOCOMOTIVE.

To face page 260.

F

FIG. 174.—TRACK LAYER AT WORK ON THE NEW TRANSCONTINENTAL LINE WEST OF EDMONTON.
(Grand Trunk Pacific Railway.)

To face page 261.

dropped to 50, 3, 2, and 0 in successive years. We now know that vitamine water-soluble B, which exists in the outer layer of the rice grain, is essential for the prevention and cure of this scourge. Funk isolated, from 200 lb. of yeast, one-twelfth of an ounce of a substance which, if not the pure vitamine is extremely rich in it, and found that one fifteen-thousandth of an ounce added to the diet cured pigeons paralysed by polyneuritis in a few hours. There is only one way to find out whether water-soluble B vitamine is present in any food, and that is to ascertain whether it will cure polyneuritis in birds or beri-beri in man.

Everyone knows of the havoc created by scurvy in mediæval armies and among Arctic explorers in modern times. It was rife during the crusades, it was known and feared by the adventurous sailors who took long voyages of discovery in the sixteenth and seventeenth centuries, and it inevitably occurs in countries stricken by famine. Men who suffer from it become dizzy and weak. The gums swell and recede from the teeth. The teeth become so loose that they can be pulled out with the fingers. The remedy for many years has been fresh fruit and vegetables—raw potato has often been used effectively. Experiments on guinea-pigs indicate that the effect of such foods is due to the presence of the third vitamine—water-soluble C.

It is found that vitamines are destroyed by heating, especially if the heating is prolonged. Thus, Miss Chick has shown that cabbage or beans heated for one hour at a temperature about that of boiling water lose 70 per cent of their value in curing scurvy. From this and other experiments it follows that canned foods are of little value unless an adequate quantity of fresh foods is taken at the same time. From the points of view of the man in the street it does not matter whether vitamines, as the physiological chemist understands them, exist or not, provided that he takes the advice given to him. Don't cook food longer than is necessary, and don't indulge too freely in preserved foods at the expense of fresh foods.

To put this into a practical form, fat-soluble A is present in milk, butter, yolk of egg, and to a less extent in beef-fat and many vegetables. There is a little in wheat, rye, and barley, but none in lard or vegetable oils. Water-soluble B is more widely distributed, and is contained in nearly all natural foods, especially milk, orange juice, and yeast, and in the outer coating of wheat, rice and other cereals. Water-soluble C is present in most fresh vegetables and fruit, especially tomatoes and oranges.[1]

How vitamines act is a puzzle. There are two views. One is that they stimulate the cells of the body into activity, and the other that they act as catalysts, facilitating those changes by which food produces energy or tissue in conformity with the requirements of the body. But, whatever the explanation, the investigations and the ideas to which they have given birth have thrown a new light on the feeding of man and of domestic animals. The mysteries of life have become deeper and more wonderful than they were before.

[1] Other vitamines with special functions have been recognized since the foregoing account was written.

CHAPTER XIII

" My son, the turkeys we eat all come out of little eggs."

" Indeed, Father, I always thought the reverse."

This controversy from *Punch* illustrates the relation between Great Britain's manufactures and her railway system. For it is certainly true that neither could have grown without the other, and the development of both owes something to geographical circumstances. The climate enables work to be done night and day all the year. The ports are never closed by ice, so that the railways can fetch and carry between the coast and the interior without interruption. The country is not mountainous, and the Great Central and Southern Plains permit of easy communication from east to west. There are few foaming torrents, or deep rifts or chasms, to be bridged, or high mountains to be tunnelled. And the childhood and youth of the railway system was fed and nourished by the increase of population, the concentration of vast numbers of people in towns, and the localization of manufactures. Wherever the railway goes there is an immediate return upon the outlay.

Contrast this with the circumstances of a railway driven across one of the great American plains. There the track had to be laid through a virgin country many times greater in extent than Great Britain, with a sparse population engaged in wringing from the reluctant earth a bare livelihood. Unable to purchase or wanting little of the luxuries of city life, with little to send away, and with no time to travel, the early settlers were able to offer small encouragement to the railway pioneers. And so the great trans-continental lines were constructed with the certain knowledge that for the first few years the venture would not pay.

The reader who has a little time to spare might employ it in an interesting way by drawing railway maps of different countries to the same scale on tracing paper, and by superposition, obtaining a comparison of the facilities for locomotion and transport. If, moreover, he compares at the same time the populations of the countries, he will obtain some idea of the connection between population and the magnitude and importance of a railway system that will illuminate many an otherwise obscure passage in national history.

To offer a description of British railways to readers who live in Great Britain seems like taking coals to Newcastle, and to describe the railways of the world would require a volume larger than the one in which this chapter appears. But it will be of interest to notice a few of the more striking features of the progress made during the life of the present generation.

Progress in railways differs according as new or old lines are being considered. In the latter it is not so much a question of invention in the ordinary sense of the word, as improvements in organization to meet new demands and increased traffic, and alterations of the track to reduce the cost of working. The past thirty years has seen a demand on the one hand for long, high-speed, non-stop journeys in England, and for through trains in which people can live for days together when crossing the great continents of Asia and America. On the other hand, the concentration of people in large cities, often within an hour's journey of one another, has required an inter-urban and suburban service of frequent trains ; and this service has had sometimes to be established wholly or partially underground. It will be convenient first to consider very briefly

TRANS-CONTINENTAL LINES

For mammoth engines, huge loads, and enormous distances, the inquirer must turn to the great continents of the old and the new world. North America was at first peopled only along the Atlantic seaboard and the Pacific slope. Until the railways came, the wide expanse of prairie and the stupendous heights of the Rocky Mountains separated the settlers in the east and west more effectually than if they had been divided by the rolling sea. The difficulties met with in the construction of these lines were enormous. On the one hand the directors, realizing that for a time the business must be conducted at a loss, demanded cheapness. On the other, great natural obstacles required an expensive scheme. In the low-lying areas there were extensive tracts of bog to be crossed, and wide streams to be bridged over. But the Rockies provided the most serious problems. Long detours had to be made to avoid tunnelling, and the trains wound backwards and forwards along the mountain sides as they

FIG. 177.—A CROSS-OVER ON THE CENTRAL LONDON RAILWAY.

To face page 264.

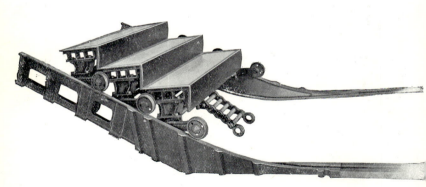

FIG. 178.—HOW THE STEP ESCALATOR WORKS.

To face page 265.

rose towards the summit. To save distance and time, steep gradients were included, and most of the Canadian and American Companies have spent millions in later years in relaying the line and driving tunnels to avoid detours and slopes that required three or four engines to mount them. In such country, indeed, it is often safer underground than on the surface ; for mile after mile has to be protected against avalanche and the peril which accompanies their fall.

In another way, too, the local conditions influenced the plans of the engineers. Timber was plentiful and cheap, and masonry and steel were scarce, remote, and expensive. So abrupt dips in the land and deep gullies through which foamed mountain torrents were spanned by wooden trestle bridges. As the country became more thickly settled and business grew, these bridges were replaced by structures of masonry and steel, allowing for heavier loads, higher speeds, and a longer life ; while as opportunity offered the single track was doubled between important stations.

Apart from the achievements of construction, perhaps no railways in the world offer so much food for thought or provide so much material for interesting speculation as the great inter-oceanic lines of Canada—the Grand Trunk, the Canadian Pacific, and, when completed, the Canadian Northern. For these owe their origin less to the needs of the local inhabitants than to the growth of the population in European manufacturing countries. Only one-fifth of the wheat consumed in England is grown at home, and Canada is one of the principal countries from which 80 per cent of our staple food supply comes. For carrying the wheat from the prairie provinces in 1912 no fewer than 190,000 cars were employed, made up into trains of from 30 to 70 each, which, placed end to end, would stretch for 1,100 miles. During that year, the Canadian Pacific Railway Company ordered 300 new locomotives and 12,500 new freight wagons. The locomotives were each 70 feet long, weigh 195 tons, and develop 1,500 horse-power (Fig. 173). The aggregate horse-power of these engines were therefore 450,000. A writer in the Special Souvenir Number of the *Liverpool Journal of Commerce*, 1913, from which this information was taken, added further that the 12,500 freight cars placed end to end would reach 92 miles. Or if they were made up into 250 trains of 50 cars each and sent off at intervals of one hour, they would take $10\frac{1}{2}$ days to dispatch.

And this order was not to establish a system but to meet the normal expansion of trade.

The construction of the track across stretches of rolling prairie has provided the American and Canadian engineers with scope for their ingenuity in mechanical track-laying, and several ingenious machines for this purpose have been used during the past ten years or so. The one illustrated in Fig. 174 was employed on the Grand Trunk Railway of Canada. A train is made up of the tracklayer, followed by half-a-dozen flat trucks carrying rails, then the engine, and lastly a number of trucks carrying sleepers. Alongside the train is a trough with rotating rollers at the bottom, and the sleepers, pitched into this from the trucks in the rear, are carried along and tumbled out at the side of the track. Here they are rapidly placed in position, while from the huge overhanging front of the tracklayer rails weighing over half a ton each are slung into place and spiked to the sleepers. The track is then ballasted and is ready for use at first by light loads and within two or three months for heavy loads at high speeds. In this way progress has been made at the rate of 5 miles per day. Supposing the sleepers are only 3 feet apart from centre to centre, and that the work could proceed continuously, this would involve fixing the astonishing number of 8,800 sleepers per day.

But the character of the land on the east presented its own special problems. There are numerous low divides from which the water drains but slowly, and when the spring sunshine disperses the winter's frost, the land becomes a swamp. It is perhaps not out of place to note here that this feature determined the original native forms of locomotion ; the birch-bark canoe enabled the water-courses which threaded the morass to serve as the highway in summer, and snowshoes and sledges became imperative in winter. Wheeled vehicles were introduced on the North American Continent by Europeans.

There is now only one of the five great continents which is not crossed by the steel road. Europe has a perfect network of railways. Asia is crossed by the Trans-Siberian Line which, though interrupted for a time by Lake Baikal, now proceeds round the southern shore of that obstacle. North America is spanned by half-a-dozen lines, and with the completion of the Trans-Andean Railway, which reaches the highest elevation of any railway in the world, the east and west coasts of South

America are now in railway communication. In 1919 the Trans-Australian Railway was completed. But though much progress has been and is being made, the broken and thickly wooded character of Central Africa has hitherto presented impassable barriers on those continents.

The enormous expense involved in trans-continental lines is an important consideration ; and unless they connect populous centres, or pass through rich mining or agricultural districts capable of early and rapid development, or serve some political purpose, they are not likely to be undertaken. The absence of water and the difficulty of feeding the workmen in the desert, or the ceaseless efforts necessary to maintain clearings in tropical forests, demand an expenditure which is only justified by some powerful motive. And the tendency is not now so much towards big projects, as towards smaller enterprises which promise a quick return.

TUBE RAILWAYS

The pioneer railway engineers soon learned to pierce their way through great mountains, and the earlier volume contains an account of the way in which the Mont Cenis and St. Gothard tunnels were constructed. A different set of problems is met with in tunnels under water, of which those under the Severn, Mersey, and Thames are examples. As compared with either of these the construction of a shallow tunnel only a few feet below the surface of dry ground is a comparatively easy matter. But the types for which the early years of the twentieth century will be famous are the spiral tunnels in the Alps and the Rockies and the tube railways driven deeply beneath the earth's surface in London. For such enterprises the way had been paved by nineteenth century experiments, and to-day they are entered upon with the confidence born of the consciousness that instruments of precision, and tools of marvellously increased dexterity and power, will enable the work to be accomplished at a fraction of the cost that would have been involved forty years ago.

The invention which banished many of the difficulties of working under water or through water-bearing ground was the Greathead Shield, named after the famous civil engineer, who converted what had previously been a suggestion into a practical device. It consisted of a large iron tube equal in diameter to the external dimensions of the tunnel to be driven. The front end had a cutting edge, and the back was closed by two parallel partitions a few feet apart, fitted with airtight doors (Fig. 175). Compressed air was driven into the front end and this held back the water. The space between the two partitions formed an air lock, so that men could enter, and material could be removed, without reducing very much the pressure in the main chamber. As the material in front was excavated the shield was forced forward by hydraulic pressure, and the tunnel was lined with iron as it proceeded.

The tunnels under the Severn and the Mersey were not constructed in this way, and the difficulty of dealing with percolating water was enormous. The first named took no less than thirteen years to complete. Both were opened in 1886. In addition to

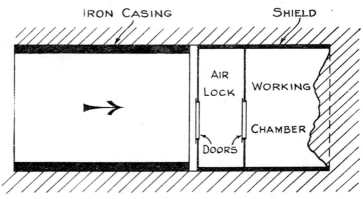

Fig. 175. GREATHEAD SHIELD AND TUBE.

the main tunnel which in each case slopes steeply towards the middle of the river, lower tunnels sloping in the opposite direction had to be made to draw off the water, see Fig. 176. Three times was the Severn tunnel flooded, the water entering at such a rate that the pumps were utterly unable to cope with it. And at one time it seemed doubtful whether that under the Mersey could be kept clear. Mr. Francis Fox describes how he and another engineer were found in the tunnel sitting under an umbrella, calculating the rate of increase in the volume of water, and poring over the conclusion that if it continued their pumps would be unable to draw it away !

FIG. 179.—CLEAT ESCALATOR.

To face page 268.

FIG. 179A.—A GARRATT ARTICULATED LOCOMOTIVE FOR THE WESTERN AUSTRALIAN STATE RAILWAYS.

The use of compressed air for shaft sinking and tunnelling had been patented by Lord Cochrane in 1830, but it was not actually used for the latter purpose until 1869, when Mr. P. W. Barlow and Mr. J. H. Greathead constructed the Tower footway under the Thames. In 1879 Mr. D. C. Haskin started to drive the Hudson River Tunnel, which runs under New York City. When, in the following year, the Northern branch had proceeded 360 feet, the compressed air in the front of the shield blew out the top, and twenty men lost their lives. For a long time work was stopped, but in 1888–91 it was extended 2,000 feet by Sir Benjamin Baker and Mr. E. W. Moir, further progress being prevented by want of funds. A final attempt was made in 1901, which proved successful, and the line was opened four years later.

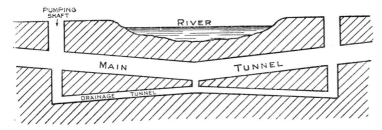

Fig. 176. TYPICAL RIVER TUNNEL—OLD PLAN.

The value of the shield was thoroughly demonstrated by the construction between 1886 and 1890 of the City and South London Railway, which, under the direction of Mr. J. H. Greathead, was driven through solid London clay, 90 feet below the surface. This was the first really deep tunnel in the world, and the nature of the material through which it was driven would have rendered any other method impossible. The original line ran from King William Street to Stockwell. It was subsequently extended to Moorgate Street, to Clapham Common, and then to Islington, forming 6 miles of double tube 10 feet 6 inches diameter. Quite recently $1\frac{1}{4}$ miles have been added, and the terminus brought to Euston. In the later work more rapid progress— nearly 12 yards a day—was attained by the use of Price's excavator. This is a large wheel fixed at the front of the shield and carrying radial cutting blades, which carve out the material and throw it behind for removal. With soft clay it performs its duty admirably, but if boulders are encountered they have to be broken up by pneumatic rock drills.

One of the most interesting of the London tubes is the Charing Cross, Euston, and Hampstead Railway, which has aided so much the development of a residential district north of London. The depth varies considerably owing to the varying level of the surface. At one of its extremities the tube emerges from the surface, but the station at Hampstead is 192 feet deep, while at a point 300 yards farther north the line is 250 feet below the level of Hampstead Heath. The tube is 11 feet 9 inches in diameter along the straight, 12 feet and 12 feet 6 inches on curves, and 21 feet 2½ inches at the stations. So great is the accuracy with which this and similar tubes are driven that it is no uncommon event for both shields (one starting from each end) to meet edge to edge. They are then left in to form part of the iron lining. In the case of the Hampstead Tube, Mr. Francis Fox states [1] that when the shields met in December, 1903, the following small inaccuracies were observed :—

Error in direction	.	.	¼ inch.
Error in level	.	.	⅛ inch.
Error in length	.	.	⅞ inch.

A completed tube is shown in Fig. 177.

These tunnels are small in diameter compared with some which are not used for railway traffic. The Blackwall tunnel, for example, is 24 feet 3 inches, and the Rotherhithe tunnel is larger—in fact, the largest in the world. The latter is 4,800 feet long, and runs for over 1,400 feet under the Thames. At one point it is only 7 feet below the bed of the stream, and it was impossible to follow the usual practice of making a " blanket " by tipping earth over the spot, owing to the inconvenience this would have caused to navigation. A railway tunnel under the East and Hudson Rivers has some similarity to that at Rotherhithe in its proximity to the bed. The principal objection is, of course, the danger of a " blow-out ", and on one occasion a workman was actually expelled right through the bottom of the river to the surface. It is remarkable that after such an experience he should have lived until the next day.

In the early days the use of compressed air was attended by some loss of life from caisson disease, so-called because it had

[1] *River, Road, and Rail.*

first appeared when compressed air was used for laying the foundations of bridge piers under water. A huge iron cylinder containing two partitions fitted with airtight trap-doors, was lowered into the water, and by its own weight (with the doors open) it sank into the bed of the river. It was then emptied of water by pumping, and air forced into the lower chamber, so that men could go down and set the masonry or concrete foundation. By allowing the door in only one partition to be open at a time the space formed an air-lock. Caisson disease has been investigated thoroughly in recent years by Dr. J. S. Haldane, Dr. Leonard Hill, and others, and has been found to be due to the fact that under increased pressure the nitrogen of the air dissolves in the blood, and is liberated in the form of bubbles when the pressure is removed. If the change from high to low pressure occurs rapidly the gas is liberated in large bubbles, which interfere dangerous with the circulation, and it is necessary for the change of pressure to take place gradually. The time required for decompression is, in fact, almost as long as the period during which the extra pressure has been experienced. All divers are liable to the same disease, and elaborate precautions are taken to avoid it. Thus a man is only drawn up part of the way at a time after working in deep water ; and Dr. Leonard Hill has devised a diving bell with a decompression chamber, into which the men go for a time before emerging into the open. But as the disease and its causes are now well understood, work in compressed air is not now more dangerous than many occupations carried on in workshops and factories.

Valuable, however, as was the Greathead Shield in enabling these subterranean corridors to be driven at a reasonable cost, not a little of their development has been due, as will appear from a perusal of Chapter XIV, to electric traction. Ventilation is at all times difficult in deep subways, and by no possibility could the smoke of locomotives have been extracted from the City and South London Railway. The use of electricity was in those days a bold experiment, but Dr. John Hopkinson's opinion has been amply justified, and all the underground railways in London and other cities are now worked by this power. Still, a stagnant atmosphere is difficult to avoid, and the air in the deeper tunnels is purified by ozone. This is a highly active form of oxygen, produced when a silent electrical discharge is passed through that gas or air. The air supplied to the London Tubes is passed through an apparatus which subjects it to a leak from metal surfaces connected to a source of high-tension electricity, and the ozone thus formed attacks and destroys the organic matter that tends to accumulate in the network of caverns below.

One of the disadvantages of tube railways is the trouble of getting to and from the platforms. Steps are very inconvenient, especially for elderly or stout people, and there are many to whom the sensation of travelling up or down in a lift is equally, if not more, objectionable. At many of the London stations—the Liverpool Street Station of the Central London Railway was an early example—escalators have been erected. These are moving stairs which work in inclined tunnels. Those at Liverpool Street rise 40 feet and the speed is 90 feet per minute. The four of them will convey as many passengers as ten of the ordinary lifts in use at the same station. The principle upon which they work is very ingenious, and by the courtesy of the Otis Elevator Company the author is able to include a description of the mechanical arrangements. The stairs are attached to a continuous chain which passes over a sprocket or toothed wheel at the top and bottom of the incline. The " tread " or standing portion of each step is supported on a wheeled truck and the two front wheels are closer together than the hind ones. Two pairs of rails are required—one pair of narrow gauge for the front wheels and one pair of wider gauge for the hind wheels. When the rails are on the level the " treads " follow one another closely, forming a moving platform, but at the foot of the incline the outer rails rise before the inner as shown in Fig. 178. In this way the tread always remains horizontal and the steps follow one another up the incline to the top, when the outer rail becomes level first. On either side is a flexible handrail, which moves at the same rate as the steps, though the motion is so steady that there is really no need for it.

The escalators usually installed have a carrying capacity of more than 10,000 people per hour. They were first exhibited at the Paris Exhibition of 1900, and have since been adopted to a considerable extent by railways, theatres, mills, and large stores in the United States. In London they are to be seen at Liverpool Street, Earl's Court, Paddington, and most of the other stations underground. These will carry over 2,000,000 passengers a day.

RAILWAYS 273

Escalators are also made without the steps, these being replaced
by a sort of chain of which the links provide a level foothold on
the slope. Such an elevator is shown loaded with passengers
in Fig. 179.

THE MODERN LOCOMOTIVE

The great railway engines of to-day—a perennial source of
interest during the period of boyhood—have not altered much
in outward form during the last fifty or sixty years, but they
have increased very considerably in size and power. A typical
engine of 1870 had a weight on the driving wheels of 15 tons ;
the Great Bear of 1908 a weight of 60 tons. The pull of the
former was over 10,000 lb. ; that of the latter was over 26,000 lb.
Sixty years ago the weight of a Great Western express, excluding
the engine and tender, was about 60 tons, whereas a modern
train will weigh anything from 200 to 350 tons, and some of the
large freight trains of America reach nearly 700 tons.

Increased power in any engine is generally obtained by
increasing the size and efficiency of the boiler and the size and
number of the cylinders, in addition to which the various devices
for preventing waste referred to in Chapter II are employed.
On a locomotive, however, the size of the boiler is limited.
If it is made higher the existing bridges would be in the way ;
if made of larger diameter the driving wheels would have to be
reduced in size, and though the largest driving wheels do not
permit of the highest speeds there is a lower limit beyond which
it is not desirable to go. With the existing gauge of 4 feet 8½ inches
larger boilers cannot well be employed, so that various methods
are adopted to increase efficiency, and these may briefly be
considered in turn.

The draught in a locomotive is caused partly by the air which
enters the front of the ash-box, owing to the motion of the
train, and partly by the discharge of exhaust steam into the
smoke-box. The earlier boilers were made with short smoke-
boxes, but it has been found that a larger space under the chimney
acts as a sort of reservoir, and the draught is much steadier. If
one of the older engines of any of the railways, or even a modern
London and South-Western, be compared with a modern London
and North-Western, for example, the extended smoke-box will
be quite noticeable.

T

A disadvantage of the locomotive type of boiler is the unequal application of the heat. From the fire-box to the smoke-box there is a rapid fall of temperature, and each square foot of heating surface at the former end is capable of evaporating far more water in a given time than a similar area at the other end. This defect is mitigated in the cone boiler of the Great Western engines of the Great Bear class, designed by Mr. Churchward. These decrease in diameter towards the smoke-box, and with a flat-topped fire-box of the Belpaire pattern, they not only hold a larger quantity of water where the heat is greatest, but they enable drier steam to be drawn off without the provision of a dome.

The special qualities of water-tube boilers have not escaped the attention of locomotive engineers, and the principle has been partially adopted on several lines. The simplest plan is that followed by Mr. Dugald Drummond on the London and South-Western engines, in which a number of inclined tubes pass through the upper portion of the fire-box from side to side. Another type which involves a much more extensive system of water tubes has been devised by Herr Brotan, and has been in use on the Austrian State Railways for the last twelve years. In this the main boiler has an upper drum or barrel, with which it is connected by two vertical wide tubes. The fire-box is really a nest of water-tubes, which discharge into the upper drum, while the burnt gases pass through ordinary fire tubes in the lower one. A third type is used on the Southern Pacific, and a fourth on the Northern Railway of France. The arrangement of water tubes in the fire-box of the last named is very similar to that of the Yarrow marine boiler illustrated on page 35.

The earliest locomotives were constructed to burn coke, and it was not until the introduction of the brick arch which prevents flames playing directly on the ends of the tubes that coal was used. On the Great Eastern Railway crude oil or creosote has been used either alone or in addition to coal. The liquid fuel is sprayed into the fire-box by means of a jet of steam. In Russia, in Mexico, and in the Far East, crude petroleum is used to a far greater extent, and more than 3,000,000 tons per annum are used in the United States for this purpose. In some cases it has been necessary to employ it for passing through tunnels on lines which ordinarily use coal, because the more perfect combustion attainable prevents the formation of smoke.

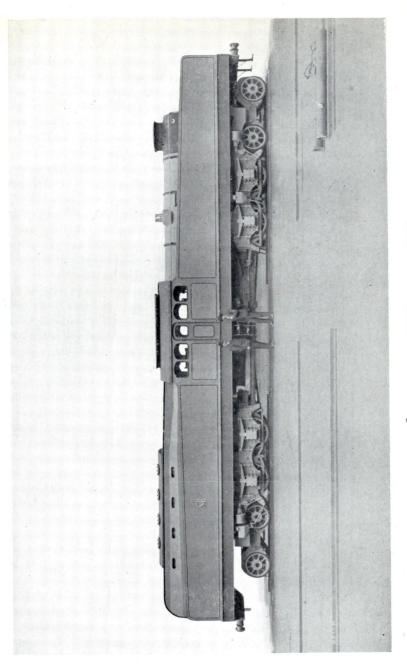

FIG. 181.—A GEARED TURBINE LOCOMOTIVE.

To face page 274.

FIG. 182.—A HUGE LOCOMOTIVE ON THE CHESAPEAKE AND OHIO RAILWAY.

To face page 275.

Perhaps no development in locomotive construction in recent years is more striking than the spread of superheating. The addition of superheaters is not new, and various forms have been introduced from time to time since 1840, but it is only during the last twenty years or so that real progress has been made, and there are now at least 30,000 locomotives in different parts of the world delivering steam to the cylinders at a higher temperature than that at which it left the boiler. The general principles upon which the superheater works and the way in which it effects economy in the engine have been described in Chapter III. On the locomotive the chief difficulty has been to arrange a considerable length of tubing in the limited space available. In some cases the smoke-box has been used, but in the latest form of the Schmidt superheater the relatively narrow tubing through which the steam passes on its way to the cylinder is contained in a number of larger flues leading from the upper part of the fire-box. The drier steam at a higher temperature leads to an economy of 10 to 15 per cent, though considerably higher figures are claimed.

The method of feed-water heating so largely adopted with stationary engines is somewhat rare in locomotives, though Mr. Drummond provided for it on some of the London and South-Western engines. There is no doubt a saving to be effected in this way, but there is a special difficulty. The locomotive boiler is fed, not by a pump, but by an injector, and readers of the earlier volume will recollect that this appliance works by condensation of a steam jet. Consequently an injector works more satisfactorily with cold water than with hot.

When the boiler has reached the largest size that the bridges and gauge will permit, and when it has been equipped with the most efficient devices for improving the draught, for heating the feed-water, and for superheating the steam, the only other direction in which greater economy can be obtained is in the utilization of the steam. The obvious method is to " compound ", and to cause the steam to expand over a greater range through two, three, or four cylinders successively. At the same time, the impossibility of mounting all the paraphernalia of condensing apparatus on a locomotive robs the system of some of its advantages. The compound engine on English railways has had a chequered career, and has been a sort of shuttlecock for successive locomotive superintendents. It has been flirted with by the Great Western, the Lancashire and Yorkshire, the Great Central, and the Great Northern ; taken up and dropped by the Great Eastern and the London and North-Western ; but is in use by the North-Eastern and to some extent by the Midland.

The Continental and American Railways are fairly unanimous on the matter, and two and four-cylinder engines of this type are the rule rather than the exception on the big lines. It is claimed on their behalf that a saving, which may amount to 20 per cent, results from the practice, and that this more than compensates for the extra cost of construction and maintenance. Some difficulty arises in starting. In the ordinary engine, steam goes direct to each cylinder, and the whole effect of the boiler pressure can be exerted to overcome the starting resistance. But in the compound engine the steam goes through the cylinders in series, and sufficient power for starting cannot always be obtained on the small high-pressure pistons. A special arrangement by which steam can be sent directly into both cylinders at first is therefore adopted, and this increases the complexity of the engine.

The apparent difference of opinion among locomotive engineers as to the merits of the compound engine is explained by a number of experiments undertaken by Mr. George Hughes, the Chief Locomotive Superintendent of the Lancashire and Yorkshire Railway. He showed that the value depended very largely upon the cost of coal. When this rose to 12s. per ton or thereabouts compound engines were a distinct advantage, but where, as in England, coal could be purchased by railway companies for about 8s. a ton, the increased cost and complexity of the compound engine were hardly worth while. During the last few years, however, the cost of coal in this country has risen materially.

In order to secure greater economy and flexibility experiments are being made in Sweden and in this country with turbine locomotives. Fig. 179A shows one built by the North-British Locomotive Co. for the L.N.E.R. The power is communicated to the axles by gearing ; there are no reciprocating parts, less fuel, water, and oil are required, and a lower maintenance cost is secured. The absence of reciprocating motion eliminates vibration.

The modern classification of locomotives is based on the number of wheels, and the extent to which they are coupled.

The front is supported by a pony truck, or bogie carriage, having two or four small wheels, while the main weight of the boiler is carried by two, four, six, eight, or ten larger wheels, which are usually coupled in order to increase the grip on the rails. The cab, too, is carried on none, two, or four bogie wheels. The usual

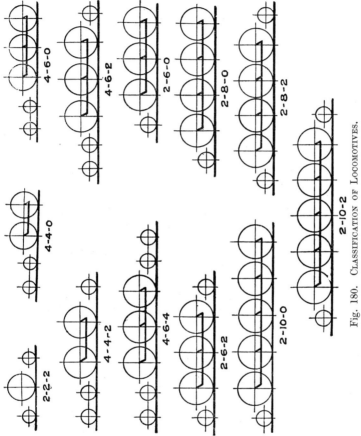

Fig. 180. CLASSIFICATION OF LOCOMOTIVES.

arrangements are shown in Fig. 180, and the description under each will render the diagram self-explanatory. The Great Western engine, the " Great Bear ", is a 4–6–2, and is a good type of a modern express, and the Northern of France large engine is a 4–6–4. An 0–6–0 and 0–8–0 would represent heavy goods engines in which a powerful grip on the rails is necessary, and 4–8–0 and 0–8–4 would be heavy shunting engines. The larger wheel-bases belong to heavy freights and steep gradients such as are not usually met with in Great Britain.

While locomotives in Great Britain exhibit no very great variety of pattern among themselves, they differ very materially from those in other parts of the world. Allusion has already been made to the powerful engines used by the Canadian Pacific Railway for drawing heavy freights. One of these has been illustrated in Fig. 173 ; it is a compound four-cylinder engine of the 4–6–4 type.

On many colonial railways the track is laid to a narrow gauge, and contains sharp curves and steep gradients necessitated by cheapness of construction. For the same reason relatively light rails are used. Of the locomotives which have been designed to meet these conditions none have been more successful than those of the famous Garratt type of articulated engines. By the courtesy of Messrs. Beyer, Peacock and Company, the author is able to illustrate one of these in Fig. 181. In November, 1906, six of these engines were delivered to the Government of Western Australia for use on the State railways, and seven more—of which one is illustrated—were delivered in May, 1913. The gauge is 3 feet 6 inches, some of the curves are only of 5 chains radius, and the gradients are as steep as 1 in 22. It was stipulated in regard to the first six that the load on each axle should not exceed 9 tons, and that the tractive force at 75 per cent of the boiler pressure should be not less than 21,000 lb.

The engine is really a double one consisting of a 2–6–0 and 0–6–2. The boiler is carried on a frame resting upon, and pivoted to, the engine frames at each end. This double joint enables the locomotive to take sharp curves without grinding, or the heavier portions overhanging so far as to endanger the stability. The distance between the pivots is 25 feet. The absence of wheels under the boiler enables it to be designed independently of the restrictions which usually hamper the locomotive engineer. The engines are not compound, but Schmidt superheaters are fitted in the last seven. The two tenders carry 2,000 gallons of water and 3 tons of coal. When full the total weight is just under 70 tons, and in no case does the load on one of the eight axles exceed 9 tons 7 cwt.

An interesting and modern type of locomotive is illustrated

FIG. 183.—AN OIL-BURNING LOCOMOTIVE ON THE SOUTHERN PACIFIC RAILWAY.

To face page 278.

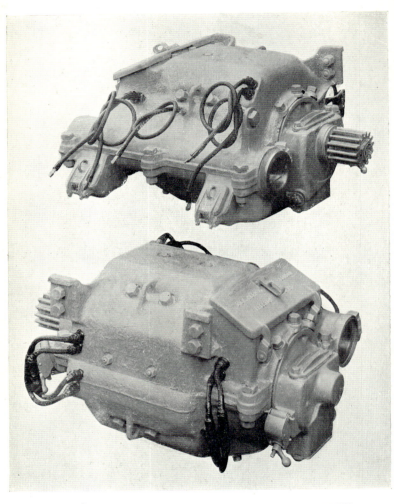

FIG. 187.—TRAMWAY MOTOR: FRONT AND BACK VIEW.

in Fig. 182. This is one of a number which have been constructed by the famous Baldwin Locomotive Works for the Chesapeake and Ohio Railway, and which are intended to haul trains weighing nearly 700 tons over the Alleghanies. The main frame is of vanadium steel and the construction is adapted for heavy work. A very good view is obtained of the Walschaert valve-gear, which is preferred on the Continent and in America to Joy's, which is used on British railways. A somewhat unusual feature is the position of the steampipe leading to the steam-chest, which in this case passes through the side of the smoke-box instead of downwards through the frame. The size of the engine may be gathered from the fact that the tractive force with 85 lb. steam pressure is 44,000 lb. That of the " Great Bear ", it will be remembered, is 26,000 lb. The tender carries 8,000 gallons of water and 14 tons of coal.

Another monster engine made by the same company is shown in Fig. 183. It was built for the Southern Pacific Company, and is a compound four-cylinder engine with an oil-fired boiler. The engine and tank-tender measure over 90 feet long, and the latter carries 10,000 gallons of water and 3,200 gallons of oil. Both this and the former engine are fitted with superheaters— the former with a Schmidt, and the latter with one of the Baldwin smoke-box type.

RAILWAY SIGNALLING

The fact that hundreds of thousands—nay, millions—of passengers are carried every year by the railways of the world with so few mishaps is a marvellous tribute to the watchfulness of engine-drivers and signalmen. The former may stand on the footplate of an engine for six or seven hours, with very few stops. He has to keep an eye on the pressure-gauge, watch the level of the water, and observe whether the signals at intervals of at most a few miles are for or against him. True, he has a fireman with him, but the driver is responsible, and though the number of matters to which he must give attention has been reduced as far as possible, the speed has to be regulated so that the scheduled time is kept. All this is sufficient by daylight, but when darkness falls there is an additional strain, which is intensified by rain, snow, or fog. In fact, some drivers will not face the responsibility, and decline promotion from the slow goods to fast passenger service.

If the engine-driver must possess clear vision, the signalman must possess a clear head ; for he must have in mind all the trains on his section of the line, and send and receive the messages that flash from box to box. At a big station like New Street, Birmingham, from which under ordinary conditions 700 trains are despatched per day, there is no time for dalliance, and no room for men who cannot concentrate themselves wholly and solely upon that section of the steel road which is under their care.

While a number of accidents arise from defective permanent-way and from culpable negligence, many have their origin in the inevitable fallibility of man. It does not seem to be realized generally that the safety of railway travelling lies in the perfection of the organization—partly human and partly mechanical—that controls the movements of the trains. Men who perform the same series of operations daily, year in and year out, act subconsciously ; they discharge their duties with a regularity that is machine-like in its precision. And this action is correct so long as the expected happens. But if the unexpected occurs ;—if by some fatal mischance a train which should be in the next section has not entered it, there is an accident. The signalman has learnt by long experience to look, not for the unexpected, but for the expected.

Whenever a railway accident occurs from the failure of a man, there is an outcry for the adoption of automatic devices ; and even as this chapter was being written the Midland Railway Company announced their intention to install an electrical apparatus to supplement the ordinary system.

Before proceeding to consider some of the plans which have been or are likely to be adopted it will be desirable to consider more exactly the object which it is desired to achieve. At present every railway line is divided into sections or " blocks " varying in length from one mile to several miles, and the problem is to prevent more than one train being on one section at the same time. This is secured by having a signalman to set the signals at clear, caution, or danger, and an engine-driver to observe them. Mr. W. H. Dammond, writing in *Cassier's Engineering Monthly* for December, 1913, contended that of sixteen recent serious accidents in England, France, and America nine would have been prevented by a signal given in the cab of the engine, and seven if the ordinary signalling arrangements had been automatic. It will be well to deal with these aspects separately.

Let us consider first the ordinary system of signalling. It may be presumed that everyone is familiar with the way in which a signal arm rises or falls when a lever is moved in the cabin, and knows that the ordinary means by which signal-arm and lever are connected is by long iron rods resting on wheels on short posts. When the signal is a long way off—and the distance must be great for fast traffic—the labour of operating the levers is considerable, and by no system of balance-weights can this be entirely avoided. Some means by which the arm can be raised or lowered by power is therefore desirable, and three systems were described by Professor W. E. Dalby in his address to the engineering section of the British Association in 1910. These three are the " all-electric", the "low-pressure pneumatic ", and the " electro-pneumatic ".

The first system is represented by the Mackenzie-Holland and Westinghouse method employed on the Metropolitan and the Great Western, the " Crewe " system on the London and North-Western Railway, and the system devised by Messrs. Siemens Brothers used on the Great Western Railway. The signal-arm in these cases is operated by an electro-magnet, the current for which is switched on or off at the signal-cabin.

Low pressure pneumatic signalling is in operation on the London and South-Western and Great Central Railways. The signal-arms are operated by compressed air at 20 lb. per square inch, and the valves are opened and closed by the usual system of levers and rods.

The electro-pneumatic is the most popular and is used on most of the other lines. The air pressure is 65 lb. per square inch and the valves are controlled by electricity. The small levers for switching on and off the electric current require less labour and occupy less space than those necessary when the signal-arms and points are operated directly. In the signal-cabin at the Central Station, Newcastle, there are no fewer than 494 levers, and in the cabin of the Central Station, Glasgow, 374. These figures give some idea of the complexity with which the modern railway engineer has to deal ; they also convey some notion of the onerous duties of the man who occupies the box.

In the remainder of this section we shall consider various additional devices which are in operation or have been proposed. Care must be exercised to distinguish between cab-signals and train-stops. Two systems of cab-signals have been in operation

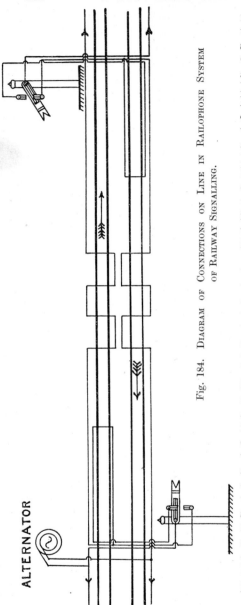

Fig. 184. DIAGRAM OF CONNECTIONS ON LINE IN RAILOPHONE SYSTEM of RAILWAY SIGNALLING.

ALTERNATOR

for some time in England—the Audible, invented by some of the staff of the Great Western Railway, and another, invented by Mr. Raven, the Chief Locomotive Superintendent of the North-Eastern Railway. They are also used on all the French railways. But so far train stops have been adopted only on electric railways, and, as we shall see, it is rather in this direction that there is the greatest likelihood of important developments.

A signal may be given in the cab of an engine in three ways ; firstly, by means of a trip lever, which stands up between or just outside the rails, and knocks over a lever on the engine as the train passes ; secondly, by means of a ramp or sliding contact standing up above the level of the rails and touching a corresponding shoe or wheel on the engine ; and thirdly by wireless communication. In all cases a mechanism on the engine is set in motion, and this may drop a small signal, blow a compressed air or steam whistle, light up an electric lamp, or even cut off steam and put on the brakes. Inattention on the part of the driver is in this case of no consequence. He is free to drive his engine, and regulate his speed, with the certainty that if the signals are against him, then snow, or fog, or darkness notwithstanding, he will know of it, and if he does not respond quickly his train may be pulled up for him.

The disadvantage of a trip lever situated near the ground (except in the case of tube railways) is that it is liable to become jammed by snow and ice. Consequently, experiments have been made with a lever mounted on a gallows, and capable of engaging another lever on the roof of the cab. In order to

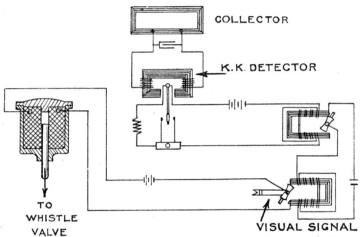

Fig. 185. Diagram of Connections on Engine in Railophone System of Railway Signalling.

avoid too great rigidity, which is undesirable for high speeds, the upper lever is made to swing, and on a wind-swept piece of line this is again a disadvantage.

The system which has been adopted by the German State railways and by the Midland Railway involves wireless communication, and is known as the " railophone ". By the courtesy of International Railophones, Limited, the author is able to give some account of the principles upon which the invention is based. A pair of insulated cables is either laid alongside the line underground or carried on poles, and along these an alternating current is sent. The current produces electro-magnetic waves through which any train on this section of a line must pass. A large coil of wire carried on a frame on the engine or tender serves to collect the waves, and the current produced in it operates a special detector which is the joint invention of Herr von Kramer, the inventor of the system, and Professor Gisbert Kapp. So long as electric waves are being received the detector keeps a battery on the engine connected up ; but immediately the waves cease, the detector stops, the current is cut off, an electro-magnet releases an armature, a small signal in the cab falls, a compressed-air whistle, or electric hooter, or bell sounds, and by suitable devices the steam may be cut off and the brakes applied. The arrangements on the line and on the engine are shown diagrammatically in Figs. 184 and 185.

Temporary interruption of the waves as the train is approaching a distant signal is effected by making what is known as a loop in the line cable. Two of these loops as shown in Fig. 184 will give rise to two short audible signals in the cab of the engine. These are merely to warn the driver that he is approaching the signal. If it is at " Clear " nothing further happens ; but if it is at danger and the driver ignores it, a prolonged audible signal is given, and the steam may be shut off and the brakes applied without further ceremony.

A more complicated form of this apparatus enables telegrams to be sent between the moving train and a station, and telephonic communication can be established in a similar way. These methods have been well tested both in England and Germany, and it is stated that satisfactory results have been obtained.

In ramp systems the apparatus on the locomotive is put into operation by the contact of a shoe or wheel with the ramp fixed at the side of the line. In this case, of course, no collecting coil or detector is required.

In all these examples the existence of a signalman has been assumed, and the object of the various contrivances has been to draw the driver's attention to the fact that he is approaching a signal, and to prevent him running past a signal standing at danger. A completely automatic system would be created by causing a train standing on a section to connect up an electric circuit and thus set in operation the current in the section behind it. A system of this kind is being tried on the London and South-Western Railway.

It will, perhaps, be of interest to describe the method, adopted on the Metropolitan Railway, which has been admirably described by Professor Dalby in the address to which reference has been made. The system is electro-pneumatic, modified so as to be automatic so far as the signalman is concerned, except at junctions and points which have to be operated. The arrangement is shown diagrammatically in Fig. 186. A_1, A_2, and A_3 show the successive positions of the same train on the line, and the same letters show the corresponding indications of the signals at the side of the track. One rail is continuous, the separate lengths being metallically " bonded " together. The other rail is " broken " about every 300 yards, this distance constituting a section. At the beginning of each of these sections is a signal. The train causes a short circuit in regard to each signal as it passes over the corresponding section, and the signal

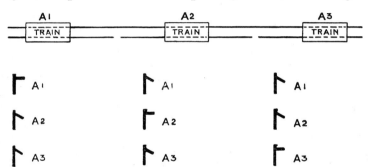

Fig. 186. DIAGRAM OF METROPOLITAN SYSTEM OF RAILWAY SIGNALLING.

behind it is raised to danger. No train may therefore enter that section until the one in front has passed out, when the signal falls to the " line clear " position, and the next one is raised to " danger ". The frequency with which trains can follow one another is remarkable. At Earl's Court no less than 40 trains per hour can pass each way, or a total of 80 trains per hour on two lines.

At junctions automatic working ceases, and the signals are controlled by a signalman. In each cabin is a small cast-iron box with 15 small spaces or windows each $1\frac{1}{2}$ inches square. These have a white background when the line is clear; at other times they show small indicators. On the approach of a train there is a click in the box, and a tablet stating the destination of the train appears in one of the windows. The signalman then presses in a plug and a similar notice appears in the next signal-cabin. As the train passes the first cabin the man presses another plug and the indicator disappears. The progress of the train is therefore notified two cabins ahead, and if the line is not clear the signalman can stop the train.

At present experiments on completely automatic arrangements are tentative, and such methods as are adopted will at first be supplementary to those already in existence ; there is no present intention of doing without signalmen altogether. But no partial provision will be of much value. There may be places where the risk of accidents is greatest—such as complicated crossings—but there are few spots on a railway where an accident due to faulty signalling or observation may not occur. The mishaps at such widely different situations as Aisgill and Liverpool St. James's testify to that.

Railway accidents are sometimes rendered more terrible by fire, which is most likely to occur when the carriages are lighted by gas ; but this illuminant is giving way to electricity, which can be generated while the train is running and be stored in accumulators in the guard's van. Another precaution is to use steel for the construction of the coaches, and some railways —notably the Hampstead Tube, and the Chicago, Milwaukee, and St. Paul—have already adopted this plan. Steel has the additional advantage that it does not splinter like timber, and is quite as capable of resisting shock. At the same time, it is easier to release an unfortunate passenger from a smashed-up wooden coach than from one of crumpled sheet-steel. The only tool that will cut steel rapidly is the oxyacetylene blowpipe (see Chapter VII) and that cannot be used in very close proximity to a person's body

So much for railways. They are purely a product of the nineteenth century, and they mark off that period in the world's history more effectively perhaps than any other results of man's handiwork. The material progress that could have been made with the horse-drawn vehicle, or even the cumbrous canal boat, might well have been great, but if these had remained the most effective means of inland communication our population, trade, food, and clothing, and many of our manners and customs would probably be still what they were in 1850.

CHAPTER XIV

THERE is nothing very humorous about an electric tramcar, and the advertisements often more nearly approach tragedy ; yet the late Professor Ayrton once said that two conductors were required—one to take the current and the other to take the current coin. Moreover, he looked forward to the time when the first of these would be unnecessary. Though this stage has not been reached, the progress has been remarkable ; for the first example of electric traction was a miniature railway laid down by Messrs. Siemens and Halske at the Berlin Exhibition of 1879. The method of conveying current to and from the motor is the same as on most electric railways to-day. A " third rail " is fixed alongside those upon which the cars run, and the current is collected by a sliding " shoe " attached to the locomotive or cars. From this shoe it passes to the motor, and back to the generating station through the ordinary rails.

The presence of a " live rail " close to the ground renders this method unsuitable for use in public streets, and at the Paris Exhibition of 1881 a railway was shown in which the current was conveyed by two overhead wires, from which it was collected by sliders attached to wires leading to the motors. The upper slider was subsequently replaced by a small wheel, and it was also found possible to have one overhead wire and to return the current through the rails—a plan which is followed on nearly all tramways to-day. In recounting some of the progress since these pioneer efforts it will be convenient to deal with tramways and railways separate.

TRAMS AND TRAMWAYS

The children and young people of to-day are hardly able to realize that trams were once small, uncomfortable vehicles drawn by horses, and in a few cases by puffing and snorting steam-engines ; and yet it was not until after 1890 that electric tramways began to make any appreciable headway in this country. A few experiments had been made in the 'eighties, such as the lines from Portrush to Giant's Causeway and from Ness to Newry, but these more nearly approached light railways than urban tramways. The causes which threw Great Britain behind in comparison with America and Continental countries were complex, and need not be discussed here. When once the initial obstacles had been overcome the rate of development was rapid, and to-day few towns of any size or importance are without electric trams. In the more thickly populated parts of the country like the Potteries, Lancashire, and Yorkshire, the services of different towns are so complete that they form a linkage along which one can travel great distances. For example, it is possible to go from Liverpool to Manchester and Manchester to Leeds by electric tram, merely by changing from one system to another at the termini.

The current for a tramway system is generated in a central station and supplied to the overhead wire at a pressure of 500 volts. As the pressure between flow and return tends to become weaker as the distance from the station increases, the wire is usually divided into sections of about half a mile in length. The reader will probably have noticed at the edge of the footpath near a trolley-wire standard a rectangular metal box or pillar about 3 or 4 feet high. This is the feeder-pillar containing the switches which enable the current from the line section which starts at that point to be cut out. A glance overhead will reveal two cables running along the bracket and connected with the trolley wires. The trolley wires before and behind this pole belong to different sections.

For very large tramway systems the current is distributed at a pressure of 5,000 volts or more to sub-stations, in which it is transformed to 500 volts pressure for feeding the overhead wire. Whether or not this system is used depends upon the distance and the power to be transmitted. It is often cheaper to erect and equip sub-stations than to put in heavy copper cables over many square miles of country.

It has already been remarked that the most suitable motor for tramway work is a series wound D.C. machine, which gives a strong torque or turning effect on starting. One of these is applied to a front and another to a back axle. The motors are fully enclosed to keep out dust. The axle passes through two holes in the casing at one side of the motor, which is suspended to the frame on the other side by a spring. This permits the toothed

FIG. 188.—TRAMWAY MOTOR : OPEN.

To face page 288.

FIG. 189.—TRAM CONTROLLER.

To face page 289.

wheel on the armature-shaft to gear with a larger wheel on the axle in spite of any jolting due to the unevenness of the road. Figs. 187 and 188 show a well-known type constructed by Messrs. Dick, Kerr and Company.

The device that usually mystifies young observers is the controller which is fixed in front of the driver and has a handle projecting from the top. Who has not sat at the front end of a car watching the jerky movements of the driver's hand and noting the readiness with which the car responds ? The principle and purpose, however, are very simple, though the details of construction are extremely complicated. The first movement of the handle switches on the current, so that it reaches the motor only after passing through a number of wire resistance coils usually placed under the seats. The second movement cuts out one of these coils and allows more current to flow through the motor. The third step cuts out further resistance, and so on until the lever is turned to full speed and the motor receives the full strength of the current. The contacts are inside the controller box, and are separated one from another by sheets of non-conducting and non-inflammable material, so that if, as is quite possible, an arc forms at one contact, the others will be uninjured. A further precaution consists of an electro-magnet which tends to blow out the arc should one be formed. The box also contains a switch for reversing the direction of the current through the motors and therefore the direction of the car. The arrangement is shown in Fig. 189.

There are one or two details connected with the overhead wires and the rails which are of some interest. The points at the junction of two overhead lines are sometimes automatic in one direction, but require to be operated by hand for a car proceeding in the other. It is rather difficult to show this by an illustration, but the reader who desires to understand it should watch the action closely as the trolley passes the points. Again, at the junction of the rails there is a tongue which opens out by the action of the wheels in one direction, but returns after the tram has passed. In some cases the tongue is quite loose and a boy is stationed at the junction to operate it for each car. A more recent plan is to operate this tongue by magnets on the car.

The overhead wire and its trolley met with no little opposition in the early days, and much was made of the unsightly

U

character of the equipment. The difficulties of providing an effective substitute, however, were so great that only two others have had a commercial trial, and these will now be described.

In London and some other places, the conduit system is in operation. A shallow tunnel or conduit, lined with concrete and of a section shown in Fig. 190, is contructed between the two rails. On either side of this tunnel are the conductors which convey the current from the central or sub-station. Through a slot in the upper portion passes an arm leading from the car, and carrying at its lower end two slippers which make contact with the conductor rails. From these the current is led by wires to the motor.

The Griffiths-Bedell (or G.B.) stud system is in operation at Lincoln. In this case current is carried by two iron conductors

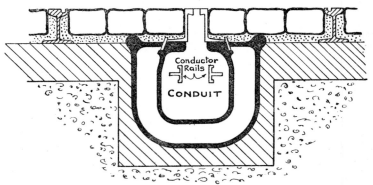

Fig. 190.   SECTION OF TRAMWAY—CONDUIT.   (After Whyte.)

in a conduit. At intervals cast-iron studs are let into the road-way, which carry at their lower ends small sliding contact pieces which are held a little above the conductor by a spring. The car has a long slipper which is always in contact with one stud, and as the car passes along an electro-magnet causes the sliding piece at the lower end of the stud to make contact with the live cable. When the car has passed the spring lifts the sliding piece of the cable and the stud becomes dead. The arrangement is shown in sections in Fig. 191.

The rapidity with which electric tramcars can follow one another without confusion is really marvellous, and on some routes a service of a minute and a half is regularly maintained.

FIG. 192.—A P.A.Y.E. CAR.

FIG. 193.—A RAILLESS CAR.

FIG. 194.—A RAILLESS CAR PASSING ANOTHER VEHICLE.

To face page 291.

Under these circumstances delay has to be avoided at all costs, and where no passengers are waiting at the optional stops both power and time can be saved if the driver keeps on. If, however, the conductor is busy collecting fares he is unable to keep that sharp look-out which is essential to quick progress. About fifteen years ago a new type of car—the pay-as-you-enter or P.A.Y.E. car (Fig. 192)—was introduced. In this the conductor remained on the platform and collected fares from passengers as they entered. Entering and leaving passengers, moreover, are separated, so that less time was required at the more important stopping-place where a considerable number board or alight from the car. This type of car has not, however, been extensively adopted.

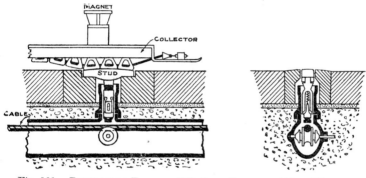

Fig. 191. DIAGRAM TO EXPLAIN G.B. STUD SYSTEM. (After Whyte.)

A powerful obstacle to the extension of a tramway system to the outskirts of a town before there is a guarantee of regular traffic is the cost of laying the track. This difficulty is being met in some towns by what is known as the trolley omnibus. This consists of a car constructed after the fashion of a motor-bus, driven by an electric motor, and collecting its current from overhead wires by means of two trolley poles, as shown in Figs. 193 and 194. No rails are necessary, the car is self-steering, and the trolley poles are so mounted that they can swing from side to side of the road without losing contact. The fact that such a car is not confined to rails renders it less of an obstruction than a car travelling on a fixed track, while it is obvious that a greater speed can be obtained. Moreover, if at any time the traffic on a route served by these cars develops sufficiently, it is quite easy to lay down the necessary rails. There are, at the time of writing, indications of a considerable extension of this plan. Trams have had to meet severe competition from motor omnibuses, and tramway undertakings are hesitating to renew the track in outlying districts. Buses have now been enormously improved ; they are quicker, less noisy, and do not obstruct ordinary traffic so much as trams. They involve smaller capital outlay, require less labour, and can adopt any route that may be desirable at a moment's notice. Many tramways are already utilizing them as feeders for, or supplementary to, their existing system. But the capital sunk in tramway undertakings will certainly prevent any violent change, and the harsh grinding roar of the trams will offend the ear for many years to come.

There is, however, another possible method by which the overhead and possibly also the surface equipment alone would fall into disuse, and that is by the use of accumulators carried on the cars and charged at the central station. The difficulty hitherto has been the great weight and fragility of the lead accumulator ; but the recent improvements made in the Edison storage battery appear to have rendered it capable of standing very rough usage. It is still heavy and expensive to manufacture, but it is very strongly made and its use on motor vehicles is extending very rapidly, especially in the United States. And if the employment of accumulators does become general, then the rumbling tram and coughing petrol omnibus will be replaced by a rubber-tyred vehicle with a silent motor, capable of high speed, and accurate steering, and possessing all the qualities which rapid urban traffic requires.

ELECTRIC RAILWAYS

The growth of enormous towns and groups of towns in close proximity reached such proportions towards the close of last century that the railway engineer found the problems which he had to solve separating into two groups. On the one hand there was the need of a quick suburban and inter-urban service with a fluctuating traffic, and on the other the need for an express service over long distances. The disadvantages of a locomotive and an ordinary train for the former are many, but one is obviously the waste in drawing a heavy engine with a tail of empty carriages during the slack period of the day. It is unnecessary to make more than a passing reference here to the establishment of rail-motor services in which a coach is fitted with a small steam-engine. This serves much the same purpose as the trackless tram already described, and is useful in dealing with traffic along a rural branch line which would not justify a large and heavy train ; and it is also used by several companies on suburban lines.

The use of electricity on railways in the early days was delayed to some extent by the notion that cheap water-power was essential and the first electric railways owe their existence to causes altogether outside the special merits of electric traction. The first real electric railway in Great Britain was the Liverpool Overhead, which runs along the whole length of the docks on an elevated platform. There were clearly objections to an ordinary locomotive 30 feet above the street level, and the promoters decided to use electricity. Again, the railway under the River Mersey was the first steam railway to be converted to the new method. Difficulties connected with pumping and ventilation, together with the steep gradients at either end, had resulted in lack of financial success, and electricity was adopted as a drastic step— a last desperate effort to convert imminent failure into ultimate success.

About this time the City and South London line was opened. It was the first really deep underground railway in the world, and in it steam was clearly quite out of the question. The difficulty of ventilating even a shallow tunnel will be realized by those who remember the Metropolitan Railway in its days of steam. Its grime and fumes were a fit inspiration for Sir Lewis Morris' poem " An Epic of Hades ", which was composed during many journeys through its poisonous atmosphere. Since then every tube railway in the world, with its paramount need for avoiding anything which would destroy the purity of the air, and many suburban lines with variable traffic and frequent stoppages, have adopted electric traction.

Among the chief advantages of electricity in locomotion is the fact that the grip on the rails need not be concentrated at one point, but may be distributed over the train so that the whole weight of the coaches will aid the adhesion of the wheels. Thus two at least of the coaches, and in most cases the first and last, are provided with electric motors, which can all be operated from either end of the train. The electric motor gives rapid acceleration on starting ; no shunting is necessary, and when the train is ready for its return journey the driver merely walks down the platform to the other end. No current is consumed when the train is not moving, and coaches can be put on or taken off to meet the variation of traffic.

The City and South London, the famous New York, New Haven, and Hartford, the Swiss, and many other railways use electric locomotives, and some enormously powerful examples of this type have been built (Fig. 195). This practice is essential where only a section of a line is electrified. Thus the difficulty of ventilating the great alpine tunnels has led to the use of electricity, and the steam locomotive hands over the train to the electrical locomotive for this part of the journey only. Generally, however, the need for rapid acceleration on starting and the absence of room for shunting lead to the use of motor-cars and trailers on suburban service above or below ground.

Pulling up suddenly and getting up speed rapidly, while achieved more easily by electricity than by any other form of power, throw a heavy strain on the equipment, and an interesting method of reducing it has been adopted on the Central London Railway. On this line each station is situated at the top of an incline, so that an approaching train slows down naturally on climbing the hill and acquires speed rapidly on descending. This natural method relieves the brakes in the one case and the electrical equipment in the other. But it is obviously not a plan that could be adopted on a line used also for fast through trains.

Most of the London tubes are operated by direct current obtained from sub-stations. At the central station alternating current is produced and transmitted over fine wires to the sub-stations at a pressure of 5,000 or 6,000 volts. Here it passes through a rotary converter (see p. 98) and is delivered to the line at 500 volts. This, in the case of the London underground system, power is produced at Lot's Road Generating Station, at Chelsea, and dispatched north, east, and west, over the whole of the area served by the tubes. This supply of a number of lines from one station equalizes the demand, and enables the generating plant to run with uniform load. Incidentally it furnishes a magnificent example of the advantages of production on a large scale, whether it be of power or manufactured goods. No set of independent stations supplying each line could be carried on so economically—there would be an inferior service, fares would be higher, and the accommodation would possess less comfort and convenience. The single station provides a means of locomotion for several millions of people to whom speed and frequency of service are essential not only for themselves, but for the business of the empire. From the farthest points of the system—Ealing, Hounslow, Highgate, Barking—the demand for more or less current is dispatched and answered in a fraction of a second. The system of wires and machines is like so many threads attached to a central elastic support, which yields, now in this direction, now in that, as occasion requires.

The character of the current produced at this station is 3-phase, and after conversion to direct current it is communicated to the train by what is known as the " third-rail " system, though, as a matter of fact, four rails are required. One rail outside those upon which the cars run conveys the current, while another in the middle of the track conducts it back to the station. Each coach carrying a motor has two shoes which slide along the conductor rails. The controller and other arrangements on the coach are very similar to those on an electric tramcar, which have already been described. There is, however, one feature which is worthy of separate description.

On a steam locomotive there are always two men, the driver and the fireman, and if one of them is taken ill or dies suddenly the other is able to act in the emergency. But the driver of an electric train is alone. True, there are conductors, but they are some distance away, the trains follow one another at high speed, and only the driver can see the signals. If anything should happen to him and the current were not shut off, there might be a serious accident. For until the train passed through a station at which it ought to stop there would be nothing to warn the conductors and signalmen that something was amiss, and some little time might elapse, therefore, before the current could be cut off from a section of the line in front of the train. Meanwhile, the cars would be rushing on to destruction with their passengers entirely oblivious of the threatening danger.

Such an event is prevented by a device known as the " dead man's handle ". On the top of the controller handle, by the movement of which the current is switched on or off, is a small button. Unless this button is pressed the handle cannot be moved from the off position. If by any chance the driver releases the pressure the handle flies back to the off position, and the train comes to a standstill. Moreover, as the signalling arrangements are operated by the train itself the section on which it stands is closed to any train behind it, and an accident is averted. This plan was adopted on the Central London Railway and a similar contrivance is now required by the Board of Trade in all cases.

To return to systems of transmission. The 3-phase system with direct-current motors was adopted on all the earlier lines, on the Metropolitan and District Railways, and on the Lancashire and Yorkshire's Liverpool and Southport Line. But about a dozen years ago a new system came into operation. High-tension single-phase current is transmitted by overhead wires all along the line, rendering sub-stations unnecessary. It is found that very large quantities of high-tension current can be collected by a slider, and as the single-phase current only requires one wire for its transmission against three for 3-phase there is an additional advantage. Practically the first line of this kind was the New York, New Haven, and Hartford Railway, but the first in Great Britain was the Lancaster, Morecambe, and Heysham section of the Midland Railway (Fig. 196). The power-house for this line is at Heysham, and uses gas from Mond gas-producers in gas-engines made by the British Westinghouse Company. The overhead equipment was put up by the Midland Railway Company, and the electrical equipment of the cars was supplied partly by the British Westinghouse and partly by Messrs. Siemens Brothers. The single-phase system has since been adopted for the Swedish State railways and for the London, Brighton, and South Coast Line on their various suburban lines. The Southern Railway still use high tension single phase for distribution, but are now using direct current for driving.

The current is conveyed along overhead wires at a pressure of 6,000 or 6,600 volts, and this part of the equipment differs materially from that in use on tramways. Not only must the insulation be more effective, but the collector must remain in contact at high speeds. The ordinary tramway overhead wire is supported rather rigidly at intervals, and everyone must have observed the " knocking " which occurs when the trolley wheel passes over these points. The electric-railway cable, on the other hand, must be flexible, so that it yields lightly but

FIG. 195.—A POWERFUL ELECTRIC LOCOMOTIVE, 2,500 HORSE-POWER, WEIGHT 108 TONS.

To face page 296.

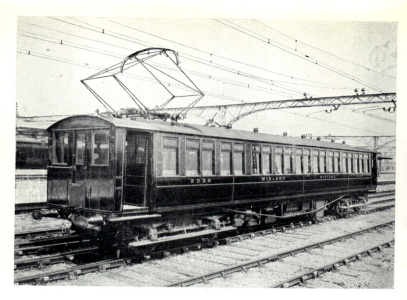

FIG. 196.—A WESTINGHOUSE CAR ON THE MIDLAND RAILWAY.

FIG. 197.—OVERHEAD EQUIPMENT AT TORRISHOLME JUNCTION
(MIDLAND RAILWAY).

To face page 297.

uniformly to the pressure of the collector. If any wire, rope, or chain is stretched between two points it sags, and the amount of sagging is greater as the distance between the two points increases. The curve which it forms is well known to mathematicians and is called a catenary, from the Latin *catena*, a chain. No amount of stretching will ever make it quite straight, and an approach to a straight line can only be obtained by supports at frequent intervals. But if the tramway method were adopted this would raise the cost of electrifying a long stretch of line enormously.

The difficulty has been overcome by hanging the conductor in stirrups from a suspended cable—as a matter of fact, two cables are used, the second being supported to the upper one by stirrups and the conductor wire being also suspended from the second by stirrups (Fig. 197). This gives a nearly level and perfectly flexible contact for the collector. The irregularities which would cause sparking and wear are, therefore, prevented. The amount of sag of the supporting cable depends upon its tension, and this is kept constant by fixing it at one end of a section and hanging a weight of 1,200 lb. at the other. A cable hanging too freely would swing with the wind, or " stagger " as it is called, and it is found that this stagger should not be more than 2 feet for a collector 7 feet across. The conductor wire is usually of figure 8 section, so that it can be clipped by stirrups without interfering with the sliding contact below.

The current in these single-phase lines is transformed on the train itself down to about 300 volts, and then conveyed, still as single-phase alternating current, to motors similar to those described on pp. 97–98. No sub-stations are required, and twenty years ago the plan seemed to be the simplest and most efficient that could be devised.

More recent practice, however, has shown a reversion to the original plan of direct current, but with a higher voltage. Thus the Manchester-Bury line opened a few years ago employs direct current at 1,200 volts. In America a higher voltage has been adopted. On the extension of the Chicago, Milwaukee and St. Paul line to Portland, Oregon, 440 miles over the Belt, the Bitter Root and the Rocky Mountain Ranges, and 211 miles over the Cascattes has been electrified. Three-phase current is obtained from hydro-electric power stations and transmitted to each 14-mile section, where it is transformed by rotary converters into direct current at 3,000 volts. In this district there is deep snow in winter, and if a steam locomotive runs into a drift it is cooled to such an extent that it is not easy to maintain a pressure of steam. But an electric locomotive can be overloaded without becoming overheated under such conditions ; and, in the winter of 1916–17, electric trains butted through drifts that would have put a steam locomotive out of action.

Before leaving the question of electric traction it is interesting to notice that electricity has in this, as in other cases, created its own demand. The old horse-car could not provide a quicker service than one every fifteen minutes. This is the *slowest* that can be economically provided by electric cars, and in many places a car starts from the terminus every five minutes or even less. Again, on the Inner Circle of the Metropolitan Railway there used to be 16 trains per hour, but with electric traction there are 40. Yet neither tram nor train run empty, and there is if anything a greater struggle for seats than ever there was in the old horse and steam days. In 1908 there were 204 miles of single-track railway worked wholly by electricity in the United Kingdom and 200 miles worked partly by steam and partly by electricity.

What London would do now if compelled to go back only fifteen years in history can hardly be imagined. The 138 miles worked by electricity carried, in 1908, 342,000,000 passengers, or one-third of the total number carried by all the railways of England and Wales in that year. In 1928 the London Electric Railways, excluding the Metropolitan, having a mileage of 142, carried no fewer than 368,367,918 passengers. The most heavily worked was the Central London Railway with a mileage of 11·06 and 42,282,585 passengers. Meanwhile the bus service in London has increased enormously.

Such means of quick transit have an important influence in extending the area of large towns. The country is healthy, and business men will live as far away from the centre of their town as they can cover in, say, an hour's journey. So London and Liverpool by their electric suburban railways, and Manchester and Birmingham by their trams, are spreading out and coalescing with places which were once distinct and isolated.

CHAPTER XV

MOTOR-CARS

WHEN a modern schoolboy was asked for the meaning of the phrase " the quick and the dead " he replied that the quick were those who got out of the way of motor-cars, and the dead were those who did not ; and there was, after all, a modicum of truth in the explanation. For the rate at which the speed of locomotion along public roads has increased in the last twenty years is amazing. The sedate and conservative citizen of the pre-Victorian period who pooh-poohed the idea of twelve miles an hour along a pair of rails could certainly not have been convinced that within a hundred years three or four times that speed would be attainable on an ordinary road surface with present-day ease. And even in the early 'nineties the members of that body which makes the law were equally sceptical, if not of the possibility, at any rate of the desirability, of such a degree of speed.

Attempts to drive vehicles on ordinary roads by mechanical power were perhaps earlier than those which led up to the railway, but were far later in coming to a successful issue. When the steam traction-engine made its appearance it was such a clumsy and terrifying contrivance that regulations were laid down to limit its speed, including the requirement that a man carrying a red flag should walk in front to give warning of its approach. During the middle of the century many steam carriages of small size were designed, but made no progress, because the weight of the boiler, water supply, and fuel were so great. Not until Daimler, after 1886, had perfected his petrol-motor was there any possibility of rapid development.

Early in the 'nineties this motor had been shown to be peculiarly suitable for propelling a " horseless carriage ", and the result of a number of trials attracted public attention, and gave a great impetus to the industry in France. No headway could be made in England until the " Red Flag Act " was repealed in 1896. Once, however, this obstable was removed, the peculiar conditions of Great Britain fostered the use, if not the manufacture, of motor-cars. No country in the world has such a good supply of excellent roads. We are a wealthy nation and people were attracted by the novelty and speed of the new mode of locomotion—people who could afford to lay out the money and spend time in mastering the idiosyncrasies of a new and occasionally recalcitrant machine. The demand for private cars and the experience of those who ran them helped in no small measure the improvement which made it possible to use them in the public service and for the purpose of commerce. At the International Road Congress, held in London in July, 1913, Mr. Lloyd George stated that there were in England more than 220,000 motor vehicles ; twice as many as in any Continental country, but only one-third as many as in the United States.

It is not proposed in this chapter to give the information that will enable anyone to select and manage a car, for such details are to be found in great fulness in the many excellent manuals devoted to the subject, and in the instructions issued by the different makers. A brief description of typical engines and of their mode of operation is contained in Chapter IV, and it will merely be necessary now to indicate the chief characters of the essential parts. The first thing, perhaps, that puzzles the uninitiated is the classification of types. Probably this is simpler in America where they have the Ford Car, the 5,000-dollar car, and the 10,000 dollar car ! But in Europe the motor-car was at first the special possession of the wealthy and the aristocratic, and a maker's catalogue read like a French *menu*. A luxurious body for two, four, or six persons is described in words which convey small meaning to one possessing only a school knowledge of the language. Thus in addition to the phaeton, there are the torpedo, the limousine, the limousine-landaulette, the coupé-landaulette, the cabriolet, the cabrio-phaeton ; and all of these in " streamline " or other appropriately named forms.

These terms refer only to the upper part or " body ". The lower part, or frame upon which the body is supported, is called the " chassis ". It is constructed of steel, generally of channel (**⊔**) section. In addition there are the engine and its subsidiary contrivances, change-speed gear, driving-gear, and steering-gear.

The general arrangement of these parts, which differs a little in different cars, will be clear from an inspection of Fig. 198, which shows a Singer chassis. The engine is in the front part

FIG. 198.—A CHASSIS FOR A 14 HORSE-POWER CAR.

FIG. 199.—THE FLEXIBILITY OF THE LANCHESTER SYSTEM
OF SUSPENSION.

FIG. 203.—EXTERNAL VIEW OF THE EPICYCLIC
GEAR ON THE LANCHESTER CAR.

To face page 301.

of the car, and just behind is the clutch, then the gear-box, and lastly the back axle. The frame is connected with the front and back axle through springs of the form which is familiar to all in the ordinary landau or " four-wheeler ". In the Lanchester car these springs are not used to brace the axles and frame together, but are free to slide at the ends. This enables them to resist shock better, and accounts for the characteristic smoothness of running. Fig. 199 illustrates the flexibility of the suspension very effectively. It will be observed that the seats are not inclined to anything like the extent that the height of the obstacle would suggest.

If the motor-car owes its existence to the petrol-engine, it owes its lightness, speed, strength, and reliability to the new varieties of steel which have been introduced during the last fifteen years. If only the older alloys had been available the car would have weighed 20 or 30 per cent more, and a corresponding increase of power would have been required to drive it. It is stated in Chapter VIII that no less than six special steels are used for the different parts, and this would certainly not be the practice unless lightness and reliability were increased thereby. The Ford car owes its lightness to the use of vanadium steel, which is close-grained, free from cavities, and extremely uniform in composition and texture. It is, however, an expensive material, and the low cost of the car is due to the fact that only one size and type is made. This admits of the extensive use of automatic machines and economy of production.

Engines for motor-cars all run at high speed—not less than 1,000 revolutions per minute, and until quite recently petrol has been their only fuel. The present scarcity of this substance is stimulating the search for substitutes, and the one most commonly employed is benzene, obtained during the distillation of coal or shale. The more usual name of this fuel among motorists is benzol or benzole. Trials (July, 1913) at Brooklands showed that nearly 30 per cent more power could be obtained from it than from an equal volume of petrol. But there are two or three disadvantages attending its use. One is that it is more difficult to ignite while the engine is cold, so that common practice is to start on petrol and then to switch on the benzol supply. A second is the liability of the substance to contain sulphur, and the products of combustion to corrode the cylinders and valves. A third is the tendency to form a carbonaceous or gummy deposit in the cylinders and ports. The second disadvantage is very largely avoided by using only the purest quality of the spirit, and the third depends very much on the design and adjustment of the carburettor. Similar troubles may attend the use of an inferior quality of petrol.

The noise made by the exhaust gases issuing from an internal-combustion engine render it necessary to use a silencer. This is a long tube pierced with holes and fixed inside one or more tubes similarly pierced. The resistance offered to the successive puffs breaks up the gases and causes them to issue quietly, in a continuous stream. The cylinders must also be cooled by water, and this has to be reduced to the smallest possible amount on account of the weight. After leaving the cylinder jackets the hot water flows through a nest of tubes having thin, waved, metal strips coiled round them on edge. These are placed in front of the car, where the strips, offering a large surface to the air, are quickly cooled. The nest of tubes is called a radiator.

The engine is placed at the front of the car and the driving-wheels grip the road at the rear. Between the head and the the tail are a series of mechanisms, the form and construction of which are as important to the smooth running, and other qualities which determine the efficiency of the car, as the engine itself. First and foremost is the clutch, by which the engine is connected or disconnected with the transmission-shaft. This may be of the cone type, but preferably of the multiple plate type described on page 133, and having either coned or flat plates. The second mechanism is an arrangement of toothed wheels which enable the speed of the car to be altered. Of this there are two forms, one depending upon a set of wheels in pairs, one of each pair sliding along a square or " castellated " shaft. There is a fixed wheel on the engine-shaft which drives a fixed wheel on the short counter-shaft. The counter-shaft has two or three wheels which gear with similar wheels on the transmission shaft. The latter is square or castellated so that it rotates when any one of the wheels is in gear. But by means of forks operated from a lever at the driver's right or left hand, only one wheel can be put into operation at once. As the pairs of wheels—one on the counter-shaft and one on the transmission-shaft—differ in size, three speeds can be obtained in this way. A fourth—and higher—speed is obtained by coupling the engine-shaft directly to the transmission-shaft. For reversal there must be an extra wheel between the counter and transmission-shafts. This will be clear from a study of the diagram of a simple gear-box,

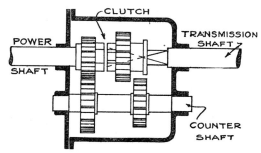

Fig. 200.   DIAGRAM OF SIMPLE GEAR-BOX GIVING TWO SPEEDS.

Fig. 200, giving only two speeds. A complete gear-box with four speeds and reverse is shown in Fig. 201.

The other form of change gearing is known as an epicyclic train, and is rather difficult for the non-mechanical reader to

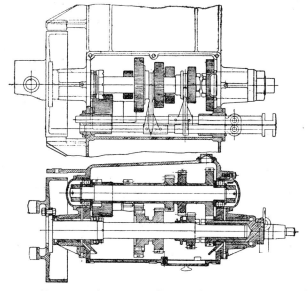

Fig. 201.   PLAN AND SECTION OF CHANGE GEARING ON ARGYLL CAR.

understand. The accompanying Fig. 202 may, however, help to make the principle clear. A and B are toothed wheels, and the arm C can rotate about the same centre as A. D is a wheel with teeth on the inner side of the rim. If the arm is fixed then the motion of A is transmitted to D through the wheel B, so that D turns more slowly in the opposite direction. If D be clamped and the arm C released, this arm rotates at a rate depending on the sizes of A and B, and in the same direction as A. If both C and D are free, power can only be transmitted through the engine-shaft. This gives two speeds ahead and one backwards, but one of the speeds ahead is that of the engine-shaft. Three sets of wheels like this are therefore necessary to give four speeds ahead and one reverse. The epicyclic train

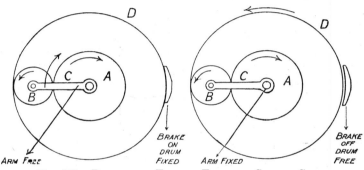

Fig. 202. DIAGRAM TO EXPLAIN EPICYCLIC CHANGE GEAR.

has the advantage of occupying smaller space than ordinary gearing, and as the wheels are always engaged there is no fear of stripping the teeth. Fig. 203 is an external view of the change gear used on the Lanchester car. The three drums are held or released by small brake-shoes which clip them on opposite sides, and the gearing is wholly immersed in oil.

The third mechanism is a flexible joint in the transmission-shaft, between the gear-box and the back axle. The engine and gear-box are fixed rigidly to the frame, and the rear axle, being attached by springs, is constantly rising and falling owing to inequalities in the road. A fixed shaft would therefore be bent, or throw undesirable strains on the bearings. The usual form of coupling is a Hooke's joint consisting of two forks fixed to a block by pins—see Fig. 204, but a smoother action is obtained

FIG. 204.—HOOKE'S JOINT
ON DRIVING SHAFT OF
LANCHESTER CAR.

FIG. 205.—SLIDING JOINT
ON DRIVING SHAFT OF
LANCHESTER CAR.

FIG. 206.—WORM AND WORM WHEEL OF LIVE AXLE
OF LANCHESTER CAR.

To face page 304.

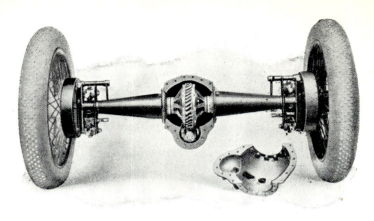

FIG. 208.—BACK AXLE OF LANCHESTER CAR, SHOWING
WORM DRIVE.

FIG. 209.—A COSY CORNER, SHOWING THE LANCHESTER
COLLAPSIBLE ARM-REST.

To face page 305.

by a sphere grooved in two directions at right angles. An additional joint, illustrated in Fig. 205, is used on the Lanchester and other cars. One end of the shaft is made square and fits into a square hole in a box on the other end. This gives flexibility in regard to length.

The fourth and the last step in the transmission of power is the arrangement for communicating the motion to the wheels and is mechanically the most interesting feature of the whole system. The interest arises from the fact that, while both the rear wheels must receive power, they must be capable of rotating at different speeds. For when the car is turning a corner the outside wheel has the greater distance to cover, and if it were not free to turn faster than the other, one of them would have to slip. The back axle is therefore cut in the middle and the ends are connected

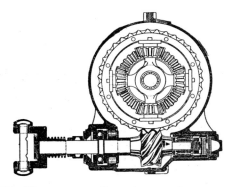

Fig. 207. MECHANISM OF LIVE AXLE OF LANCHESTER CAR.

through gearing which constitutes the " live axle " first applied to a full-sized motor-car by Mr. F. W. Lanchester in 1896.

Notice first that each half of the axle has a bevel wheel fixed rigidly on the end. Between these two bevel wheels are two or four small bevel wheels in gear with the larger ones. The small wheels have their short axles mounted in a ring (see Fig. 207) or the interior of a circular box, so that the wheels point inward. If this ring or box is rotated then the small wheels carry the larger bevels round with them, and do not themselves rotate, so long as both the large bevels turn at the same rate. But the two bevels are quite free to rotate at different rates, and in that case the small wheels also turn. One of the bevels and the

x

four small wheels are shown clearly in Fig. 207. Here one shaft and bevel has been removed. By comparing this Figure with Fig. 206 the way in which the small wheels fit the inner surface of the ring will also be clear. The ring in this case is operated by a worm fixed to the end of the transmission-shaft, and the whole arrangement is shown in Fig. 208.

This worm-drive is adopted on the Lanchester and other cars, and is noiseless and admirably efficient. Most other makers drive the ring or box by a small bevel pinion on the shaft, and this is in line with the back axle. The Lanchester worm is below the axle. The former method ensures adequate lubrication, and when once the case has been filled with oil it will run without attention for 1,000 miles.

The various speeds and the clutch are operated by a lever at the driver's right or left hand, and steering is effected through a wheel in front of him. When this is turned it causes the two front wheels to swing round in the desired direction. These front wheels are mounted upon short axles attached to brackets which are fixed to the end of the front axle by pins about which they are free to turn. The so-called front axle is therefore part of the frame, to which it is attached by springs. The short axles upon which the front wheels are mounted are as a rule not horizontal, but inclined downward. This causes the lower portion of the rim to lie just under the pin about which the wheel is turned when steering. Any other arrangement would cause the wheel to roll forward or drag ; it would be harder to work and less sensitive in action.

The brakes are usually operated by a foot-lever, and consist of a broad, flat ring or band of metal like a short drum on each rear wheel. Inside these is a split ring which ordinarily does not touch the outer one, but which is expanded by the movement of the foot-lever. It is usual now for all four wheels to be braked in this way.

On the top of the steering-wheel the reader will have noticed one or two small quadrants with a radial arm or arms. The purpose of one is always to regulate the petrol supply, of the other to regulate the time of the spark. When the engine is running rapidly the time which the mixture of petrol and air takes to burn causes the explosion to lag behind the piston. The spark is then caused to take place a little earlier, so that the piston receives the full pressure of the burning gases.

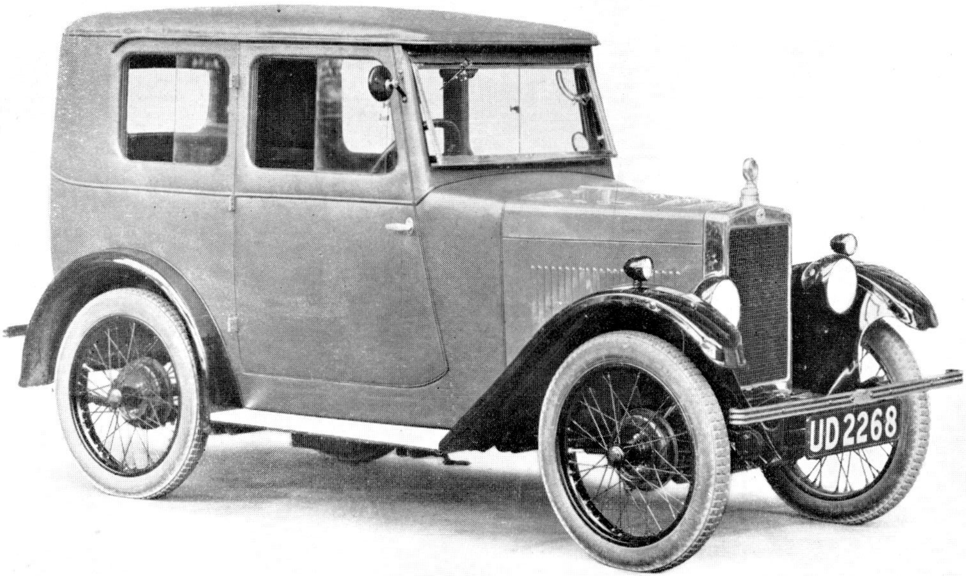

FIG. 209A.—A TYPICAL SMALL CAR OF TO-DAY : THE MORRIS MINOR.

UD 2268

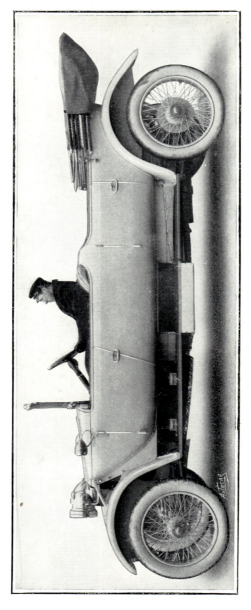

FIG. 210.—31 HORSE-POWER 6-CYLINDER TOURING CAR DE LUXE.

To face page 307.

The car was originally, and still is in many cases, started by giving a half-turn or so to the engine shaft, but for this purpose the driver has to leave his seat, and this journey round to the front of the car is to be avoided if possible. The earlier attempts followed the pattern of large engines and employed a small air pump worked from the engine, which automatically charged a reservoir with compressed air, or a cylinder of compressed acetylene. It was then easy to arrange for this to be used again in starting the engine. A very satisfactory starting device has now been fitted to many cars in connection with the electric-lighting equipment. It consists of a small dynamo driven from the engine and charging a set of accumulators which operate the lamps. To start the car a switch is employed to connect the accumulators up in such a way as to drive the dynamo as a motor, and this drives the engine for a few seconds until the explosions begin. The device is simple, reliable, and effective.

The modern motor-car is in all cases a comfortable conveyance, and in the more expensive types it embodies a greater degree of luxury than any other medium of locomotion, except perhaps the Atlantic liner. Less smooth in movement than the aeroplane, but without the monotonous roll that characterizes a ship, the irregularities of the road are softened and toned down by the most resilient upholstery that man has yet devised. The amount of room in a limousine body is surprising, and no cosy corner in a lady's boudoir is more inviting that that shown in Fig. 209. Here is a complete protection from the weather, warmth, a gentle oscillation, and a gliding panorama of scenery outside the window. For those who prefer the open air the touring body provides all the comfort that, and as much air as, anyone has a right to enjoy, for a little smaller cost.

A substantial well-built modern car costs, with all accessories, from £300 to £600, though it is easy to run the price up to £1,500 or £2,000. Some of this value represents the expensive system of advertising and trials adopted by the industry, and a good deal more to the fact that in catering for a wealthy body of customers the number of types has been needlessly increased. A cheap car can only be produced when the pattern is standardized and the whole of the machinery and organization of a large works is concentrated upon its production. The Ford car, first manufactured in America, which can be purchased for £125, is a case in point. It was a light car, and English makers preferred to turn out a heavier vehicle, which was more durable, and which, they considered, would give the best results in the long run. The few light cars which they had produced could not compete with the Ford in price, and the cycle-car, built upon a very light framework, did not catch the popular fancy.

English manufacturers, however, gained a vast experience of mass production during the war, and during the last ten years a number of light cars has been put upon the market. Of these the Morris Cowley, costing about £180, was, perhaps, the most successful. It has a better appearance than the Ford, and was an excellent car for the man of small means. Then Humber and Austin cars, smaller, and costing about £160, were produced, and now the Morris Minor costing about £120 has appeared. It must be borne in mind that a large number of the cars one sees upon the road are secondhand, and cost their present owners considerably less than £100. The extension of motoring is due, in considerable measure, to this transfer from people in more fortunate to people in less fortunate circumstances.

To those who can remember the old high bicycle and its displacement by the safety, the extraordinary development of the motor cycle is merely a repetition of history. Here, again, 230,000 miles of the best roads in the world, coupled with vast wealth from manufacturing industry, enable thousands of people to lay out the £50 or £60 that is required, and this in its turn enables business to be done in less time, and holidays to be taken more frequently and at less expense, than if the railway and the horsed vehicle were the only means of locomotion.

It has already been remarked that the first use of the motor-car was for pleasure. It was expensive and not very reliable, and possessed many of those qualities upon which the spirit of adventure feeds. The first commercial use was probably made by medical men, who found that speed was of considerable value in enabling a larger practice to be built up without assistance. But when through the tribulation of private owners and the efforts of manufacturers a reasonable degree of reliability had been secured, a public service was established first in the form of taxi-cabs, and later by tradesmen's delivery vans. Perhaps no change in the appearance of the streets of large towns has ever been so rapid as that which has resulted in the partial displacement of the horse. At the inaugural meeting of the International Road Congress in London in July, 1913,

Mr. Lloyd George gave the results of observation carried out on a Sunday morning on one of the secondary arteries leading out of London. There were 100 bicycles, 50 motor-bicycles, 30 motor-buses, 300 motor-cars, and 15 horsed vehicles. Another investigation undertaken in 1926 on another road showed a traffic consisting of 687 motor-cars and cycles, 232 heavy motors and trailers, and 56 horse-drawn vehicles per day. And this was not on a main road !

This enormous traffic has introduced a new problem for the civil engineer. On the Canterbury Road, Willesden, in 1923, the average daily weight of traffic was 12,583 tons. The old macadamized road, which has served the purpose for 100 years, has had its day. Under the endless succession of vehicles that flash from point to point with frequent stoppages, the surface crumbles up, and it has been necessary to use the hardest material, such as broken granite with bitumen or tar to bind the separate fragments together and form an elastic matrix which, by yielding to pressure, reduces wear.

The motor-van or dray is now an essential part of military equipment. No country in the world has such a close network of railways as Great Britain, yet there are many parts specially adapted for military operations which are not readily approached by train. But the enormous amount of transport required by a large army would render it a very expensive matter to purchase and maintain a sufficient number of motor vehicles, and the plan adopted by European countries before the war was to subsidize private owners for the use of their vans in time of war. Very stringent regulations as to weight, speed, and hill-climbing power were laid down, and this tended to standardize the type of vehicle used.

The remarkable improvement which has been made during the last ten years in the motor omnibus is threatening, if not the existence, at least the extension, of the tramways. With equal speed, and very little smaller capacity, it has the advantage of not requiring an enormous expenditure on track and overhead equipment, and of being able to thread its way through crowded thoroughfares with far less interruption to the traffic than an ordinary tramcar. The struggle between tramways and railways for the suburban traffic in large towns is now complicated by the motor vehicle, and the petrol omnibus is making rapid headway.

It is interesting to note that a long time must elapse before competing systems of this kind arrive at a steady state in which one or the other is the victor. The extraordinarily rapid growth of towns during the past thirty or forty years has enabled new forms of mechanical transport to gain a footing without as a rule appreciably affecting the old. There has been enough an d to spare, for both.   The new methods have simply encouraged more people to take advantage of the facilities, and the new supply has created a new demand.

CHAPTER XVI

MODERN SHIPS

PROBABLY no field of invention has been more startling in its results than that connected with ocean transport. The old vessels in which the adventurous spirits of the sixteenth and seventeenth centuries sallied forth across the waste of waters and founded the British Empire, possess only a superficial resemblance to the magnificent vessels of the present day. And when one looks at the comfort and convenience of the modern steamer the imagination is exercised to picture the daring and hardihood that planned and executed those early voyages. In the volume dealing with the nineteenth century an account is given of the advent of the iron and steel ship of the growth in power and speed which had taken place by 1890. It may safely be said that the progress during the past twenty-five years has been as remarkable as anything that preceded it. At the same time, there is probably a good deal of popular misconception in regard to size. The newspapers have dealt so generously with the giant Cunarders, the *Mauretania* and *Aquitania*, and with the still larger White Star *Olympic*, that these huge liners are regarded as representative. But as a matter of fact, they are only engaged in the Atlantic trade. Together with the equally large vessels of the Hamburg-America Line, they owe their existence to, and represent in full measure, the commercial interests existing between Europe and America.

Thus, in 1912, if steamers of less than 500 tons are excluded, the average size of the ships launched in Great Britain was 4,000 tons. Only 16 vessels were over 10,000 tons, and 54 were between 6,000 and 10,000 tons. The general cargo boat is not as a rule more than 6,000 tons, because beyond this size difficulties arise in making up and breaking cargo. The finest passenger and cargo vessels, with the exception of those already named, are under 20,000 tons—thus the White Star vessel *Ceramic*, of 18,000 tons, is the largest vessel sailing to Australia, and the *Laurentic*, 14,500 tons, of the same company is the largest in the Canadian trade. The tonnage of the ships of this company shows in a striking manner the suddenness of the increase

311

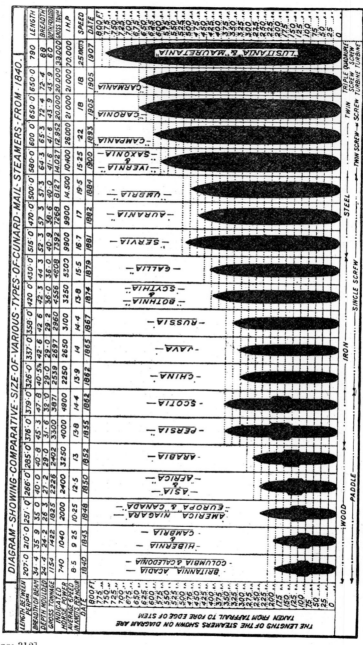

Fig. 211.

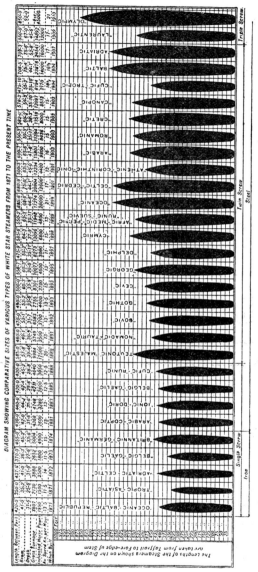

DIAGRAM SHOWING COMPARATIVE SIZES OF VARIOUS TYPES OF WHITE STAR STEAMERS FROM 1871 TO THE PRESENT TIME

The Lengths of the Steamers shown on the Diagram are taken from Taffrail to Fore-edge of Stem

in size. Only six vessels of the fleet are more than 20,000 tons.
They are :—

Celtic .	.	.	.	21,000 tons.	Built in 1901.
Cedric	.	.	.	21,000 ,,	,, 1901.
Baltic .	.	.	.	24,000 ,,	,, 1904.
Adriatic	.	.	.	24,500 ,,	,, 1907.
Olympic	.	.	.	46,000 ,,	,, 1910.
Britannic	.	.	.	50,000 ,,	,, 1919.

Similarly the first Cunard vessel over 20,000 tons, the *Caronia*,
was launched in 1905, and was followed two years later by
the *Mauretania*, of 32,000 tons. The tables given on pages
312 and 313 will show how the fleet of each company has grown
in size, speed, and means of propulsion, since it was first
established.

CONSTRUCTION

The problem of constructing a ship of adequate strength is
a very interesting one. The chief forces that have to be considered in an ocean-going boat are those which arise from the
uneven surface of the water. Fig. 213 shows how at one moment

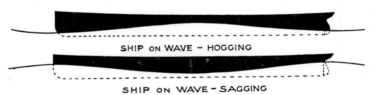

SHIP ON WAVE – HOGGING

SHIP ON WAVE – SAGGING

Fig. 213. THE NEED FOR LONGITUDINAL STRENGTH.

she may be supported at both ends on the crests of two waves,
and at another supported in the middle on the crest of one wave,
with both ends free. In both cases the forces called into play
are the same as those in a beam supported in a similar way.
And it is clear that a form of construction similar to that of a
box-girder shown in Fig. 214 is essential. The effect is obviously
more serious as the length of the ship increases.

The old wooden ship was of no great length, and no great
strength was required in a longitudinal direction. All the
heavy timbering was concentrated in the ribs running from
deck to keel. If she was strong enough to escape being battered
in, her bottom, sides, and deck were strong enough to prevent

FIG. 215.—A SHIP IN COURSE OF CONSTRUCTION ON THE ISHERWOOD SYSTEM.

To face page 314.

FIG. 216.—THE RUDDER AND PROPELLER SHAFTS OF THE "OLYMPIC" JUST BEFORE THE LAUNCH.

FIG. 217.—NO. 4 FUNNEL OF THE "OLYMPIC" READY FOR FIXING IN POSITION.

To face page 315.

her back being broken. But the advent of iron and steel ships brought a great increase of length, particularly when it was found that an increase in carrying power could be effected in this way without a corresponding increase in the horse-power required. As the length increased the transverse system became no longer permissible, and methods were devised by Scott-Russell and others to stiffen the frame in a longitudinal direction. At the present time a good many ships are being constructed on the Isherwood system, of which an illustration is given in Fig. 215. The transverse frames or ribs are slotted at intervals

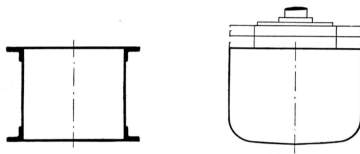

Fig. 214. Comparison of Sections of Box-girder and Atlantic Liner.

on the outer edge and longitudinal girders are let into them. Both transverse and longitudinal framework are therefore flush on the outer surface, and the plating is fixed rigidly to both. In addition to increased strength arising from a better distribution of material, the inventor claims that it gives an increase of space for bale goods, or a greater dead-weight carrying capacity, while the ventilation is simplified. It appears to be particularly suitable for oil-tank vessels, but is being adopted for all classes of ships. Though only introduced in 1908 there were 248 built or being built on this system in 1913. Forty-eight shipyards were engaged in the work, and the tonnage amounted to over a million. No fewer than 86 were oil-tankers, with a combined capacity of half a million tons.

A more recent method is the Monitor system, in which a double curve of corrugation is arranged very nearly from end to end of the ship. It is claimed for this system that less power is required for a given speed, and that the ship is less prone to roll. At the time of writing, 1913, only two ships—the *Monitoria* and the

Hyltonia—have been constructed on this plan, but Messrs. Furness, Withy & Company, the owners, had ordered another to be built in the same way.

There is a very general tendency nowadays to look closely into the quantity and distribution of the metal in the hull, and it is quite possible that a considerable saving of weight will be effected, and result in a corresponding increase of carrying capacity. The shipbuilder benefits by the improvements in the manufacture of steel described in Chapter VIII. Not only is the material more reliable, but it is supplied in larger pieces, so that the amount of labour involved is less. When the *Great Eastern* was built in 1858 the plates used in her " skin " were 10 feet long and 2 feet 9 inches wide ; the plates used on a large modern vessel are 30 feet long and 5 feet wide. The area of the old plate was therefore $27\frac{1}{2}$ square feet ; of the modern one 150 square feet. In the White Star Liner *Olympic* most of the plates are 30 feet by 6 feet and weigh between $2\frac{1}{2}$ and 3 tons, while the largest shell plates are 36 feet long and weigh $4\frac{1}{4}$ tons. The stern frame, which carries the rudder and the bearings for the propeller-shafts, is a steel casting. That of the *Aquitania* weights 62 tons. The *Olympic's* stern frame is cast in portions weighing together 70 tons ; there are in addition an after bracket weighing over 73 tons, a forward bracket weighing 45 tons, while the rudder, cast in six pieces, weighs $101\frac{1}{4}$ tons !

Some idea of the meaning of these figures will be given by the accompanying illustrations. Fig 216 is a view of the stern of the *Olympic* just before she was launched. The size will be gathered from a comparison of the men with the propeller-shaft. Fig. 217 is a photograph of No. 4 funnel on the same vessel, ready for fixing in position. It is oval in form, the longest diameter being 24 feet 6 inches and the shortest 19 feet. They reach to an average height above the furnace bars of 150 feet. These four funnels have to serve 24 double-ended and 5 single-ended boilers. The former are 15 feet 9 inches diameter and 20 feet long with 6 furnaces each. The latter are of the same diameter, but only 11 feet 9 inches long with 3 furnaces. There are consequently 159 furnaces.

The largest vessel hitherto launched in Great Britain is the Cunard quadruple-screw turbine steamship *Aquitania*. She is 901 feet long, 97 feet beam, and with a depth to the boat-deck of 92 feet 6 inches. Her gross tonnage is 47,000 tons, her speed

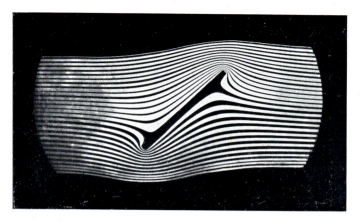

FIG. 218.—STREAM LINES IN GLYCERINE PASSING ROUND
AN OBSTACLE.

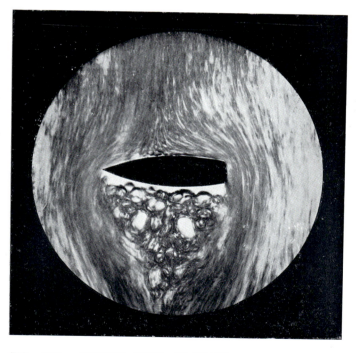

FIG. 219.—SINUOUS FLOW IN WATER, SHOWING EDDIES
BEHIND AN OBSTACLE.

To face page 316.

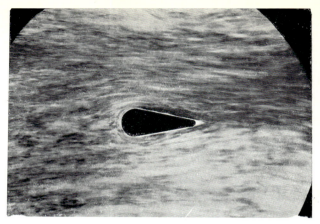

FIG. 220.—SINUOUS FLOW IN WATER, SHOWING RELA-
TIVELY SMALL IMPORTANCE OF A BLUNT STEM IN
A SHIP IN THE PRODUCTION OF EDDIES.

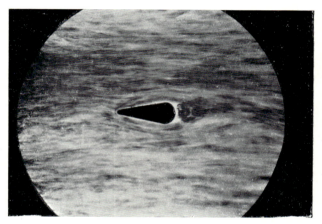

FIG. 221.—SINUOUS FLOW IN WATER, SHOWING
INFLUENCE OF A BLUNT STERN IN PRODUCING
EDDIES.

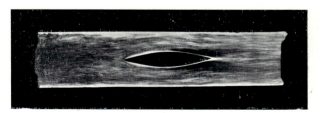

FIG. 222.—SINUOUS FLOW IN WATER, SHOWING
HOW EDDIES ARE AVOIDED BY GIVING A SHIP
FINE LINES FORE AND AFT.

To face page 317.

23 knots, and she has accommodation for 3,250 passengers and a crew of nearly 1,000 men. There is a double skin with an average distance between the two of 15 feet. This space is subdivided by numerous bulkheads at short intervals, and the whole of the ship has 16 bulkheads passing right through from port to starboard. Frahm's anti-rolling tanks, which have been thoroughly tested on the *Laconia*, have been installed, and the boats are sufficient to carry every member of the passengers and crew. The accommodation for passengers covers eight decks, and, in addition to the rooms to which travellers by this line have become accustomed, all the comforts and conveniences of more recent vessels have been incorporated.

SPEED

Another feature in which the giants of the White Star and Cunard Lines are not representative is the speed. The usual speed of a cargo boat is 10 to 12 knots, though the newer ships which bring chilled beef from the Argentine and frozen mutton from Australia and New Zealand are capable of steaming at 15 knots. Some of the most popular passenger vessels make from 17 to 20 knots. Thus the Cunard Liners *Umbria* and *Etruria*, sister ships of 8,000 tons each, which were launched in 1884, had a speed of 19 knots and held the Atlantic record for many years. They were replaced by the *Campania* and *Lucania* of 13,000 tons and 22 knots. The *Mauretania*, launched in 1907, held the record until the week before these pages were revised, when the honour passed to the German liner *Bremen*. The *Mauretania* regularly makes 25½ knots, and has done over 26. The new German liners make only 23, and the expense of attaining the extra two or three knots is so great that no company, unless heavily subsidized, would find it worth their while to undertake it. The money for this ship was advanced to the Cunard Company by the Government, and an annual sum paid for its upkeep. In return for this the Government have a right to its use in time of war. The ship carries two guns, and is specially strengthened for this purpose.

The practice of stating the speed of a ship in knots is somewhat confusing to a landsman, who often fails to realize exactly what the figures mean. A knot is 6,080 feet, and that is nearly 1⅕ land miles. A speed, then, of 20 knots is, in terms which the landsman understands, a speed of 24 miles an hour, and the *Mauretania* travels at 30 miles an hour. On many sections of British railways the speed does not exceed this figure. The reader who has not reflected upon this matter before hould estimate the velocity of the train on his next railway journey, and notice how the trees and hedgerows appear to fly past. He can then imagine a ship like the *Mauretania* racing through the waves hour after hour with never a stop for four days and four nights until she comes in sight of land. It will be possible, too, to realize how little time there is in which to avoid a collision. The *Mauretania* covers a mile in two minutes. A ship or an iceberg sighted five miles away is reached in ten minutes, and a vessel of this size cannot be reversed in a hurry without fear of damage to her engines.

In order to secure speed with a minimum of power it is important that a ship should have such an outline as will enable her to move through the water with the least resistance, and the power required to drive a vessel of given displacement and form at any particular speed can be determined beforehand with considerable accuracy.

If any object is held at rest in a stream the water will divide on meeting it in a particular way which depends mainly upon its shape. Any small portion of the water meeting the object will be deflected from its course, and will pursue a curved path round it. If, now, the water be relatively at rest and the object is forced through it, a corresponding effect will be observed. The object will be continually forcing the water out of the way and the particles displaced will describe curved paths until they regain their former relative position in the stream.

It will perhaps assist the reader to realize the kind of movement that goes on when a stream of fluid meets an obstacle if a short description is given of the method of study devised by Dr. H. S. Hele-Shaw, F.R.S., some thirty years ago. In Dr. Hele-Shaw's experiments alternate holes in the end of a small glass tank are fed with coloured and uncoloured glycerine and produce a series of parallel bands. If an obstacle, in this case representing a ship's rudder, is placed in their path these bands divide and curve round its surface, as shown in Fig. 218, which is from a photograph shown by Dr. Hele-Shaw at the Institution of Naval Architects in 1900. It was found that in a few cases in which calculation was possible, the stream lines corresponded exactly in form to those which would be produced

FIG. 223.—THE WILLIAM FROUDE MEMORIAL TANK : MODEL READY TO BE TOWED.

To face page 318.

FIG. 224.—SHAPING THE MODEL.

To face page 319.

in a perfect fluid ; and with the co-operation of the famous mathematician, Sir George Gabriel Stokes, it was shown that this was generally true if the liquid was viscous, distributed in a thin film, and the motion slow.

Water, however, is not viscous—at any rate in comparison with glycerine. Moreover, a thin film can only represent the influence on a floating body at a particular depth, and the speed of a vessel is usually greater than that at which the glycerine experiment fulfils the ideal conditions. Another illustration (Fig. 219) exhibited by Dr. Hele-Shaw will make the problem clearer. The liquid used in this case was water, and the thickness of the film was increased. The stream lines in front of the obstacle were replaced by sinuous motion, and the space behind was filled with " dead water " and eddies, in which all steadiness of motion disappeared.

While these experiments are not conducted under the actual conditions of a ship moving through the water they serve to convey some general ideas as to the most suitable form which a vessel should take in order to reduce resistance to a minimum. Thus the object of having a sharp prow is to effect the gradual displacement sideways of the water. But it is equally important to provide a sharp stern. For the stream lines tend to close in gradually as the ship moves ; and a blunt stern would tend to cause cavities behind, which would act as a drag on the ship's progress. These facts are well illustrated in Figs. 220, 221, and 222. In a screw steamer there is an additional reason for this form of construction. The propellers are continually forcing the water backwards, and unless it can flow in freely in front of them the maximum push on the water cannot be obtained. A blunt stern would act as a shield.

When a vessel is to be constructed the purchaser stipulates a certain tonnage and speed ; and the shipbuilder must decide what horse-power will be necessary to attain this. But the power required will depend very considerably upon the " lines "— that is upon the change of shape of the submerged portion from stem to stern. And though experience allows a very good result to be achieved, there is a method which enables the best lines to be determined, and the necessary power to be ascertained, with great accuracy.

It was in 1871 that the late Mr. William Froude designed for the Admiralty, at Torquay, a long tank in which scale models of ships could be towed and the power required for any given speed could be measured. Between the size, power, and speed of the model, and the size, power, and speed of a large vessel of the same shape, there is a definite relation, which enables the naval architect to draw his plans with the certainty that the result will be satisfactory.

The Admiralty tank was moved to Haslar, and was utilized by Mr. Froude and his son, Mr. R. E. Froude, for many valuable investigations in ship design. For a number of years it was the only one in the country, until another was built by Messrs. Denny of Dumbarton. Afterwards tanks were established by Messrs. John Brown and Co. of Clydebank, Messrs. Vickers, Ltd., of Barrow, and at the National Physical Laboratory.

The William Froude National Tank at the National Physical Laboratory, which owes its origin to the generosity of Mr. (now Sir) A. F. Yarrow, is built of concrete, and is 550 feet long, 30 feet wide, and $12\frac{1}{2}$ feet deep. These dimensions, with the models used, are equivalent to open water for a large ship. A false bottom can be put in so as to permit of trials in shallow water. It is spanned by a bridge running on rails and driven by four electric motors. This bridge serves to tow the models, and is equipped with delicate measuring instruments for recording the pull and speed. These arrangements are shown in Fig. 223.

The models are made in paraffin wax, from 12 to 20 feet long, with sides and bottom about two inches in thickness. They are cut out in a sort of milling machine in which the cutter is actuated in accordance with the motion of a pointer, which is made to travel along the lines of the drawing, which rests on a table at the side of the machine (Fig. 224). The tool thus shapes the wax to the exact form intended by the designer. The marks of the cutter are removed by scraping so as to produce a smooth body, and ballast is then added until the model floats at the required depth, see Fig. 225.

Until the last eight or ten years it was not usual to check the designs of cargo boats by a model, but the installation of the National Physical Laboratory tank has enabled many shipbuilders to adopt the precaution. Mr. Baker, the superintendent of the tank, stated in 1920 that more than two-thirds of the designs submitted for tests have been improved by at least 2 per cent. In the 23 test examples the average reduction of indicated horse-power, at service speed has been $8\frac{1}{2}$ per cent. Assuming that

FIG. 225.—WEIGHING THE MODEL PREPARATORY TO BALLASTING.

To face page 320.

FIG. 226.—FIXING THE INNER SKIN ON THE " OLYMPIC."

FIG. 227.—ONE OF THE 29 BOILERS OF THE WHITE STAR STEAMER " BRITANNIC ": WEIGHT OF EACH BOILER, 105 TONS.

To face page 321.

only one ship is built to each of these 23 designs, the net saving of coal per year for the 23 ships amounts to 15,000 tons, on a basis of 200 steaming days per year. With bunker coal at £3 per ton, the average saving per year per ship in these 23 test cases amounts to £1,950. All classes of ships can be dealt with in this way.

SAFETY AND COMFORT

The growth of passenger traffic has necessitated greater attention, not only to internal comfort, but also to the prevention of rolling, which at times increases to an alarming extent the unpleasantness of a voyage. The earliest effective device was the provision of bilge keels. These are thin fins fixed on either side of the bottom just where it begins to curve upwards, and running all the way along the wider portion of the ship. When the ship rolls they meet the water at right angles and offer resistance to the motion. Within the last few years passenger vessels have been fitted with Frahm's anti-rolling tanks, the first British vessel to be so fitted being the Cunard Liner *Laconia*, a vessel of 18,000 tons, launched in 1912. They are placed on either side of the ship and are open to one another by a narrow passage at the bottom. They thus constitute a sort of U tube and when full of water possess properties which are several times referred to in this book. In order to understand how they act it must first be clear that when the ship rolls the water falls relatively on the higher side and rises in the tank on the lower side, and that once the water is set oscillating in this way it will continue to do so for some little time. Suppose, now, a ship rolls so that it rises on the starboard side first, the water flows from the starboard to the port tank, so that when the port side of the ship rises, the quantity of water in the tank on that side is larger than it would be when the ship is at rest in still water. The roll from port to starboard, therefore, is prevented to some extent by the extra weight on the port side. In the same way the water flows back into the starboard tank to compensate the starboard to port roll.

This simple explanation, unfortunately, does not state the whole of the case. The cause of rolling is a succession of wave crests, meeting the vessel broadside. Their distance apart and their velocity will determine the number of impulses on the ship in a given time. Again, the ship itself has a definite period of roll. If this is very different from the period of the impulses there will be very little rolling. But if the period be the same then resonance occurs ; the ship receives a series of impulses which tend to increase the motion, and what was originally a mere discomfort now becomes a positive danger. In fact, the utter and complete disappearance of some vessels in recent years may have been due to their capsizing in mid-ocean. They go singly and leave no trace. A collision is practically out of the question ; the rockbound coasts are so well watched and the ocean highways are so well patrolled that shipwreck or fire could hardly have occurred without some trace remaining. And certainly a probable view is that the ships met with a wave motion that synchronized with their own natural period, and turned turtle without warning.

This shows that under special circumstances, which are fortunately rare, the tanks would have to meet unusual conditions. For the water in them has itself a natural period of oscillation, depending on their dimensions, and if this does not coincide with the rolling motion of the ship the water will not have accumulated at the right moment on the side on which it is needed. In fact, it may reach its highest level, for example, in the port tank at the moment when the port side is at its lowest point, and thus increase the tendency of the ship to overturn.

Experiments are being made and mathematical investigations are being carried out with tanks connected in various ways—some with a restricted connection, some with a connection having the same sectional area of the tanks themselves. The subject is complex, and the scientific solution in its infancy. But in the meantime, the limits of periodicity of the average waves met with are well known, the period of the ship can be calculated, and the tanks can be, and are, constructed of such dimensions that, while they add much to the steadiness of the ship, they are extremely unlikely to throw their weight on to the side of catastrophe.

Another plan which has been the subject of experiment for some years is to use a gyroscope. This apparatus is more fully explained on pp. 341–342. Here it may merely be remarked that when a heavy wheel is rotating at high speed, it resists, in an extraordinary way, any tendency to alter its plane of rotation. Its application to prevent the rolling of ships is mainly due to Mr. D. Sperry, an American, who has installed it

FIG. 228.—THE BRIDGE OF THE "MAURETANIA."

FIG. 229.——CLEAR-VIEW SCREEN.

To face page 323

recently on a 10,000 ton vessel. Two gyroscopes were used, each having a wheel or disc weighing 25 tons. A small pilot gyroscope is used to control the motions of the main ones. During one experiment the ship had a roll of 22°. When the gyroscope was put into operation the roll was reduced to 2°. On stopping the gyroscope the roll became 31°, which was immediately reduced to 3° by the gyroscope. The prevention of rolling by this means renders bilge keels unnecessary and saves power to the extent of 6 per cent. The vessel is also drier, simply rising and falling on the waves.

Safety at sea is secured partly in the construction of the ship, and partly by the use of subsidiary appliances. Thus the vessel is divided into a number of watertight compartments separated by partitions or bulkheads, and covered by a watertight steel deck. Communication from one compartment to another and through the deck is obtained by sliding doors which fit in watertight grooves. These can all be closed when necessary from the bridge. They are operated by hydraulic pressure, and the force is so great that any obstruction, such as a lump of coal, is cut through during the closure. The control is fixed on the bridge, and immediately behind the lever which operates the doors is a model of the ship with an electric lamp corresponding to the position of each compartment. Should one of the doors fail to act when the lever is set to close, a lamp lights up corresponding to the compartment with the open door.

It has generally been assumed that a modern ship will continue to float with any two of her compartments full of water, but the naval architect now makes assurance doubly sure. The *Olympic* is provided with a complete inner skin and the same plan has been followed in the *Aquitania*. Fig. 226 shows the inner skin of the *Olympic* being fitted as an additional precaution after one or two voyages had been made. In the case of the *Aquitania*, in addition to sixteen bulkheads right across the ship from port to starboard, there is a lining about 15 feet inside the outer hull, so that there are practically two ships, one nearly 70 feet and the other nearly 100 feet beam, and both provided with watertight compartments. It is extremely unlikely that any sharp object such as a rock or a jutting ledge of an iceberg will penetrate the inner skin, and safety is secured against collision or a glancing blow.

But if a ship runs full tilt against an obstruction big enough to stop her, no system of stiffening, or bulkheads, or inner skins can prevent her crumpling up like a paper bag. When one compares the thickness of the skin and longitudinal bulkheads with the whole width of beam, it is clear that the great ship is a frail thing indeed, and no precaution that will keep her clear of icebergs or a rockbound coast can be safely neglected. During the last dozen years an " ice-scout " has been employed to watch the movements of ice in the North Atlantic and to report its presence and position to all ships on the track.

The proximity of ice can sometimes be inferred by a sudden fall in the temperature of the water, and an instrument known as McNab's frigidometer enables the officer on duty on the bridge to detect any striking change of this character immediately it occurs. It consists of a special thermometer near the forward end of the ship, immersed in a vessel through which the sea-water is kept constantly circulating. An indicator on the bridge, which can be set to give an indication at any temperature of the occurrence of which the officer desires to be warned, shows a red light and rings an electric gong whenever that temperature is reached. The instrument registers in the same way the temperature of the air, and the officer uses his judgment as to whether it is desirable to alter the course of the ship.

An ingenious device fitted on the *Mauretania* and other vessels notifies the officer on duty of a fire in any one of the holds. On the bridge (Fig. 228) are a number of tubes fitted with caps, the removal of which enables the officer to tell whether fire has broken out. Every half-hour a bell rings, and this can only be stopped by removal of the caps—an action which is equivalent to an inspection of the hold.

A further precaution against fire is taken by the replacement of wood and other combustible material by steel. Even while this chapter was being written two large vessels have caught fire at sea and the lives of hundreds of people have been endangered. In these cases it was the cargo, but the advantage of reducing the quantity of inflammable material used in construction is obvious. Messrs. Roneo, Limited, have devised a system of thin steel partitions and doors, together with steel furniture such as is now used on warships. It is found that two thin plates about $\frac{1}{20}$ inch in thickness with an intervening air space are more effective than even a $\frac{1}{2}$-inch or 1-inch solid plate. The system was tested by the Cunard Company in refitting the *Carmania*,

FIG. 229A.—CLEAR-VIEW SCREEN IN USE ON A MOTOR-CAR: UNTOUCHED PHOTOGRAPH.

To face page 324.

H

FIG. 230.—ONE OF THE KITCHENS ON THE "MAURETANIA."

and has been adopted for the *Aquitania* and other vessels which may be built in the near future.

Sometimes the course of a vessel has been altered, and a ship has been wrecked when the captain believed himself to be clear of any rock or coast. The recording compass enables him to ascertain whether such an alteration has taken place. It consists of a roll of paper on a rotating clockwork drum, upon which a line is traced by a pen. If the course of the ship alters the change in direction and the exact time at which this took place are indicated by a bend in the line on the paper.

An additional method of avoiding dangerous coasts has been introduced during the last ten years in submarine signalling. Ever since the famous experiments of Colladon and Sturm on the Lake of Geneva in 1826 it has been known that sound travels through water with a velocity four times greater than through air ; but it was left to Mr. H. T. Mumby, an American, to realize that this method of transmitting sound signals was free from the disturbances that occur when sound passes through the atmosphere. The transmitting apparatus made by the Submarine Signal Company is now provided on many dangerous coasts, and fitted to lightships and buoys ; and the receiving apparatus is installed on passenger and cross-Channel steamers, and the vessels of the Royal Navy. The sending apparatus, which is always a bell, is made in four forms, the special uses of which may be briefly described.

The electric *shore station* consists of a bell weighing about 2 cwt. hanging from a tripod 21 feet high, resting on the sea bottom, and operated from a shore station or a lightship by electricity. It is stated to be reliable up to 15 miles, and signals have been reported up to 20 miles. The *lightship equipment* is suspended from a lightship and worked by compressed air. The number of, and interval between, successive strokes enable the mariner to identify the ship just in the same way as he is able to recognize a lighthouse by the number of, and interval between, its flashes. The plan has been largely adopted by the United States Government, and by the Brethren of Trinity House, the body charged with the management of lighthouses and lightships round the British coasts. When attached to a *buoy* the bell hangs about 16 feet below the surface, and the movement of the buoy on the waves operates the mechanism. For cross-Channel traffic a handbell is suspended from the *pier or jetty,* and worked by hand.   The sound of this in foggy weather enables the boat to steer for the pier even when the lights cannot be seen.

The receiving apparatus consists of two shallow tanks, about 22 inches square, fixed to the outside of the ship below the water-line, and on the port and starboard bows respectively.   Each tank contains a microphone, from which wires are carried to a telephone placed in the pilot house.   By moving a switch the observer can tell whether the sound is coming from port or starboard.   Even though it has only been in operation a few years this apparatus has been instrumental in saving hundreds of lives and thousands of pounds' worth of property from being lost in the greedy ocean.   The captain in the wheel-house with his eye on the chart and the telephone at his ear recognizes the tinkle of a bell, and is able to steer his ship in a direction that will carry him clear of the treacherous coast that lies in his path hidden from view by the fog.

Even if there is no fog, rain or snow may render it impossible to see a yard away from the bridge.   In order to overcome this disadvantage, the Clear View Screen made by Messrs. George Kent & Company is an admittedly simple device.   It consists (see Figs. 229 and 229A) of a glass disc about 11 inches in diameter rotated by means of an electric motor.   Any rain or snow which falls upon it is immediately flung off by centrifugal force.   In the figure it is mounted in front of a hood which gives protection against the weather, but it is also made to fit in a window of the wheel-house.

THE LUXURY OF A MODERN LINER

Perhaps no feature of a large liner is more interesting than the mechanical devices that enable 3,000 to 4,000 people to be fed with regularity, and served with such luxury as cannot be surpassed in the best hotels on shore.   Let us glance for a moment at the food required for a round voyage of the *Mauretania*.   The list includes :—

45,000 lb. Beef.	100 lb. Cavaire.
17,000 ,, Mutton.	2,000 Chickens.
3,000 ,, Lamb.	600 Fowls.
2,500 ,, Pork.	300 Ducklings.
1,500 ,, Veal.	150 Turkeys.
1,200 ,, Assorted Fresh Fish.	60 Geese.
750 ,, English Salmon	1,500 Various small Birds.
20 barrels Oysters.	150 brace each Pheasants,
3 Live Turtles.	Partridge, and Grouse.
200 boxes Dried Fish.	5,500 lb. Butter.

FIG. 231.—RECEPTION-ROOM TO A CAFÉ ON THE "OLYMPIC."

FIG. 232.—SECOND CABIN DRAWING-ROOM: R.M.S. "AQUITANIA."

To face page 326.

FIG. 233.—A STATE-ROOM ON THE "OLYMPIC."

To face page 327.

28 tons Potatoes.
1,500 bricks Ice Cream.
6,000 jars Cream.
3,000 gallons Milk.
1,000 lb. Tea.
1,800 lb. Coffee.
10,000 ,, Sugar.
720 quarts Pickles.
2,800 lb. Dried Fruits.
80 boxes Oranges.
230 ,, Apples.
800 lb. Grapes.
1,500 Peaches, Nectarines, etc.
40 boxes Pears.
150 English Melons.

20 bunches Bananas.
30 boxes Grape Fruit.
1,000 lb. English Tomatoes.
20 boxes Lemons.
300 bottles Sauces (assorted).
2,600 lb. Jams and Marmalade.
450 tins Biscuits.
8,000 lb. Cereals.
210 barrels Flour.
2 tons Salt.
1,400 lb. Ham.
4,000 ,, Bacon.
1,600 ,, Cheese.
40,000 Eggs.

All this food is prepared for the table in a series of kitchens (Fig. 230), each serving a special portion of the ship's company, and equipped with steam ovens and electrical heating devices. Roast meat, at one time unobtainable at sea, is now cooked to perfection in electric ovens, and chops and steaks are grilled over charcoal heated to redness by an electric current. Bread is kneaded in an electrically driven dough-mixing machine, and baked in an oven which completes the process in a definite time without any attention. Ice-cream, whisking, and cake-making machines are all driven by electricity, and boiling is carried on by steam. If an egg is to be boiled, it is placed in a wire basket, and lowered into boiling water, and an index on a graduated rod set to determine the number of minutes the egg must cook. When the time is up a bell rings and the egg-basket rises out of the water. And this is done for 1,000 eggs a day.

Apart from preparing the food many of the other domestic duties are performed by electrical power. The 40,000 pieces of crockery that are used daily are washed and dried in a " Vortex " machine ; the knives are cleaned as fast as a man can feed them in ; and boots are polished by electrically actuated brushes at the rate of 1,500 pairs a day. The fifty or more clocks on the ship have no independent works. They are all electrically driven from the one on the captain's bridge, which is altered nightly to suit the easterly or westerly change of longitude. Telephones are fitted all over the ship, and while it is at the landing-stage a connection is made with the trunk lines. The air-supply to the cabins and corridors is kept at uniform temperature on the *Mauretania* by 53 tanks which pass 212,000 cubic feet per minute. The electrical cables contain over 100 tons of copper, and placed end to end would span a distance of 250 miles.

The newer and larger liners such as the *Olympic*, the *Aquitania*, and the German *Imperator* rejoice in everything possessed by the *Mauretania* except her speed ; and the comforts and conveniences are on a grander scale. To the luxurious drawing and dining-rooms, smoke-rooms, reading and writing-rooms, lounges and verandah cafés, the newer boats add gymnasia and swimming-baths. The gymnasium is equipped with all the usual apparatus for physical exercise, together with special machines for exercising the same muscles as are required in rowing and cycling. Figures 231, 232, 233, and 234 will show that in beauty of decoration, spaciousness, and service, these boats are unsurpassable, until experiments on land shall have shown that there exists a degree of luxury greater than that which has yet been attained.

Perhaps the increase in space will be most effectively emphasized by comparison with the accommodation on an earlier ship. The dining-room of the *British Queen*, which sailed for New York in 1842, was 60 feet long and 30 feet wide. That on the *Olympic* is 114 feet long and 92 feet wide, with a reception-room which measures 92 feet by 54 feet.

Lest the quantity and variety of food described on pp. 326–7 be considered exceptional, it will be well to give for comparison that required for the *Olympic*, which is larger than the *Mauretania*. Here is a list of the stores required for one voyage :—

75,000 lb. Fresh Meats.	180 boxes (36,000) Oranges.
11,000 ,, ,, Fish.	50 ,, (16,000) Lemons.
4,000 ,, Salt and Dried Fish.	1,000 lb. Hothouse Grapes.
7,500 ,, Bacon and Ham.	1,500 gallons Fresh Milk.
8,000 head Poultry and Game.	600 ,, Condensed Milk.
6,000 lb. Fresh Butter.	50 boxes Grape Fruit.
40,000 Fresh Eggs.	7,000 head Lettuce.
2,500 lb. Sausages.	1,000 quarts Cream.
1,000 Sweetbreads.	800 bundles Fresh Asparagus.
1,750 quarts Ice Cream.	3,500 lb. Onions.
2,200 lb. Coffee.	1¼ tons Fresh Green Peas.
800 ,, Tea.	2¾ ,, Tomatoes.
10,000 ,, Peas, Rice, etc.	20,000 bottles Beer and Stout.
10,000 ,, Sugar.	15,000 ,, Mineral Waters.
1,120 lb. Jams.	1,500 bottles Wines.
200 barrels Flour.	850 ,, Spirits.
40 tons Potatoes.	8,000 Cigars.
180 boxes Apples.	

The enormous cost of the outfit, and the extent to which the construction of a passenger vessel provides employment in

FIG. 234.—A VERANDAH CAFÉ ON THE " MAURETANIA."

FIG. 235.—STERN OF SHALLOW-DRAFT VESSEL, SHOWING SCREW TUNNEL WITH YARROW HINGED FLAP.

FIG. 236.—MODEL TO SHOW ACTION OF PROPELLER WORKING IN A TUNNEL.

To face page 329.

industries at first sight widely remote from engineering and shipbuilding, are illustrated by the list given below of the linen, crockery and glass, and cutlery and silver on the *Olympic*.

Linen.

4,000 Aprons.
7,500 Blankets.
6,000 Tablecloths.
2,000 Glass Cloths.
3,500 Cook's ,,
3,000 Counterpanes.
3,600 Bed-covers.
3,600 Beds.
800 Eiderdown Quilts.

15,000 Single Sheets.
3,000 Double Sheets.
15,000 Pillow-slips.
45,000 Table Napkins.
7,500 Bath Towels.
25,000 Fine ,,
8,000 Lavatory Towels.
3,500 Roller ,,
6,500 Pantry ,,
40,000 Miscellaneous.

Crockery and Glass.

4,500 Breakfast Cups.
3,000 Tea ,,
1,500 Coffee ,,
3,000 Beef Tea ,,
1,000 Cream Jugs.
2,500 Breakfast Plates.
2,000 Dessert ,,
12,000 Dinner ,,
4,500 Soup ,,
1,200 Coffee Pots.
1,200 Tea ,,
4,500 Breakfast Saucers.
3,000 Tea ,,
1,500 Coffee ,,
3,000 Beef Tea ,,
1,200 Pie Dishes.
1,000 Meat ,,
8,000 Cut Tumblers.
2,500 Water Bottles.
1,500 Crystal Dishes.
300 Celery Glasses.
500 Flower Vases.
5,500 Ice-cream Plates.
2,000 Wine Glasses.
2,500 Champagne Glasses.
1,500 Cocktail ,,
1,200 Liqueur ,,
300 Claret Jugs.
2,000 Salt Cellars.
500 Salad Bowls.

1,500 Soufflé Dishes.
1,200 Pudding ,,

Silver and Cutlery.

400 Sugar Basins.
400 Fruit Dishes.
1,000 Finger Bowls.
400 Butter Dishes.
400 Vegetable ,,
400 Entrée ,,
400 Meat ,,
8,000 Dinner Forks.
1,500 Fruit ,,
1,500 Fish ,,
1,000 Oysters ,,
400 Cream Jugs.
400 Butter Knives.
1,500 Fruit ,,
1,500 Fish ,,
8,000 Table and Dessert Knives.
300 Nut Crackers.
400 Toast Racks.
5,000 Dinner Spoons.
3,000 Dessert ,,
2,000 Egg ,,
6,000 Tea ,,
1,500 Salt ,,
1,500 Mustard ,,
100 Grape Scissors.
400 Asparagus Tongs.
400 Sugar Tongs.

The Cunard, the White Star, the Allan, and other lines produce a daily paper during the voyage, containing the news flashed by wireless telegraphy from either shore. Concerts are held, games can be played, and an ocean voyage is now no longer an interruption of life, but a period into which all its pleasures can be concentrated to an extent not one whit inferior to the same time spent on land. The contrast with the conditions of fifty years ago brings out vividly not only the enormous increase in the variety of comforts and conveniences, but also the increase in the spending power of the great industrial nations of to-day. In fact, the provision of comfort is not more striking than the power of thousands of people to enjoy it.

In contrasting present-day types of ships with those which were to be seen in the 'eighties nothing is more remarkable than the development of special forms to meet the changing needs of industry and commerce. For example, although in 1886 petroleum ranked fourth in the list of American exports, nearly all of it was shipped in iron casks or wooden cases lined with tinplate. Mr. J. Montgomerie, M.I.N.A., writing in the special marine number of *Cassier's Magazine* in 1911, stated that the vessels engaged in the trade were mostly wooden sailing ships belonging to foreign owners ; and he attributes the growth of the modern oil-carrying vessel to the enterprise of British shipbuilders. By 1893 there were eighty vessels of an average tonnage of 2,500 engaged in the trade, and at the present time there are nearly 300 with an average tonnage of 3,000. Five-sixths of these, carrying nine-tenths of the total quantity, are steamers.

The modern oil-carrying ship is called an oil-tanker, because the oil is contained in tanks which occupy the bulk of the ship. She is loaded by pumping the oil into her, and unloaded by pumping it out, but whatever simplicity attends this method, the design involves special problems which require skill and judgment to overcome them. It is probable that the reader may ask why the vessel is divided up into tanks—why not utilize the whole of the hold in one or at most two or three compartments ? The main reason is that oil is not a fixed and immovable cargo. Any motion of the vessel would set it oscillating from side to side, or surging fore and aft in the hold. Moreover, any motion given to the oil might coincide with its natural period of vibration, and the force exerted by several thousand tons of oil would burst the decks or capsize the ship.

In a modern oil-tanker the tanks occupy nearly the whole length of the ship. They are about 28 feet long, and each one is divided by a longitudinal bulkhead. Each half of a tank is provided with a sort of neck in the upper portion to allow for expansion. Vacant spaces are left between the fore and aft end tanks and the cargo hold and engine-room for safety. They are known as coffer-dams and serve to isolate the oil from any part of the ship in which it might become ignited. The engine-room is in most vessels placed at the after end of the ship, partly because the cost of constructing a tunnel for the shaft through after tanks is thus avoided, and partly because this arrangement increases the amount of space for oil. The lighter varieties of oil give off a highly inflammable vapour, and exceptional precautions have to be taken to prevent the cargo catching fire. But in spite of these the captain, officers, and crew of such a ship must possess unusual courage or a profound contempt for danger, and the fact that ten thousand or so are engaged in the navigation of petroleum-carrying ships gives some idea of the nature and extent of the personal qualities which lie at the back of industrial progress.

Not content with merely supplementing the work of the railway engineer by carrying goods from rail to rail across the trackless ocean, the shipbuilder has been pressed into service to carry whole trains with their passengers and luggage across rivers, lakes, or even narrow seas. There are cases where for various reasons bridge-building is impossible ; the wide detour which would be necessary to avoid a stretch of water is prohibitive in cost of construction or maintenance, and it has been preferable to build large steamers upon which the train is run, and conveyed across the obstruction. Perhaps the most striking service of this character is the one connecting the terminus of the German State Railway at Sassnitz with the terminus of the Swedish State Railway at Trelleborg, across 65 miles of the stormy and treacherous Baltic. Four steamers, two provided by the German and two by the Swedish Government, are employed. The former were constructed in Germany. One of the Swedish boats was constructed in Sweden, and the other, the *Drottning Victoria*, in England, by Messrs. Swan, Hunter & Wigham Richardson. They are all similar, being about 370 feet long, 53 feet beam, and over 4,000 tons displacement, with a speed of $16\frac{1}{2}$ knots. The main deck has two lines of rail each nearly 300 feet long, and capable of receiving four coaches. The vessel has large tanks into or out of which water can be pumped to alter her depth of immersion or " trim " so that the train can be run directly on to her rails over a bridge or gangway. The wheel frames of the coach are chained tightly to the deck to prevent movement on the journey, and hydraulic jacks between the rails are used to lift the bodies of the coaches off the springs. The journey is made at night, and there is no need for passengers to leave their sleeping-berth in the train.[1]

The *Lake Baikal*, constructed by Messrs. Armstrong, Whitworth & Company for the Russian Government, carries trains on the Trans-Siberian Railway across Lake Baikal. As the lake is frozen over in winter the vessel is an ice-breaker as well. Her dimensions are 290 feet long, and 57 feet wide, and she displaces over 4,000 tons. Steel plating 9 feet wide and 1 inch thick protects the hull at the water-line, and the prow is so constructed that it tends to rise on top of the ice in its path. Twin screws are provided for propulsion, and an additional screw in front which disturbs the water under the ice, removes its support and assists the vessel to break through. Owing to the difficulties of transport the *Lake Baikal* had to be made in sections, sent to St. Petersburg by sea, and conveyed by rail and sledge to the lake side. For nearly thirty years the service has been regularly maintained, though there is now an alternative route by rail along the southern shore of the lake.

The *Saratovskaia Pereprava* is another train ferry employed in crossing the River Volga, under extraordinary disadvantages. The difference between the average summer and winter level is about 45 feet, and though separate landing-stages are used these have to be supplemented by hydraulic lifts, capable of raising or lowering the coaches through no less than 25 feet. Other examples exist in Denmark, the United States, and Canada.

The British firm which constructed the special steamers for Lake Baikal and the Volga have also built a number of ice-breaking vessels, the largest of which is the *Ermack*. This not only breaks up the ice in the southern Baltic, but has even done good work within the Arctic Circle, though for this purpose the forward screw had to be removed, and reliance placed on the three screws astern. She has been instrumental in making a way into port for over 400 vessels whose value is estimated to be £4,000,000.

The development of traffic on the great Canadian lakes, which has been enormously increased since they were connected by

---

[1] A similar service was established between England and France during the war.

FIG. 237.—THE GEARING ON THE ISLE OF MAN STEAMER, "KING ORRY."

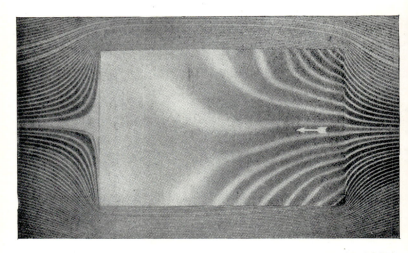

FIG. 240.—FLOW OF LUBRICANT BETWEEN TWO PARALLEL SURFACES.

canals, such as the famous Sault Ste. Marie between Lake Superior and Lake Huron, has demanded a special type of steamer. Here the water is comparatively still, and the material to be carried is ore and corn in bulk. Ships can therefore be employed which have an enormous carrying capacity, but of a form which would render them extremely unseaworthy in rough water. Their speed through the canals is limited to 4 miles per hour, but in view of the short distance between the locks they must be able to start and stop very quickly. A screw of special form is used, having wide blades. The horse-power required is only 150. On the lakes a speed of about 10 miles per hour is usual, and 750 horse-power is required.

The navigation of rivers presents special problems to the shipbuilder. The small size of Great Britain, the shortness of her rivers, the possibility of having ports at or near their mouths, and the excellence of her railway system, render it difficult to realize the importance of natural waterways in large continents. Only a vague conception exists of the enormous traffic on rivers like the Danube and the Mississippi. If rivers like these are important in highly civilized countries possessing a not inconsiderable railway system, how much more vital must they be for example in Africa where the forest resents even the narrow clearing demanded by a railway line. Since the British railway companies bought up the canals and permitted them to fall into disuse the Englishman has grown up with no tradition of the value of the narrow waterway as an alternative to the macadamized road or the steel track.

Generally speaking, the rivers which lend themselves to navigation are slow-flowing, sluggish streams, which amble along shallow depressions, and do not carve out for themselves the deep channels that the ordinary ship demands. Even in the case of ports which are situated at the mouths of rivers dredgers have to be kept constantly at work to remove the silt which the river deposits on its way to the sea ; and Glasgow is a standing example of a port that owes its growth and existence to extensive and persistent dredging. The mouth of the Clyde has been literally scooped out of the earth during the past hundred years.

The characteristic of most river steamers, then, is shallow draft, and many of them must not draw more than 18 inches of water. They are more like flat-bottomed houseboats, with great breadth of beam, and all their accommodation for cargo and passengers above the water-line. A common form of propulsion is a single paddle wheel mounted over the stern, but Mr. H. F. Yarrow constructs river boats with a screw working in a tunnel with a hinged flap at the after end (Figs. 235–6). A screw having a diameter more than twice as great as the draught of the boat can be used, because, once it has started rotating, it throws up the water until the tunnel is completely full. For high efficiency the upper surface of the tunnel should be nearly horizontal, and yet, especially at starting, the opening should be wholly beneath the surface. If the latter condition be fulfilled when the boat is loaded it will not be fulfilled when she is light. Mr. Yarrow therefore attaches the upper surface of the tunnel, from the screw aft, to a hinge, so that it can be adjusted with the outer end a few inches below the surface whatever be the load carried. Increased speed is obtained without increase of power, and the engines work with maximum efficiency under all conditions of load.

MARINE PROPULSION

The means of propelling ships is at the present time undergoing a remarkable upheaval, and the result of the extensive experiments which are being carried out will in all probability be half a dozen different forms, each specially suited to some particular service. Considering first steam-power it may be remarkable that the triple or quadruple expansion engine has held sway for more than thirty years. It is efficient, gives a large power at a reasonably low speed, and is thoroughly understood by the present generation of sea-going engineers. When the turbine was first introduced it used a large quantity of steam—about 16 lb.—per horse-power, but this has been reduced to 10 lb. or less, and this is quite as small as can be shown by any reciprocating engine working under similar conditions. Moreover, it occupies a much smaller space and leaves more room for cargo. As compared with the reciprocating engine it has, however, at least two disadvantages—non-reversibility and high speed. Large engines of any type whatever run at slower speeds than small ones, yet the turbines of the *Mauretania* make 700 revolutions per minute. The non-reversibility has been overcome by fitting " astern " turbines on each propeller-shaft, which are usually capable of giving half the power. The high speeds were at first met by reducing the diameter and altering the pitch of the propeller.

Within the last fifteen years three other methods have been tried, and each seems likely to have an extended use in particular circumstances. One is to connect the turbine-shaft to the propeller-shaft by means of gearing, see Fig. 237, and has been rendered possible by the improvements of Sir Charles Parsons in the cutting of toothed wheels, to which reference has been made in a previous chapter.

The second method is to use electricity. The steam-turbine is at its best when driving a dynamo, and the current is used to drive electromotors mounted on the propeller-shaft. Reversal is then effected by means of a reversing switch. Electrical drive was adopted extensively in the American Navy, and possesses great flexibility. It has been adopted for the huge new liner to be built at Belfast for the White Star Line, but the electric generators will probably be driven by Diesel engines.

The third system has been devised by Professor Fottinger. In this case the turbine drives a high-speed turbine pump, which delivers water to a low-speed turbine on the propeller-shaft. Actually there are two water-turbines in the same casing, one used for driving the propeller ahead and one astern. The same water circulates round and round, through pump and turbine, and the heat produced in it by friction is utilized in raising the temperature of the feed water for the boiler. A test of a 10,000 horse-power plant in Germany in 1913 showed an efficiency of over 90 per cent.

The supremacy of the steam-engine has been challenged during the last twenty years by the Diesel heavy oil-engine. In Chapter IV some account is given of the saving in space, and Chapter II contains a statement of the special value of oil-fuel. To the points there enumerated may be added the reduction in the amount of auxiliary machinery. Those who have not actually seen the engine-room of a steamship can have no conception of the complicated mass of machinery it contains. Apart from the engines which turn the screws, there are condensers, air-pumps, feed-water purifiers, and a host of indispensable appliances which use power, take up space, and materially increase the possibilities of breakdown. By comparison the oil-engine is far simpler, but at present there appears to be some difficulty in building engines of large power. About

6,000 horse-power was the maximum for any one " set ", or from 1,000 to 1,500 horse-power per cylinder. For cargo boats the Diesel engine had achieved a rapid and extraordinary popularity, and more than half the world's tonnage launched in recent years is dependent upon oil fuel. But there will still be a question of fuel sufficiency, and many believe that coal will always, so far as can be seen at present, be the primary fuel for marine propulsion.

It must be borne in mind that all this work is of very recent growth. The first large ships to be equipped with steam-turbines were the Allan liners *Victorian* and *Virginian*, and the Cunarder *Campania*—all three in 1905. Combined reciprocating engines and low-pressure steam-turbines were first used in the *Otaki*, belonging to the New Zealand Steamship Company and showed a fuel economy 12 per cent over a sister ship fitted with reciprocating engines only. The same plan has been adopted for the *Laurentic* of the White Star Line, a vessel of 15,000 tons, and the largest engaged in the Canadian trade, and the giant liner *Olympic*.

The geared turbine was introduced in 1910, and a cargo steamer of the Caira Line built by Messrs. Doxford of Sunderland showed 15 per cent economy over reciprocating engines.

It is interesting to note that the combination of the reciprocating engine and low-pressure turbine, and the geared turbine were both introduced less than twenty years ago, and both appear to give about the same increase of efficiency. It is probable that the former method will be largely adopted for the large fast-passenger boats, and the latter for the smaller cargo vessels. To understand what such an increase of efficiency means it is necessary to look at the enormous amount of coal required. The Canadian Pacific boats, for example, burn 3,000 tons per day regularly, year in and year out, and the *Mauretania* burns nearly 1,000 tons per day. For a trip to New York and back the coal of this would require 22 trains of 30 ten-ton trucks to convey it to the stage. A saving of 10 per cent at 10s. per ton reckoning 30 trips per year would amount to nearly £10,000 a year.

During the last sixteen years the geared turbine has outstripped its competitors for fast passenger ships, and its success has been due very largely to the invention of the Michell Thrust Block. When the propeller is rotating it is forcing the water backward and the " reaction " which causes the ship to move forward is communicated to the ship through the propeller shaft. In order to take this thrust it was customary for the propeller shaft to be furnished with a number of collars or flanges, which pressed upon a number of " horse-shoes " fixed to the frame of the ship. These plates were faced with anti-friction metal, and kept cool by circulation of water when high-powers were involved. This multicollar thrust block (Fig. 238) was elaborate and costly,

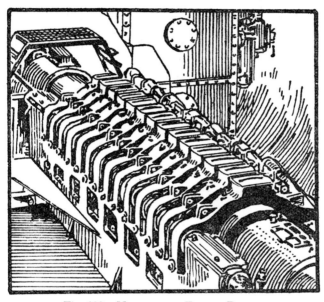

Fig. 238. MULTICOLLAR THRUST BLOCK.

it occupied a great deal of space, and it required very careful and accurate adjustment to equalize the pressure on the horse-shoes. But once adjusted it worked well with reciprocating engines.

When turbines were introduced they were mounted on the propeller-shaft, and as the steam acted on the turbine blades in such a way as to oppose the thrust on the main shaft, a thrust block on the main shaft was not necessary. But in a geared turbine the thrust block had to be reintroduced, because the steam no longer acted on the propeller shaft. Then the multicollar thrust block broke down for reasons which will be apparent

z

after we have devoted a little attention to the mechanics of lubrication.

The theory of lubrication was worked out many years ago, experimentally by Mr. Beauchamp Tower and mathematically by Professor Osborne Reynolds. If a "block W (Fig. 239) is loaded and moves over a lubricated surface in the direction of the arrow, the edge *b* must lift to allow the lubricant adhering to the stationary surface to enter the film space. *The block mounts over the oil*"! Consequently surfaces are most effectively lubricated when the film of oil is wedge-shaped, tapering in thickness towards the rear. In order to fill this wedge-shaped space, the oil exerts pressure, and will begin to flow into such a

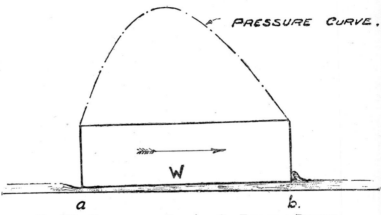

Fig. 239. FORMATION OF PRESSURE OIL FILM IN A BEARING.

space as soon as the rubbing speed exceeds 7 ft. or 8 ft. per minute. If the surfaces are both stationary and oil is forced between them, the lines of flow between the surfaces are not parallel. The oil enters at the front edge and leaves at the sides as well as the back as shown in Fig. 240. Unless the front is lifted the lubricant cannot enter with sufficient rapidity, cavitation or thinning out occurs, and friction with consequent overheating result.

With a reciprocating engine it was not difficult to keep a multi-collar thrust block lubricated, because the unevenness of motion caused a slight rocking which destroyed the parallelism of the surfaces in contact. But the motions of the turbine and its

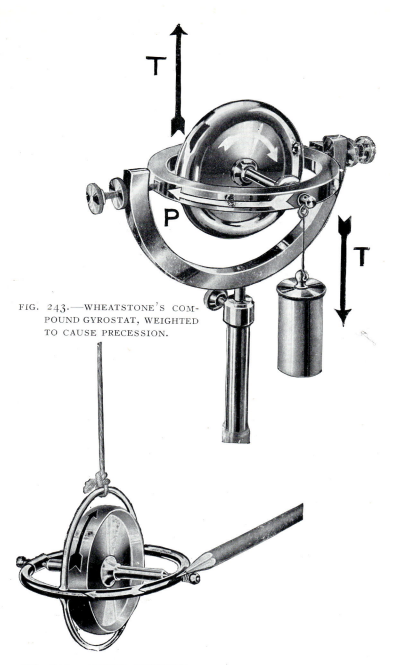

FIG. 243.—WHEATSTONE'S COM-
POUND GYROSTAT, WEIGHTED
TO CAUSE PRECESSION.

FIG. 242.—SIMPLE GYROSTAT.

To face page 338.

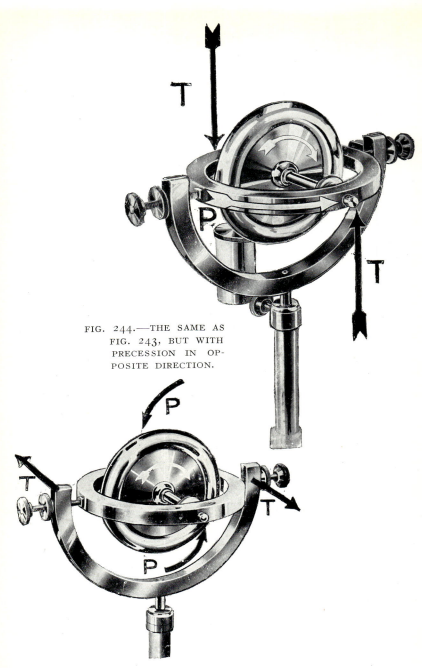

FIG. 244.—THE SAME AS
FIG. 243, BUT WITH
PRECESSION IN OP-
POSITE DIRECTION.

FIG. 245.—THE SAME AS FIG. 242, BUT COUPLE
APPLIED TO HORIZONTAL AXIS.

To face page 339.

geared shaft are so uniform that no wedge is formed, any oil present was squeezed out, and the bearing " ran hot ".

Some years before this problem arose, Mr. A. G. M. Michell, of Melbourne, had invented a thrust block which had small blocks,

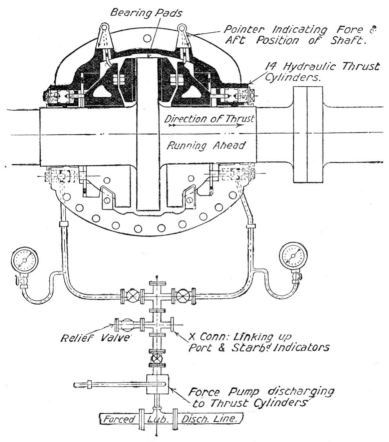

Fig. 241. MICHELL THRUST BLOCK, FITTED WITH APPARATUS FOR INDICATING THE THRUST.

capable of rocking about a point or a line, on the face of the horse-shoe. As soon as the shaft begins to rotate the blocks rock and form wedges into which the oil flows, and the friction under similar conditions is only one-twentieth of that which occurs between fixed flat surfaces. Fig. 241 is a half-section of a marine type thrust block fitted with an arrangement for indicating the magnitude of the thrust. The block is only just coming into use for merchant vessels, but Mr. J. Hamilton Gibson stated in a paper read before the Institution of Naval Architects in April, 1919, that up to 8th March of that year, the Admiralty had in service or under construction Michell Thrust Blocks for over 9,750,000 shaft-horse-power.

A Michell Thrust Block seven feet long replaces a multi-collar thrust block 25 feet long and three times its weight. A single collar only is necessary, and owing to the rocking blocks the thrust may safely be 200 lb. or 300 lb. per square inch compared with 20 lb. or 30 lb. per square inch for the older type, and there are blocks in use transmitting no less than 25,000 horse-power through a single shaft.

THE GYRO-COMPASS

For a thousand years the mariner has navigated the ocean by the magnetic compass. A small needle or needles attached to the under surface of a graduated card have enabled him to plot his course from hour to hour and from day to day. When the sun and stars were obscured by fog or cloud, the small instrument in the brass case has enabled him to steer his ship with the certainty and confidence that come of long experience. He has discovered new lands, brought North and South, East and West into communication and made the whole world kin. Definite ocean highways have been established, and sea voyages are carried out with a punctuality that depends upon the navigator and his instruments no less than upon the engineer and the powerful forces he controls.

The use of iron and steel in place of wood for ships conferred size and safety, but led to special difficulties of navigation. Any mass of iron or steel influences and is influenced by a magnetic needle ; and the enormous masses of magnetic metal in modern ships are liable to exercise an effect upon the direction of the compass needle which entirely overshadows that of the earth. Special adjustments are necessary, and the readings have to be checked from time to time.

But with the dawn of the new century experiments were undertaken which have resulted in an instrument that will

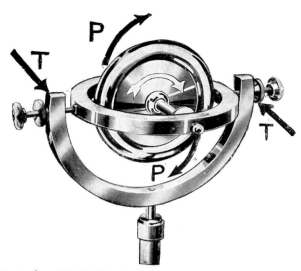

FIG. 246.—THE SAME AS FIG. 244, BUT COUPLE APPLIED
TO HORIZONTAL AXIS IN OPPOSITE DIRECTION.

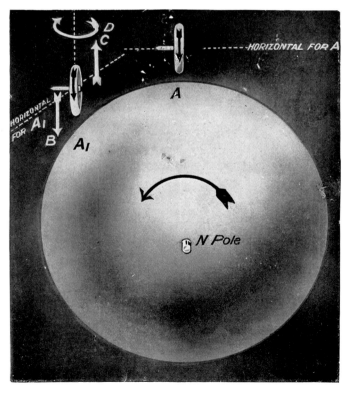

FIG. 247.—RELATION BETWEEN A GYROSTAT AND THE
AXIS OF ROTATION OF THE EARTH.

To face page 340

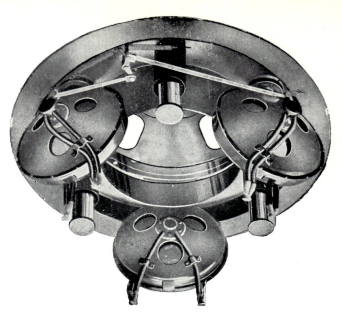

FIG. 250.—GYRO-COMPASS, 1912 PATTERN, SEEN FROM BELOW.

FIG. 251.—GYRO-COMPASS, 1912 PATTERN, SEEN FROM ABOVE.

To face page 341.

point a north and south direction quite independently of the nature of the material of which the ship is made, and the gyroscope, which has for years been a popular scientific toy and had found a single permanent application in the torpedo, seems destined to guide the world's shipping with a certainty that the frail compass needle under the new conditions could never achieve.

A gyrostat is simply a heavy wheel, the axle of which is mounted in a ring (see Fig. 242). When the wheel is set rotating at high speed, either by means of a piece of string or by pressing the pulley wheel of a small electromotor against the axle, it resists strongly any attempt to twist the wheel so as to alter its plane of rotation. Few things are more striking than the way in which any attempt to move the frame in any direction except one in which the axis remains parallel with itself is met by a vicious " kick " which, if the wheel is a heavy one rotating at high speed, almost throws the apparatus out of one's hand.

This kicking propensity of the instrument is really the source of its usefulness, and it will be interesting to observe the exact effect of the twisting force upon it. If the simple form already illustrated is suspended by a string, as in Fig. 242, and pressure is applied to one end of the axis by a pencil for example, the wheel tends to turn in the direction of the arrow marked on the horizontal ring. The wheel and its axle turn in a direction at right angles to the force which is applied, and the rotation of the axis is known as *precession*. If the pencil is applied to the other end of the axis, the rotation is in the opposite direction.

These results are more easily observed in Wheatstone's Compound Gyrostat, in which the wheel is mounted in two rings capable of rotating about axes at right angles to one another. Such an instrument is illustrated in Figs. 243 and 244. The force is applied by hanging a small weight to one end of the axis, and so long as it remains the precession is continuous, while immediately it is removed the precession stops.

If the axis is caused to rotate, then a force is produced at its ends, and a " kick " is produced in a direction at right angles to that about which the turning takes place. This reverse effect is illustrated in Figs. 245 and 246. Gyroscopes or gyrostats mounted in this way—so that they are capable of rotation about three axes at right angles—are said to have three degrees of freedom. If one of the possible rotations is prevented, then the rotating wheel will have two degrees of freedom, and it is a gyrostat with two degrees of freedom that is suitable for use in navigation.

In order to understand how this result has been achieved it is necessary to recall the pendulum experiments of the famous French physicist Foucault, conducted about the middle of last century. He showed that if a pendulum were set swinging and were subject to no disturbing influences, it would maintain its original plane of vibration throughout ; and though the earth might be turning beneath it, the pendulum would still swing to and fro in the same absolute direction as that in which it was started.

This, in fact, provides one of the most beautiful methods of proving that the earth itself rotates. Foucault set up a long pendulum carrying a small pointer beneath the weight or bob. This pointer traced a line in sand as the bob passed through the lower part of its path, and as the earth rotated on its axis the line in the sand showed more and more deviation from the original trace.

The rotation of a heavy wheel at high speed produces a more powerful tendency to maintain the original direction of motion than does the to-and-fro motion of the pendulum bob ; and Foucault concluded that any gyrostat with three degrees of freedom would indicate the rotation of the earth in the same way. In other words, such a gyrostat would maintain its original direction independently of the movement of the body to which it was attached. Moreover, he stated that a gyrostat with only two degrees of freedom would, at any place on the earth's surface except the two poles, tend to set itself with its axis of rotation parallel to the axis of the earth. For consider the cases presented by Fig. 247, in which a gyrostat at A, with its axis horizontal, has three degrees of freedom. When, owing to the earth's rotation the gyrostat has moved to A_1, having maintained its original direction the axis is not now horizontal, but the black end dips downwards. If the gyrostat is suspended by a thread as a pendulum, or supported by means of a float, in such a way as to keep the axis in the horizontal, this constraint gives rise to precession in the direction indicated by the curved arrow D. The ultimate result is to turn the gyrostat so that the axis points true north and south.

At the time when Foucault arrived at his conclusions mechanical science and accuracy of workmanship were insufficient

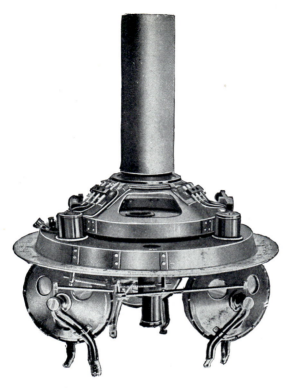

FIG. 252.—GYRO-COMPASS, 1912 PATTERN,
SUSPENDED FOR TESTING.

FIG. 253.—THE DIAL OF THE GYRO-COMPASS.

To face page 342.

FIG. 255.—THE FARMAN BIPLANE.

To face page 343.

to enable a practical demonstration to be made. It was not until the use of steel for ships, and particularly ships of war and submarines, had enormously increased the difficulties of compass adjustment that the need became great, and even then the theoretical and practical obstacles effectively prevented a solution. But in 1900 Dr. Anschütz began a series of experiments which six years later were crowned with success.

Once the initial difficulties were overcome, simpler methods of obtaining the results presented themselves, and in 1908 an exhaustive series of trials extending over four weeks was carried out on the German battleship *Deutschland*. These were so successful that the instrument has now been adopted by practically every navy in the world.

The earlier form, while ordinarily giving good results, was liable to error owing to the pitching of the ship when on a quadrantal course, and a new form was introduced in 1912 which is independent of any kind of motion to which the vessel may be subject. It will be desirable to describe both types, because the earlier one is the simpler, and will form a stepping-stone to the comprehension of the other.

First then, as to the gyrostat itself. The wheel is mounted on a long flexible shaft [1] and has rigidly attached to it a small squirrel-cage rotor the stator of which carries the windings. The two constitute a small 3-phase motor, and the whole is mounted inside a metal case. The motor requires 120 volts and about 1·1 amperes with 333 alternations per second, and drives the wheel at 20,000 revolutions per minute. The wheel and spindle are constructed from one solid piece of special nickel steel, and the stress in the rim produced by such an enormous speed amounts to 10 tons per square inch. The velocity of a point on the rim is 500 feet per second, or 340 miles per hour ! Ball bearings are employed, and 95 per cent of the power is used in overcoming the friction of the air. When the wheel has run for a few thousand hours its surface has a perceptibly higher polish than it had on leaving the grinding machine in which the finishing process was conducted.

The construction of the motor is in itself no mean achievement. When the compass was first invented no machine of such small size and great power was obtainable, and many experiments had to be carried out before success was attained.

[1] See Chapter IV for the reason of flexibility.

The general arrangement of the 1908 type of compass is shown in Fig. 248. In this figure K is a bowl of annular form containing mercury, Q. The gyrostat A is contained in a casing B which is suspended from the under side of a bell. This bell has along its lower edge a hollow steel ring S which floats in the mercury, and gives sufficient buoyancy to support the gyrostat.

The compass card R is fixed to the upper portion of the bell, and the glass top G excludes dust and air currents. Of the

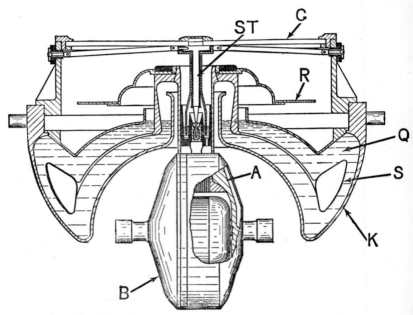

Fig. 248. SECTION OF GYRO-COMPASS—1908 PATTERN.

three wires conveying the current, one is attached to the casing, and the other two are attached to the insulated rod S and tube T, the lower ends of which dip into mercury cups. From thence the current is led to the motor.

The 1912 model is shown in section in Fig. 249. The mercury trough is in the centre, and the bell supported by the hollow floating steel vessel carries three gyrostats about 6 inches in diameter at 120° apart. One gyrostat is set with its axis under the north and south line of the card. The appearance of the actual instrument is shown in Figs. 250, 251, and 252. If these are compared with the section previously given a fairly clear notion of the instrument will be obtained.

A discussion of the theory of the instrument would carry us beyond the range of a popular book, and it must suffice to say

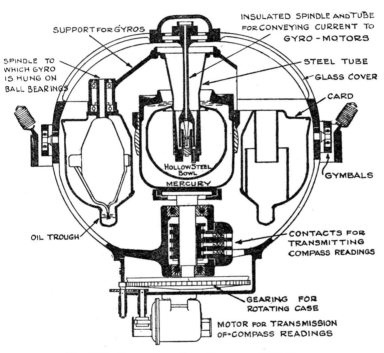

Fig. 249. Section of Gyro-Compass—1912 Pattern.

that accurate indications are given, and that very few corrections are necessary. Moreover, the readings are transmitted electrically to any part of the ship and indicated on dials (Fig. 253) in the upper and lower conning towers or in any other place that may be desired. It is practically unaffected by the vibrations which result from the discharge of big guns ; it is independent of the material of which the ship is made ; and it is uninfluenced by magnetic storms.   No instrument designed in recent years involves greater delicacy of craftsmanship in its manufacture, or more reliability in the materials of its construction.   It is difficult to believe that within its silent casing there are three wheels making 20,000 revolutions a minute, involving a linear velocity only one-fifth of that of a projectile as it leaves the muzzle of a gun.

CHAPTER XVII

THE CONQUEST OF THE AIR

JUST as the beginning of the nineteenth century saw the achievements of the railway and the steamship, so the beginning of the twentieth century has witnessed navigation of the ocean of air. The aeroplane and the dirigible are no sudden advances in man's struggle with nature, but rather the final yielding of defences which have withstood his attacks for a hundred years. From the time when the French physicist Charles in 1784 explained the action of Montgolfier's balloon, and constructed the first balloon to be filled with the light gas, hydrogen, discovered eight years before by Black of Edinburgh, one of the methods of aerial navigation merely awaited a motor. In 1852 Giffard, the inventor of the injector which bears his name, constructed a balloon fitted with a steam-engine and propeller, and succeeded in driving it at the rate of 5 or 6 miles an hour. In regard to another of the main problems, Sir George Cayley in 1809 stated the essential principles of aerial locomotion with a machine weighing more than the air it displaced.

Before passing to a consideration of the achievements of the last twenty years it will be convenient to glance briefly at the history of aeroplanes. Probably everyone is familiar with the way in which a kite is flown. When it is drawn through the air against the wind it rises, and will then go higher and higher as the string is paid out. There are now three forces acting on it :—

(a) The weight which tends to make it fall ;
(b) The pressure of the wind on its surface ;
(c) The pull of the string.

As the force in the string and the weight of the kite act in a downward direction the wind tends to lift it. In fact, the wind can be regarded as having an effect in two directions, one tending to move the kite along in the direction of its own motion, and the other tending to lift the kite vertically. These two effects vary with the angle of the kite and speed of the wind. If the kite has not too heavy a tail the lower end gives before the wind and the kite rises. If the effective speed of the wind is increased by the boy who holds the string running against it, the kite goes higher. If the boy runs with the wind the kite sinks lower and tends to fall.

Ever since kites have been flown it has been known that they are capable of raising considerable weights,[1] and it was this fact that led to the proposal to drive a plane or thin sheet of material through the air with such a velocity that it would support a man. About 1871 a German named Otto Lilienthal commenced to study the flight of birds—more particularly the position and shape of their wings—when gliding near the surface of water, and the construction of kites. Six or seven years later he constructed a frame carrying a pair of wings, and commenced to make gliding flights for the purpose of learning how to balance himself in the air. The wings of the machine were 27 feet from tip to tip, and had an area of 100 square feet. By seizing the frame between the wings and launching himself from the top of a hill he was able to glide several hundred feet, and to alter his direction by swinging his legs. Three years later he constructed a glider with two planes one above the other in order to obtain greater lifting power. Lilienthal met with a fatal accident in 1896, and Percy Pilcher, who started similar experiments in England in 1895, was killed in the same way four years later.

The evolution of the aeroplane on scientific lines was aided by the work of Professor S. P. Langley of the Smithsonian Institution, Washington. He made a great number of experiments on the power necessary to drive a plane of given size through the air with given velocity, by fixing the planes at the end of a long rotating arm ; and he followed this up by constructing models of gradually increasing size, and studying their flight when launched through the air. Having calculated exactly the power necessary, he succeeded in constructing a steam-engine which propelled the model for a minute and a half—the limit which the amount of fuel and water allowed.

Some time before 1900 gliding experiments with a biplane were made in America by the brothers Wilbur and Orville Wright. They increased the area of surface of 160 square feet finally employed by Lilienthal to 305 square feet, and in 1901 succeeded in making flights more than 600 feet long. They reduced the air

---

[1] Many experiments were made with man-lifting kites in the 'nineties by the British Army, and Colonel Cody, who was appointed instructor by the War Office, once crossed the English Channel in a small boat drawn by a kite.

resistance by lying flat on the lower plane instead of hanging from the framework as Lilienthal had done. In 1903 they constructed a motor and made flights lasting about a minute. The following year this was increased to over 5 minutes, and a year later to 38 minutes.

Meantime, progress was being made in France. In 1906 Santos Dumont flew over 200 yards, in 1908 Farman covered over 300 yards, and in April, 1908, Delagrange remained in the air more than 9 minutes. The Wrights had put away their machine and were negotiating with several governments for its sale ; but the development in France brought Wilbur Wright across the Atlantic. After some delay in getting his machine to work he effectively abolished all criticism by flying for more than two hours, and by carrying passengers at a height of 400 feet. The experimental stage was now passed. The building of both monoplanes and biplanes was started in real earnest, and the following year saw the first aviation meeting at Rheims when Glenn H. Curtiss won the speed race on his biplane, making 47 miles per hour, and Latham won the height test by attaining an altitude of 500 feet.

THEORY AND CONSTRUCTION OF AEROPLANES

Let us now turn to the theory and construction of aeroplanes, and consider the monoplane as being theoretically the simpler, though actually the more difficult to construct. If a plane is held horizontally with its edge to the wind, the only resistance which it offers will be that due to the sliding of the air over its surface, or skin friction as it is called. But if it is tipped so that its front edge rises the wind flowing underneath will be deflected, and will exert an upward pressure and a resistance to the forward motion of the plant. These two forces are known as " lift " and " drift " and the shape of the plane should be such that the lift is as high and the drift as low as possible. The forcible passage of a body through the air is liable to give rise to whirls and eddies, but these are avoided if the surface of the moving body coincides as nearly as possible with the paths of the particles displaced. Thus the section of the planes is generally of the form shown in Fig. 254. Here the curve of the under surface causes the pressure to be increased gradually, and tends to reduce the disturbance at the advancing edge of the plane.

The planes are more or less rectangular, oblong in form, and with a long edge facing the wind. If they meet an upward or downward current more or less pressure will be produced on the under surface, and considerable rocking may take place. Similarly, a side wind will spend most of its force on the windward planes and tend to overturn the machine. It is therefore

Fig. 254. Section of Aeroplane Wing in Relation to Horizontal Stream Lines.

necessary to provide some means of securing stability in a fore and aft or longitudinal direction, and lateral or transverse direction.

Longitudinal stability is secured by one or two horizontal or nearly horizontal planes at the rear, the condition being that the inclination of this tail to the horizontal is less than that of the main planes. If the machine pitches forward the pressure under the tail is reduced more rapidly than that under the planes and may even fall on the upper surface. The tail therefore tends to fall. Conversely, any tilt backwards brings more pressure to bear under the tail, causing it to rise and right the machine. A contrivance of this kind has, in fact, what is known as inherent stability, because it operates independently of the pilot.

Lateral stability is secured in one of four ways. The first is the use of ailerons. These are small supplementary planes attached to the extremities (generally the rear edge, see Fig. 255) of the main planes, and capable of being rotated slightly from the pilot's seat. If the machine tends to tilt downwards to the left it is clear that the wind pressure under the wing on the right is greater than that under the wing on the left. The aileron on the left is then lowered, and that on the right raised. The lifting force under the left or lower wing is thus increased, that under the right or upper wing is decreased, and the machine

FIG. 256.—THE VICKERS MONOPLANE : BACK VIEW.

To face page 350.

FIG. 257.—THE VICKERS MONOPLANE: FRONT VIEW.

To face page 351.

returns to its correct level. These were employed on Cody's biplane. A second method, used on the original Wright biplane, consisted of hanging flaps at the ends of the main planes, which could be raised or lowered. Their action can be easily understood.

A method more frequently used on monoplanes was to warp the wings. In this case the rear edge of each wing or plane can be bent up or down and the pressure under them adjusted. The fourth method was used on biplanes. It consisted of a supplementary plane on either side, the inclination of which can be varied. It is never used now.

A typical monoplane and a typical biplane (1913) are shown in Figs. 256, 257, and 255. The former had a long, narrow boat-shaped body, containing the engine, pilot, and passengers, with the screw at the front end. The wings consisted of strong, light frames fixed rigidly to the body or fuselage, and held in place by tightly stretched wires. The tail and rudder are shown at the back, fixed by means of outriggers to the fusilage. The rudder was generally operated by wires attached to a lever, upon which the feet of the pilot rested, and the warping of the wings or raising or lowering of the flaps or ailerons by similar wires connected with a lever worked by hand.

The biplane consists of a strong framework connecting the upper and lower main planes. This consists of a series of struts holding the two planes apart, and wires bracing them together. The tail consists of two horizontal planes fixed on outriggers, with one or two vertical planes to act as rudders. There used to be one or two movable horizontal planes in front which acted as elevators, but these have now been discarded.

The framework was made of wood or less usually of steel. Of nine machines built up to 1913 two only were of steel and seven were of wood, ash and silver spruce being the principal varieties used. The wings are covered with textile material treated with rubber or some form of varnish, and weighing from 2 to $6\frac{1}{2}$ oz. per square yard. In the early machines the pilot sat on a wooden seat and was fully exposed to the weather. Some of the later machines had the body encased in a very light material impervious to wind. A type of biplane made by A. V. Roe & Company had a small cabin fitted with celluloid windows, which was entered by a door in the side.

One of the most important details of construction is the

chassis or landing carriage. It requires a high degree of skill to settle on the ground without a bump, and from the very beginning some means had to be adopted other than the bare skids. A very effective plan was devised by Farman. The wheels were fixed on castors, so that they would turn readily in the proper direction. The rod which carried each wheel passed through the skids and was held down by rubber bands. On landing these bands were stretched, and their resilience broke the shock of impact.

The engines used for aeroplanes are practically motor-car engines, working on petrol and constructed in the lightest possible way. It is noteworthy that the brothers Wright started their machine by drawing it along a short length of tramway rail by means of a rope which was released when it rose from the ground. As the engine had nothing to do but drive the machine through the air, one of 24 horse-power was found to be sufficient. The early French aeroplanes, however, were started by running along level ground, and for this purpose a more powerful engine was required. At the first flying meeting at Rheims in 1909 engines of 50 horse-power were common. It was at this meeting that Farman and Delagrange used the Gnome engine on a biplane and a monoplane respectively.

Perhaps the real merit of a light engine is only realized when the weight of the petrol and lubricating oil is taken into account. Thus the 100 horse-power Gnome engine consumed 0·87 [1] pints of fuel per horse-power per hour, and the amount required for an hour's journey would be 22 gallons, weighing about 150 lb. Several gallons of lubricating oil would also be necessary for the same time. This will readily explain why even a small machine fully loaded for flight weighs nearly a ton, and why flights of more than four or five hours' duration were not at first usually attempted.

Propellers are generally two-bladed, and were formerly made either of metal or of wood, but the latter is considered the more reliable. They are built up in layers and then shaped. The diameter is from 5 to 10 feet, and the pitch from 3 feet to 6 feet. The Wright biplane had two propellers, placed behind the machine. Modern biplanes have either single or double propellers placed either in front or behind, but as most machines are

---

[1] The improvements in this engine, described on p. 70, resulted in a considerable decrease in the consumption of petrol.

FIG. 258.—CONSTRUCTING THE WINGS OF THE VICKERS MONOPLANE.

To face page 352.

I

FIG. 259.—THE HANDLEY PAGE PASSENGER AEROPLANE.

To face page 353.

constructed it is safer to have them in front. Some accidents have in all probability been due to a wire breaking and becoming entangled in the propeller. For the purpose of determining the best size and pitch, Messrs. Vickers, Ltd., erected at their works at Barrow a huge rotating arm 110 feet long. The engine and propeller are fixed on the outer end, and the rate at which the arm is rotated for a given power of engine can be observed.

TYPICAL AEROPLANES

One of the earliest types of monoplanes was the Bleriot, whose designer was the first to fly across the Channel. It was made to carry one or two persons. The span of the wings from tip to tip was just under 30 feet in the smaller machine and 36 feet in the larger, and the lengths were 25 ft. 6 in. and 27 ft. 6 in. respectively. The areas of the main planes were 187 square feet and 263 square feet respectively. The single-seater has a lifting and the two seater a fixed tail ; the latter is therefore provided with elevators, which are unnecessary with a movable tail. The weights are 550 and 700 lb., and each is driven by a 50 horse-power Gnome engine. Just in front of the pilot's seat is fixed a lever which, moved to and fro in a fore-and-aft direction, warps the wings, and moved sideways governs the lifting tail or the elevator. The rudder is manipulated by a foot-rest.

The monoplane constructed by Messrs. Vickers, Ltd. (Figs. 256, 257, and 258), has a frame of weldless steel tubes which carries the fusilage and wings. The rudder is actuated by a foot-bar, and a universal lever enables the wings to be warped and the elevator to be raised or lowered. The wings are made of ash, and are covered with an extremely strong light material with a smooth surface which is impervious to water. All metal parts are tinned to prevent rusting. The machine is very strong, and with the great resources at the firm's command the materials are thoroughly tested before use. With an 80 horse-power Gnome engine a speed of 70 miles per hour is attained when carrying a passenger, and fuel for 3 hours, and under these circumstances it will climb 400 feet per minute. For transport it can be dismantled and placed in a case 25 ft. 6 in. long, 9 ft. 6 in. high, and 5 feet wide, and taken out and erected ready for flying in 45 minutes.

The Deperdussin monoplane was the invention of M. Bichereau, and was noteworthy as holding for some time the record for speed.

With it at Chicago in 1912 Vedrines won the Gordon Bennett Cup with a velocity of 105 miles per hour, and a greater speed than this was said to have been attained. This, however, was on a machine specially built for racing, having only 97 square feet of wing surface and a 140 horse-power engine. The ordinary machine embodies a good deal of attention to detail in its construction, no less than three different kinds of timber being used in the longitudinal span of the wings. The fusilage is not fixed rigidly to the landing chassis, but is slung to it by two flexible belts. The rudder is controlled as usual by the foot, and the wings are warped by turning a wheel mounted on a bridge. Movement of the bridge backwards or forwards operates the elevator.

Having dealt so fully with the early types of monoplanes the biplane will require less explanation. It is a heavier and slower machine but capable of carrying a greater weight for the same spread of wing. There may be one or two engines and the propellers may be placed in front of, or behind, the planes. The original machines had an open fusilage, composed merely of a framework, but the general practice arose of covering this with canvas or thin wood, partly to reduce air resistance and partly to afford protection to the aviator from the intense cold of the upper regions of the atmosphere.

Before the war machines were constructed to rise from or alight upon the water, the landing carriage being replaced by floats. These were satisfactory when the water was still, but decidedly unsafe in a rough sea. Still if a sea-plane fell into the water, the airman had a better chance of remaining afloat until help arrived.

THE PROGRESS OF AVIATION

Turning now to the progress in aerial navigation since the historic meeting at Rheims in August, 1909, it is hardly possible to realize its magnitude and rapidity. Before the end of the year Wilbur Wright had flown with a passenger for an hour and a half, and Henri Farman had actually remained in the air for 4 hours and 17 minutes. Nor had Englishmen been idle. Mr. S. F. Cody, whose lamentable death occurred less than a week before these lines were written (in 1913) had made short flights on Laffan's Plain. Mr. J. T. Moore Brabazon flew both in England and on the Continent in a Voisin biplane, and Mr. A. V. Roe flew

Interior of new Handley Page. W.8.

FIG. 260.—INTERIOR OF NEW HANDLEY PAGE PASSENGER AEROPLANE.

To face page 354.

FIG. 261.—THE BRISTOL PULLMAN TRIPLANE AND THE BRISTOL BABE.

To face page 355.

in a triplane propelled by an engine of only 9 horse-power—the smallest power with which flight has ever been accomplished.

It is not without significance that many of these pioneers were expert motorists—men who understood the petrol motor, were accustomed to high speeds, and possessed the nerve and the delicacy of balance and touch that made the control of their machines almost automatic. These qualities enabled them to acquire rapidly a degree of skill which the Wrights and others had previously developed by long-continued experiments in gliding.

In 1910 aviators began to make long-distance flights across country. The race from London to Manchester for the *Daily Mail* prize of £10,000 will be freshly remembered by all British readers. Twice was Graham White pulled up by treacherous winds in the Trent Valley, while Paulhan managed to cover the 183 miles with one stop. In America Curtiss flew from Albany to New York, a distance of 150 miles, with one stop. Again, in the same year the Hon. C. S. Rolls flew from Dover to Calais and back without descending, while Loraine flew from Holyhead to Ireland across 52 miles of sea. With more confidence airmen began to attain greater heights, and within a few months a succession of records was made and broken. Armstrong Drexel began by rising to 6,000 feet at Lanark, and before the year was over he and half a dozen others had beaten that by more than 60 per cent. First Morane attained a height of 8,469 feet, then Chavez 8,790 feet, Wynmalen 9,174 feet, Drexel, again, 9,450 feet, Johnstone 9,714 feet, and finally Legagneux 10,746 feet— a height which has since been surpassed on several occasions. The record is now over 20,000 feet, or nearly 4 miles.

The year 1911 was again a year of long flights. The circuit round Great Britain, a distance of 1,010 miles, the Paris to Rome race of 815 miles, and the European circuit of 1,030 miles were won by Lieutenant Conneau ; and the Paris to Madrid race of 874 miles was won by Vedrines, who afterwards flew for the Deperdussin Company. An experiment was also made in the carrying of mails from Hendon to Windsor. The following year was notable for the War Office trials on Salisbury Plain.

But the pioneers of aviation were by no means satisfied with past results, and Vedrines completed a flight from Paris to Cairo ; received an invitation to continue his journey as far as the Cape of Good Hope ; and talked lightly of flying round the

world. Nor did the fascination of spectacular flying decrease.
During the year 1913 Pegoud and Chantaloupe, and later B. C.
Hucks and G. Hamel, showed how to " loop-the-loop " and set the
nerves of timid persons tingling with the reckless daring of their
feats. All this went to show that with the necessary coolness,
judgment, and skill, the aeroplane was absolutely under the
control of its pilot, and had simply added a new dimension to
locomotion. Freedom of movement in a horizontal plane on
land and water has been within man's power for long ages, but
movement in a vertical plane had hitherto been possible only in
a laboured way and to a limited extent.

The early achievements were secured on machines that were
for the most part flimsy contrivances, but ill-adapted for
exploring the unknown currents and vortices of the atmosphere.
Accidents have occurred by sudden gusts of wind which tipped
the machines over to such an angle that they could slip sideways
to the ground. Then, again, an " air-pocket ", the nature of
which is at present little understood, may be entered without
warning. These are not cavities, but regions in which the
pressure or movement of the air is such that it offers but little
support to the machine, which may thus fall a thousand feet or
more. They are generally met with over valleys, and constitute
a danger which cannot be anticipated either by the construction
of the machine or by skilful pilotage.

But the manufacture of aeroplanes was now in the hands of men
whose experience enables them to provide strength where strength
is needed, and retain the lightness which is so necessary. The
age of indefinite experiment has been succeeded by a period
of definite and progressive design. Stronger material, thoroughly
tested, is being used, and the general proportions which give the
most satisfactory results are known to a nicety. With the
great increase in the number of airmen, there has been a decrease
in the proportion of reckless spirits, and catastrophe is not
invited so freely as in the early days. If a proper standard of
care is maintained, then the causes of accident are reduced to
two, viz. treacherous winds, and breakage ; and the first of these
is the only one which cannot be controlled. There is a tendency
to exaggerate the number of accidents, because they, and not
the ordinary successful flights, are recorded in the newspapers.
An investigation by the Aero Club of France showed that during
1912 only one fatal accident occurred for every 92,000 miles
flown. Since then the ratio has become still lower.

FIG. 262.—THE AVRO BABY.

FIG. 263.—THE VICKERS VIKING : AN AMPHIBIAN TYPE.

To face page 357.

PROGRESS SINCE THE WAR

The important facts of aviation since the war may be summed up in the development of civil aircraft, the establishment of regular aerial services, and the spectacular triumph of the Atlantic crossing.

So long as aeroplanes were required only for military purposes they had to be heavily engined, either for low-speed long-flight bombing, or high-speed, quick-moving fighting machines. They had to be capable of rising with the greatest rapidity from the ground, and of ascending to high altitudes in order to escape observation or be beyond the range of gun fire. When hostilities came to an end some of the heavy bombing machines were converted into machines for carrying passengers and cargo ; and when the Air Navigation Act was passed in 1919, Civil Aviation was at once accorded recognition and placed under regulation.

On 25th August, 1919, the Air Transport and Travel Company began a regular service between London and Paris, using DH4 machines fitted with Rolls-Royce engines. Within a few days Handley Page instituted a service on the same route, and on 25th September, a service between London and Brussels. By 3rd January, 1920—19 weeks—Air Transport and Travel had made 200 journeys out of a possible 272, and the machines had covered 55,520 miles. Between 2nd September and 1st January, 618 passengers and 16,982 lb. of goods were carried over the same route, covering 34,660 miles, while on the London–Brussels line they had carried 251 passengers and 25,888 lb. of goods 25,895 miles. In July, 1920, Handley Page started a service to Amsterdam, the Instone Air Line had begun operations, and there are now 400 or 500 people per week making use of the new form of transportation.

The converted bombing machines have gradually been replaced by aeroplanes specially designed for the work. The heavily-engined bomber required 700 to 750 horse-power for its two passengers, while the modern commercial aeroplane, with two 450 horse-power Napier " Lion " engines, will carry 16 or 20 people. It does not rise so rapidly from the ground ; its normal altitude is less than 5,000 ft., it has a normal speed of 115 to 130 and an average speed of 85 to 100 miles per hour. Moreover, while it answers quickly enough to the helm, it has none of the

" liveliness " which characterized some of the fighting machines during the war. The cabin is fully enclosed, provided with comfortable seats, and electrically heated. Typical examples are the machines designed by Captain de Havilland for Air Travel and Transport, the giant Handley Page craft (Figs. 259–60), the Bristol Pullman Triplane (Fig. 261), and the vessels of the Instone Air Line. The Bristol Pullman carries 14 passengers, has a radius of 500 miles at 100 miles per hour, and is driven by four 410 horse-power Liberty engines. Its great size is indicated by contrast with the Bristol Babe which is shown in Fig. 261 crouching for protection under the tip of one wing. The " Babe " has a wing span of 19 ft. 8 in., a length of 14 ft. 11 in., a height of 5 ft. 9 in. The wing area is 107·8 sq. ft., and the load is 6·34 lb. per sq. ft. The maximum speed at 5,000 ft. is 80 miles per hour, best cruising speed 65 miles per hour, and landing speed 40 miles per hour.

This aeroplane illustrates a movement which arose after the war to produce a small cheap machine for the " owner-driver ". The first was the Avro Baby (Fig. 262), which has a span of 25 ft., length of 17 ft., height of 7 ft. 7 in. When empty it weighs 610 lb., and when loaded 825 lb. The wing area is 180 square ft. and the loading 4·58 lb. per sq. ft. It carries a 45 horse-power Green engine, 6 gallons of petrol and one gallon of oil. Its speed at the sea level is 80, at 10,000 ft. 71, and on landing 33 miles per hour. It will rise 5,000 ft. in 11 minutes, and 10,000 in 25 minutes. In the Aerial Derby on 21st June, 1919, Captain Hammersley, on this machine, won the handicap with an average speed of 70·8 miles per hour, and at the corresponding race in 1920 the same competitor won the handicap again with a speed of 78·89 miles per hour. On 1st May, 1920, Lieut. Hinkler, D.S.M., on an Avro Baby, made a non-stop flight to Turin— 9½ hours—on 20 gallons of petrol.

In most respects the machines are operated in the same way as the earlier ones. Changes have been made, but they are in detail rather than in principle. Thus the steering lever which formerly acted on the rudder through cables, now acts through rods, because the cables tended to become frayed. In regard to construction much has been done to ascertain the shape of cross-section of struts which offers the least resistance to the wind and wires of flattened section with sharp edges are used. Modern aeroplanes are constructed almost entirely of the light aluminium

FIG. 262A.—NEAR VIEW OF UPPER PORT WING OF HANDLEY PAGE "HARROW"
TORPEDO CARRIER: SLOT IN OPEN POSITION.

FIG. 262C.—THE SLOT AND INTERCEPTOR FITTED TO A MOTH AEROPLANE.

To face page 359.

alloy *duralumin* and steel. An enormous amount of investigation, both mathematical and experimental, has been undertaken in regard to design. The experimental study of the best form for the wings, or " aerofoils ", struts, etc., is carried out in tunnels specially constructed so that a current of air can be driven through at various velocities. The areofoil or strut is placed in the tunnel and the pressure of the air current is measured.

On the mathematical side a great deal of work has been carried out on the stability. Before the war Lieut. Dunn designed a machine that was extremely stable. It had wings which projected backwards, forming an obtuse Vee. But a compromise has to be made between stability and manageability. If a machine is too stable it does not respond readily to the control of the pilot.

One of the most interesting developments of the last few years is the Handley Page slotted wing. The weight that an aeroplane will carry is dependent upon the wing surface, upon what is called the lift-coefficient, and upon its velocity. The wings are normally fixed in area and have a definite angle to the wind when the aeroplane is on even keel. If the velocity of a given machine falls below a certain value, the machine falls. This fact is of special importance when landing, because if the velocity falls below the minimum when the machine is near the ground, it bumps. In the early days landing was particularly dangerous. What was required was some means of varying the lifting power, and this could be accomplished in a practical way by altering the section of the wing. Many experiments have been made, and a number of machines have been built with a hinged flap along the front and back edges of the wings. In 1921 Handley Page first described his slotted wing, which achieves the same purpose. The front edge of the wing can be moved forwards— see Fig. 262A—leaving a slot behind it. The underside and back of this front piece is so shaped that the air flows over it in stream-lines, without forming eddies, and the effect is to increase the wing area. By this means the lifting power may be increased 60 per cent. The minimum velocity for landing is thus reduced, landing is easier, and the aeroplane can rise more quickly from the ground. These advantages enable a smaller landing ground to be used.

Within the last two years the operation of the slotted wing has been made automatic in a way that is best explained by the aid of Fig. 262B. The movable portion of the wing is connected to the corresponding aileron.   If, when stalled, the machine tends to roll the pilot pulls down the aileron on the lower wing and pulls it up on the upper wing.   As soon as the aileron is

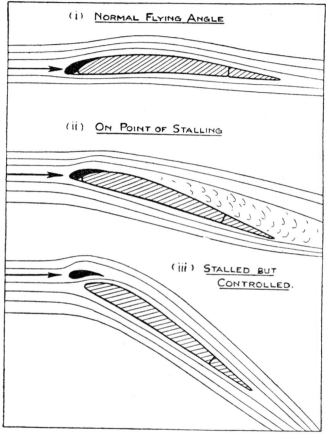

Fig. 262B. DIAGRAM TO EXPLAIN ACTION OF SLOTTED WING.

pulled up the slot on that side closes, and the lift on that upper is reduced. The machine then comes down to an even keel. Still greater control is secured by the use of an interceptor. This is a small plate lying in a recess behind and parallel to the front spar, which forms the back of the slot, see Fig. 262c. When turning very sharply so that the machine tilts at a high angle both slots remain open very nearly to the same extent. But if, for example, the left wing is depressed, the aileron on the right is raised by the pilot, and this causes the interceptor to rise as a vertical plane behind the slot. Then, though the right slot be open, its lifting power is destroyed by the interceptor, and the machine tends to right itself.

The engines have been continuously improved. Many of the Vee type used during the war had twelve cylinders arranged in six pairs. Both this and the radial or star type are still used. Lightness has been secured firstly by making the crank cases of aluminium, instead of cast-iron. Subsequently aluminium pistons with steel piston-heads to protect the more fusible metal from the action of the hot gases effected a further reduction in weight. Some engines have aluminium cylinders with steel liners in which the pistons run. One difficulty is to maintain the power at high altitudes where, owing to the low barometric pressure, the weight of air drawn into the cylinder at each suction stroke is smaller than at lower levels. A plan which is being adopted to overcome this disadvantage is " supercharging ". The engine does not draw air directly from the atmosphere, but receives it from a pump, which delivers it at approximately the sea-level pressures. For driving this pump Rateau has designed a small turbine which is operated by the exhaust gases.

The design of aeroplanes to meet conditions other than those which ordinarily obtain in civil aviation has by no means ceased. The seaplane with floats which enable it to rise from or settle upon water has been succeeded during the war by the flying-boat, largely developed by Curtiss in America. A famous example of this type is the Vickers Viking (Fig. 263), which is an amphibian machine, adapted for use on either land or water. The landing wheels can be drawn up out of the way, or lowered as required. The machine will alight upon water, it can then be steered towards shore, and will climb out of the water under its own power.

This machine combines the special features of the aeroplane, the seaplane, and the flying boat. The Viking Mark I was built in 1919 and in 1920 was placed first at the Antwerp Exhibition for the shortest time in rising from the water, fastest time over a given circuit, shortest time in rising to 1,000 metres, and highest " ceiling " of altitude with full load. The Viking Mark III gained the Air Ministry's prize of £10,000 for speed of flight, low speed of landing, rapidity in rising, reliability in flight, economy, sea-worthiness, and general excellence. The later features of this type of machine are embodied in the Mark IV. Like the others, this has wings which fold forwards, and it then occupies a space of 32 ft. by 35 ft. The overall length is 34 ft. 2 in., height 14 ft., and span 50 ft. The ailerons are operated by wheels, the elevators by the fore and aft rotation of the control lever, and the rudders by foot levers. The engine is a 450 horse-power " Lion " driving a " pusher " air-screw, and is mounted above the hull. Near the sea level the full speed is 115 miles per hour and the landing speed 50 miles per hour, while the machine has a radius of about 350 miles at full speed and 390 miles at 80 miles an hour. The weight fully loaded, with crew, is 5,790 lb.

With lighter, more efficient, and reliable engines, the length and duration of flight increased rapidly. The *Daily Mail* prize of £10,000 offered in 1913 for a journey across the Atlantic came more and more within the range of possibility. It might have been done earlier but for accidents and the war. Hamel was preparing for an attempt from Ireland on a Handyside machine, and Lieut. Porte proposed to cross from America on a Curtiss biplane. But Hamel was lost somewhere over the English Channel and Lieut. Porte was obliged to postpone his effort. After the Armistice preparations were renewed. Fresh candidates appeared, with larger and more powerful machines. A number of them gathered in Newfoundland and waited for favourable weather. Among the machines were four American Naval Aeroplanes which were not strictly competitors, because their pilots had no intention of attempting a non-stop journey. They started on 16th May, 1919, and only one, with Lieut.-Commander Read, completed the journey. He flew from Trepassey in Newfoundland to Hosta in the Azores, a distance of 1,381 miles, and thence by Porta Delgado, Lisbon, and Ferros to Plymouth, arriving on 31st May.

On 18th May H. G. Hawker and Lieut.-Commander Grieve left St. John's and for a week all trace of them was lost. They had been obliged to descend when about 750 miles from the Irish coast, had been picked up by the steamer *Mary*, which was not equipped with wireless, and until they were transferred to H.M.S. *Revenge* on 25th May, it was impossible for them to make their whereabouts known. Meantime, there were others who retained

FIG. 264.—THE VICKERS VIMY TRANSATLANTIC AEROPLANE : THE MACHINE THAT CROSSED THE ATLANTIC.

Photo: Topical Press.

FIG. 266.—A ZEPPELIN AIRSHIP OVER BERLIN.

To face page 363.

their faith in the possibility of the task. Captain John Alcock
and Lieut. John Whitton Brown had been unable to start with
Hawker and Grieve. They had reached Newfoundland later than
the others, and their machine, a Vickers-Vimy bombing machine,
(Fig. 264) fitted with Rolls-Royce engines and hurriedly adapted
for the purpose, took some time to get ready. Furnished with
865 gallons of petrol, sufficient for a journey of 2,440 miles, they
left St. John's at 5.28 a.m. on 14th June, and landed at Clifden,
Ireland, at 9.25 a.m. Allowing for the difference of longitude,
the actual time of flying was 15 hours 57 minutes. For nearly
16 hours they had been out of sight of land, often completely
enveloped in fog, rarely able to see the sun, moon, or stars, and
frequently unable to see the water below them. Half an hour
after they started the wireless apparatus broke down so that
they were unable to signal their position to any passing ship
or to announce the imminence of their arrival. For four hours
the machine had been encased in ice and when, finally, they
made a nose-dive into a bog both aviators were dazed. But
the task had been accomplished. The wide Atlantic had been
crossed.

In the same year Captain Ross Smith and his brother Keith,
accompanied by Sergeants Bennet and Shiers, won the £10,000
prize offered by the Australian Government for a flight from
England to Australia in 30 days. They accomplished the task in
27 days 20 hours. In 1925 the Marquis de Pinedo flew alone
from Italy to Japan via Australia and back again, a distance of
35,000 miles. A year later Major (now Sir) Arthur Cobham
flew from London to Capetown and back, a distance of 17,000
miles. Again, in 1927, Lindbergh flew alone and unannounced
from New York to Paris, and a short time afterwards
Chamberlin and Levine flew from New York to Eisleben, in
Germany. Within the last few months a non-stop flight has
been made from England to India. It is only twenty-four
years since Wilbur and Orville Wright made flights of 24 miles
and remained in the air for 36 minutes. To-day the distance
record is nearly 6,000 miles, and the duration record more
than 52 hours. The maximum speed now exceeds 300 miles
an hour.

But progress has not been confined to long distance flights
or attempts to break records. Commercial aviation has made
rapid progress. Imperial Airways, for example, run a daily

service between London and Paris, and London, Brussels, Cologne ; and a weekly service from London to Cairo, Gaza, Baghdad, and Basra, besides several special services during the summer months. In the three years ending March, 1927, their machines had covered 2,502,626 miles, they had carried 39,705 paying passengers, and had transported 1,855½ tons of freight, mail, and excess baggage. The aerial mail service is extending rapidly both in the old world and the new.

AIRSHIPS

It is a little curious that the earlier and more obvious method of navigating the air should have been overshadowed by a later and less obvious one. Ever since balloons filled with hydrogen or coal-gas were constructed it has seemed to be clear that sooner or later they would be fitted with some means of propelling them against the wind. It has already been mentioned that Giffard did, in 1852, use a steam-engine for the purpose, and though he bent the chimney twice at right angles and covered the stoke-holes with wire gauze in order to prevent sparks reaching the inflammable gas above, there are too many risks attached to such an experiment to encourage its extension. His gas-bag was round in section and pointed at each end (Fig. 265). Its largest diameter was 60 feet, length 140 feet, and it held 90,000 cubic feet of gas. To the network over the bag was slung a pole about 60 feet long, from which the car was suspended. This is not a satisfactory method, as it permits the car to oscillate independently of the bag. The steam-engine was of 3 horse-power, and the three-bladed propeller made 110 revolutions per minute. Three years later a larger vessel was constructed, and after making a trial flight it came to grief on landing, thus creating precedent which has been followed by so many of its kind.

An interesting balloon was constructed in 1872. The engine was driven by gas obtained from the balloon itself, and air was pumped into a smaller bag called a ballonnet contained inside the larger one, in order to compensate for the gas consumed. This attained a velocity of 3 miles an hour, but the plan was not proceeded with on account of lack of funds. The French Army used a balloon in the Franco-German War in which the screw was worked by eight men, and at a later date one was constructed with an electric motor driven by a battery.

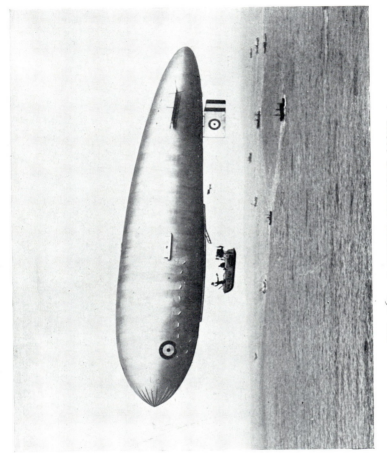

FIG. 267.—A BRITISH SUBMARINE SCOUT.

To face page 364.

FIG. 268.—A BRITISH AIR SCOUT: " S.S.Z. " CLASS.

The first definite achievement, however, is due to Renard and Krebs, two captains in the French Army. They constructed a balloon of which the envelope containing the gas was shaped something like a fish. The car was very long—over 100 feet—and about $4\frac{1}{2}$ feet wide, and 6 feet high. It contained a sliding weight which could be moved fore and aft to secure balance. The motive power was an electric motor of 9 horse-power driven by a chromic acid battery. In 1885 the " La France " as the balloon was called, flew successfully over Paris and returned to its starting-point, attaining a maximum velocity of 14 miles per hour.

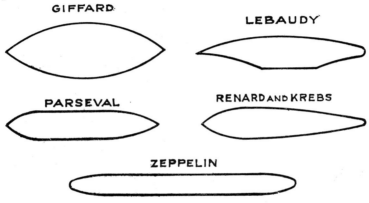

GIFFARD

LEBAUDY

PARSEVAL

RENARD AND KREBS

ZEPPELIN

Fig. 265. SHAPES OF AIRSHIP ENVELOPES.

Modern airships were developed before the war principally in Germany and France, and are of three types—rigid, non-rigid, and semi-rigid. The first of these is exemplified by the famous airships constructed by Count Zeppelin. The bag consisted of a framework of aluminium, or one of its alloys, divided by partitions into separate cells, and covered with fabric. Each cell contained a gasbag filled with hydrogen, which gave the necessary buoyancy. Underneath the envelope was a lattice-work keel, triangular in section, and covered with cloth, running nearly the whole length. Two boat-shaped cars were placed in gaps in this keel, and between them was a weight running on rails, by which the fore and aft balance could be adjusted. There were four three-bladed propellers 10 feet in diameter, driven by two motors, one in each car, of 110 horse-power.

At the rear end of the envelope were two planes set out at right angles to its surface, and making an angle of $22\frac{1}{2}°$ with one another. These are called stabilizing planes, and their use is an interesting example of the value of mathematical investigation. It has been proved by Captain Renard, and more fully by Lieutenant Crocco, that as the velocity of an airship increases it becomes unstable, and that this critical velocity is much higher when stabilizing planes are used. Ascent or descent or vertical steering can be affected by means of elevating planes fixed fore and aft, and the horizontal direction is controlled by a triple rudder at the rear.

This huge vessel (Fig. 266) was as large as an Atlantic liner, being 550 feet long. It weighs nearly 10 tons, was capable of carrying nearly 6 tons of fuel, water, crew, passengers, and cargo, and attained a speed of 30 miles an hour. It maintained a regular passenger service, and in seven months made 183 journeys and carried nearly 4,000 persons. In the later forms the envelope was about 525 feet long, and 54 feet in diameter. There were three cars, and eight engines, giving altogether 820 horse-power. A speed of 75 miles per hour has been attained, and the airship could remain aloft for four days and nights.

The French showed a preference for the non-rigid type in which the shape is maintained by the internal pressure of the gas ; and the semi-rigid, in which buckling is prevented by a long girder or keel, which may be inside or outside the envelope. Such an envelope is divided into compartments, and the hydrogen is contained in a ballonnet or little balloon in each. In some cases the gas is contained in the compartment and the ballonnet contains air. One of these methods is necessary because gas is lost if the airship rises to a considerable altitude, and the envelope would be flabby on returning to the original pressure. Consequently, air was forced into the ballonnet or compartment to maintain the shape. The blowers originally used for this purpose have been replaced by a metal scoop placed so as to collect air from the slip-stream of the propellers.

Among the ships of the former type was the Clement-Bayard which was acquired by the British Government and, after a short and chequered career, came to grief by colliding with a house ; and the Lebaudy airship purchased for this country by readers of the *Morning Post*. The envelope of the latter was 340 ft. long, and 39 ft. in diameter, and had a capacity of nearly 350,000 ft.

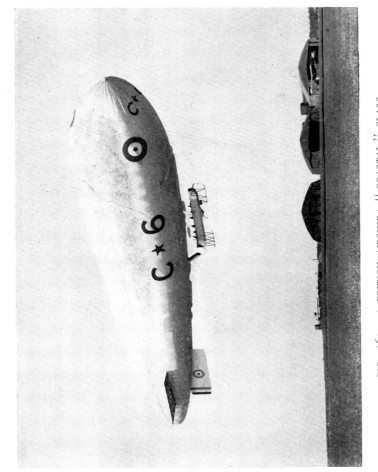

FIG. 269.—A BRITISH AIRSHIP : " COASTAL " CLASS.

To face page 366.

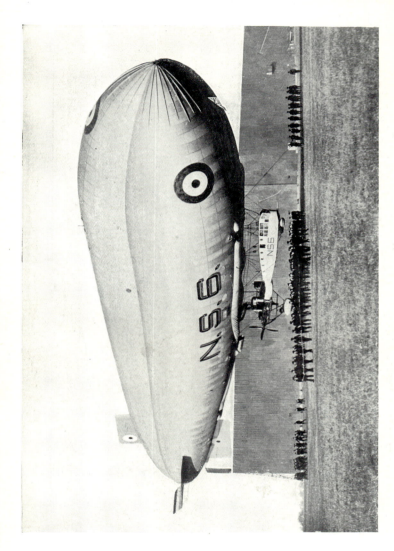

FIG. 270.—A BRITISH AIRSHIP: "NORTH SEA CLASS."

Underneath was a long frame which served as a stabilizer, and the shape was maintained by three internal ballonnets. There were two propellers driven by motors of 135 horse-power.

Most of the experimental work in Great Britain before the war was done by Mr. Willows. His first vessel, of only 12,500 cubic feet capacity, was completed in 1905 ; the third, which was the second ship rebuilt, flew across the Channel ; the fourth, in 1912, was a semi-rigid, with a diameter of 20 ft., and a capacity of 24,000 cubic feet was acquired by the British Army. Meantime, the Balloon Section of the Army began experiments in 1902, and resumed them after an interruption in 1907, when the *Nulli Secundus* was built. In September of that year she flew over London, remaining in the air for 3 hours 25 minutes, a longer time than any other airship up to that time. The next year she was rebuilt without success, but between 1909 and 1912 the *Alpha, Beta, Delta,* and *Eta* were built, and then handed over to the Navy because the Army Airship service was disbanded.

When war broke out, these vessels, together with the deflated envelope of Willow's No. 3, were all that Great Britain possessed against Germany's Zeppelins ; though six non-rigids and one rigid were under construction. Three of the non-rigids were of the Parseval type and the one that was delivered before August, 1914, did yeoman service. The envelope was of streamline shape with a pointed tail and a capacity of 300,000 cubic feet. The suspension was so attached and arranged that the vertical load fell mainly on the top of the envelope, the pull from the ends fell mainly on the lower half. The open car was of duralumin (an aluminium alloy about 5 times as strong as that metal, discovered in 1909) and contained two 170 horse-power Maybach engines. It carried two officers and a crew of seven.

From February, 1915, Great Britain devoted herself mainly to the development of non-rigid airships for scouting and coastal patrol, and by 1918 had a fleet of several hundred surpassing those of any country in the world. The first of the S.S. (Submarine Scout) class (Fig. 267) was made in a fortnight by slinging an aeroplane fuselage to the envelope of the Willows' airship which was lying deflated in a shed at Farnborough. Beginning with a capacity of 20,000 cubic feet, they were increased rapidly to 60,000 or 70,000 cubic feet. They had engines of 75 horse-power, and a wireless outfit with a range of 50 or 60 miles. The armament comprised a Lewis gun and bombs, the crew consisted of two men, and on one occasion a ship patrolled for more than 18 hours.

This type was followed by the S.S. Z class (Fig. 268), which had a similar capacity and engine-power but a more comfortable car which carried three men—gunner and signaller, engineer, and pilot. It was built rather like a boat and with 3-ply wood for " planking ". They had a speed of 45 miles an hour, and though 17 hours was the recognized limit, flights of 25 to 50 hours were accomplished.

For more extended patrols, the " Coastal " (Fig. 269) and " C Star " class were designed. The envelope was 170,000–210,000 cubic feet capacity, and was of the Astra-Torres shape, pointed at each end and having a cross-section with three lobes.[1] This had been designed by a Spaniard named Torres and manufactured by the French Astra Company ; it enables the car to be slung closer to the envelope than the circular cross-section. The car was covered and provided accommodation for a crew of five men. The engines were of 100 and 220 or 260 horse-power, and the petrol tanks were placed inside the envelope in the " C Star " and in various positions on the Coastal. One Coastal airship flew more than 66,000 miles in the 2 years and 75 days she was in service.

The N.S. (North Sea) class of airship (Fig. 270) was built from 1916 onwards in order to enable long patrols to be undertaken. The envelope had a capacity of 360,000 cubic feet and contained six ballonnets. The car which had a length of 35 ft. and a breadth of 6 ft. was covered in, and provided accommodation for a crew of ten. This enabled journeys of more than 24 hours' duration to be made. There were two screws driven by engines of 240 or 250 horse-power each. A gun is mounted on the top of the envelope and other guns and bombs are mounted on the car. These ships have regularly made journeys of 30 hours' duration, and in some cases have remained in the air for 49, 61, and 101 hours—more than 4 days and nights.

Though a rigid airship had been built in this country between 1909 and 1911, it had been a failure, and it was 1916 before the next one—R 9—was ready. A number of others, called the R 23 and R 26 class, were constructed, but owing to the long experience which the Germans had in designing this kind of vessel they were inferior to the Zeppelins which made unwelcome

[1] The cross-section is an equilateral triangle with a semicircle on each side.

visits to our shores. It was not until the autumn of 1916 that British engineers had an opportunity of examining the latest results of German science and German ingenuity ; but the L 33 which was damaged by gunfire and descended in Essex was a super-Zeppelin, and though the crew set her on fire it was easy to examine the metal portions in detail. From the information thus obtained, the R 34 and the R 36 were built. The length of these vessels is 643 ft., diameter 79 ft., and extreme height 92 ft. The envelope contains about 2,000,000 cubic feet of gas. Inside this runs a hollow triangular keel which contains sleeping accommodation for the crew, water ballast, petrol tanks, and other paraphernalia. There are five cars—one navigating and control car forward, with a wireless cabin attached, a car just behind this containing an engine, two wing cars amidships containing the engines, and a large car aft which contains two more engines. The five engines are all 250 horse-power Sunbeams.

It was the R 34 (Fig. 271) which made the famous flight to America and back in 1919. Leaving East Fortune at 2.38 a.m. on 2nd July, she arrived at Mineola, New York, at 9.55 a.m. on 6th July, after a journey which was rendered difficult by fog and electrical storms off Newfoundland. She had travelled 3,130 miles in just over 108 hours. The return journey was accomplished between 5.57 a.m. on 10th July and reached Pulham in Norfolk at 7.57 a.m. on 13th July, having been in the air 75 hours, and with one engine out of action for the last two-thirds of the time.

In spite of their frailty, airships have shown a wonderful power of withstanding high winds, and most of the accidents have occurred on landing. They are awkward things to get into a shed, and experiments have been tried of mooring them by the nose to a tall mast in the open so that they can swing with the wind. Though less speedy than aeroplanes, they have a longer range, owing to their buoyancy, and they can carry much heavier loads. Confidence in their future suffered a rude shock by the deplorable accident to the R 38 over the Humber in 1920. The loss of life was mainly due to fire and an explosion, and of the former there will always be some danger, since petrol is the source of power. The danger of both fire and explosion would be reduced as the non-inflammable gas helium becomes cheap enough to be used in place of hydrogen for the envelope.

So much for the navigation of the air. The time will arrive

B b

when on any clear, still day a glance overhead will reveal aircraft of many sizes and varied type travelling in straight lines from origin to destination with a velocity never less, and frequently greater, than that which is possible on land to-day.

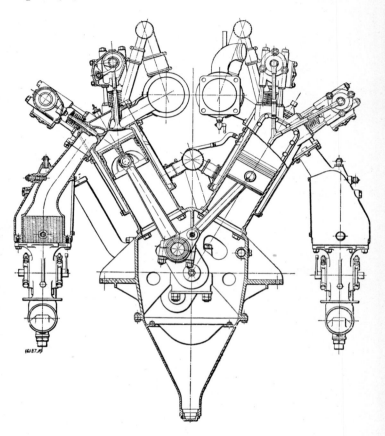

Fig. 271A.   Section of the Sunbeam-Coatalen "Maori" Engine installed in the Airship R 34.

Their paths will generally be at a great height, in order to avoid the gusts and eddies that occur near the ground, and they will float serenely through the calm airs of high altitudes.

But as the boy of the past was able to recognize, by their size and rig, the ships which traded regularly with the port in which he lived, and as the boy of the present recognizes the steamship by the colour of its funnel, so the boy of the future, in land as well as on the coast, will take a delight in identifying the speedy vessels which flit across the azure dome above him.

CHAPTER XVIII

WIRELESS TELEGRAPHY AND TELEPHONY

THOUGH one or more means of transmitting messages by electricity have been known now for eighty years, the mechanisms by which they are accomplished are understood only by those who take a general interest in physical science, and the few to whom electrical communication is a profession. So far as theory and details of working are concerned, there are a good many people still in the same shadowy frame of mind as the old Aberdeen postmaster, of whom the well-known story is told. When asked to explain the working of his instrument he said, " Look at that sheep-dog. Suppose we hold his hind-quarters here and stretch him out until his head reaches Glasgow. Then if we tread on his tail here he will bark in Glasgow." As it is not convenient to stretch a dog, we stretch a wire, and that serves the purpose.

As the name implies, " stretching a wire " is unnecessary in Wireless Telegraphy, though in order to understand the finer points of theory one needs to stretch the imagination a little. That, however, is not so much because there is any inherent obscurity or difficulty in the underlying principles, as because the mechanism of all electrical effects is more or less intangible. Electricity and magnetism operate across apparently empty space, and the links which connect cause and effect have to be pictured and supplied by imagination.

Before Marconi arrived, Sir (then plain Mr.) Oliver Lodge and Sir William Preece both succeeded, independently, in transmitting messages between two stations quite unconnected by wires ; but they employed the induction currents discovered by Michael Faraday in 1832. As long ago as 1888 Professor Heinrich Hertz had succeeded in producing and examining the properties of electric waves, but their interest for investigators lay rather in their similarity to waves of light than to their use as a means of communication. The idea first occurred to a young

Italian, Guglielmo Marconi, who came to England and laid his plans before Sir William Preece, from whom he received no little encouragement and assistance, and to whom some credit for the earlier successes is due.

Soon after Marconi applied for his British Patent in 1896, signals were sent across Salisbury Plain over $1\frac{3}{4}$ miles. Next year the distance was increased to 4 miles in March, 8 miles in May, 10 miles in July, $14\frac{1}{2}$ miles in November, and 18 miles in December. During the naval manœuvres in July, 1899, messages were exchanged between three vessels up to 85 miles apart, and in August signals were transmitted across the Channel. By 1901 the distance at which signals were possible had risen to 1,800 miles. The two stations were St. John's, Newfoundland, and Poldhu in Cornwall. The vast stretch of space between England and America had been bridged, and a new link was forged between the mother country and her sons across the sea—a fitting and auspicious event at the dawn of the twentieth century.

Before proceeding to examine the methods by which in such an incredibly short time it has been possible to link up every country in the world by " wireless " it will be profitable to consider what wave motion is, and to review the manner in which electric waves can be produced.

WAVE MOTION

When a stone is dropped into still water little ripples spread over the surface in ever-expanding circles, and communicate to any small floating object in the vicinity a vibrating motion about its position of rest. The water itself does not move with the ripple, and there is no appreciable tendency for the floating object to be translated in any direction—not even in that taken by the waves. As the stone reaches the surface it first makes a depression, then forces the water out of the way and breaks through. After it has disappeared, the water returns inward and upward to its former position with a swing, and even becomes heaped up where it was originally depressed. This swing is repeated a number of times, but to a gradually decreasing extent, and the ripples become smaller and smaller until they cease to be formed at all. Further, as each ripple spreads out from the centre of disturbance it gets fainter and fainter until it fades away.

This method of producing ripples is accompanied by a splash, and a good deal of the energy of the falling stone which might go to form waves is thus wasted. A more effective method is to float a small block of wood—a wooden ball, for example— and then to tap it slightly with a hammer. There is here no splash, and nearly the whole energy of the vibrating ball is utilized in forming waves.

The waves produced by both these methods are in short trains, each consisting of a few vibrations which soon die away. They can, however, be made continuous if the ball is lightly tapped every time it rises to its highest point. Each ripple then may be as large as, and may even become more powerful than, its predecessor, and a persistent stream of uniform ripples will extend over the surface of the water.

These waves consist of alternate crests and depressions or troughs, and the distance from crest to crest or trough to trough is called one wave-length. It represents the distance the wave travels while the object which caused it makes one complete vibration. If the object makes 10 vibrations a second, then 10 waves will be produced per second, and the first wave will have travelled 10 times the wave-length in one second. Or if N is the number of vibrations per second, L is the wave-length, and V is the velocity of the wave, then

$$V = N \times L$$

The experiments described show waves along the surface of the water only ; but if the vibrating body were immersed in a block of jelly the waves would spread out in all directions and not merely in one plane. The reader who is familiar with the way in which sound is propagated will have no difficulty in realizing how a disturbance can cause waves to extend outwards, not in circles, but in spheres or other solid shapes depending on the shape of the vibrating body. The medium, however, in which electric waves are produced is not water, nor air, nor any kind of matter as we know it. They can be produced in, and will travel through, a vessel which has been deprived of its air by the most powerful and effective pumps yet constructed. But inasmuch as they are modified considerably by the matter— air, water, earth—through which they pass, it cannot be said that matter is not concerned in their transmission.

THE PRODUCTION AND TRANSMISSION OF ELECTRIC
WAVES

An electric wave requires for its source an electrical vibration, and this was obtained by Marconi in the following way.   In Fig. 272, a battery or dynamo supplies current to the primary wire of an induction coil or transformer.[1]   The secondary coil of the transformer has one end connected to the aerial wire (a long wire suspended from a tall mast), which terminates in a small knob, and the other end connected to a similar knob below.   When the operator closes the switch the current flows round the primary

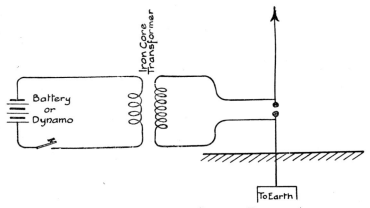

Fig. 272.   TRANSFORMER, SPARK GAP, AND EARTHED AERIAL.

circuit, and tends to include a similar current in that connected to the aerial.   Electricity may be regarded as running into the aerial and the earth, until the former can hold no more, when a discharge takes place between the knobs and the current flows down to earth.   When the switch is opened the stoppage of the current in the primary causes a similar momentary current in the secondary, but in the opposite direction.

Under proper conditions the spark discharge is oscillatory, the electricity rushing backwards and forwards from one end of the wire to the other many times a second.   Each spark discharge, therefore, is accompanied by a short train of waves produced by the to-and-fro motion of the current in the wire.

[1] For an explanation of a transformer see Chapter V.

This may be illustrated by considering a rope, one end of which is held in the hand and the other attached to a point so far away that the rope is fairly straight. If now the end held in the hand be moved up and down smartly a few times a wave will travel along the rope. The rope then represents a direction in which a wave can be sent, while the movement up and down of the hand represents the upward and downward flow of the current in the aerial wire. The *earthed* aerial represents the essential features of Marconi's original invention.

If the switch is replaced by a Morse tapper—a lever for making and breaking contact, and the coil is worked continuously by a trembler such as is used on an ordinary induction coil, then the emission of waves can be broken up into " dots and dashes " and

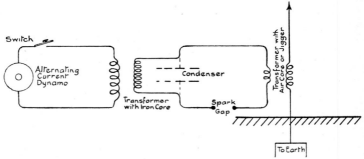

Fig. 273. Transmitting Aerial with Coupled Circuits.

signals sent in the Morse Code. For though each spark only produces a short train of waves, yet with 200 or more sparks per second and fifty or more waves per spark, there will always be some waves radiating from the aerial. That is to say, the duration of contact even for a Morse dot is so long that it cannot be completed in the interval between the death of one train of waves and the birth of the next.

Unfortunately, this delightfully simple arrangement is similar to the splash method of sending out water waves, and is not very effective over long distances. A large amount of energy is wasted in the spark, and a series of violent impulses or splashes is obtained instead of a persistent, penetrating wave motion such as is desired. It was accordingly replaced by the apparatus shown diagrammatically in Fig. 273. The alternating current is supplied by a dynamo to a transformer, which increases the electromotive force to a considerable extent. The primary circuit may therefore be regarded merely as providing a suitable supply of electricity. The secondary circuit, in which for the moment the dotted lines showing the condenser may be disregarded, consists of two coils and a spark gap. The left-hand coil belongs to the transformer, and every time the current given by the alternator reverses its direction a current in the opposite direction is induced in this coil. The flow of this current produces a spark across the gap and, flowing through the right-hand coil, induces a current in the aerial coil which is wound on the same axis.

The condenser, shown in dotted lines, consists of a series of metal plates separated by air or paraffin. The even numbers are connected with one wire—say the lower in the diagram, and the odd numbers to the other. They are thus equivalent to one pair of plates of many times the area. Their purpose is to absorb electricity until sufficient has accumulated to make a powerful spark across the gap ; in other words, the condenser increases the *capacity* of the circuit.

One other important property of these circuits and the aerial must be mentioned. They all have the same natural frequency. When an electrical discharge completes a circuit as at the spark gap in Fig. 273, there is a natural period of oscillation which depends on the form and dimensions of the circuit and the size of the condenser, and this period of oscillation can be varied by altering any one of these factors. To illustrate this, suppose the U tube shown on page 62 to be filled with water, and the level in one limb to be first depressed by blowing into the end of the tube and then released, the water will swing backwards and forwards, occupying a time for each oscillation which depends on the form and dimension of the tube and the quantity of water in it. To go back to the " coupled circuits " in Fig. 273, the dynamo will produce alternating current of a certain definite period, and the circuits and the aerial are arranged to have the same natural period. They are then said to be in " syntony " or to be in tune with one another. The period of vibration of any circuit is usually altered by varying the effective size of the condenser—cutting part of it out of action—and the operation is called " tuning ". The methods of securing syntony have been worked out mainly by Sir Oliver Lodge, and the Lodge-Muirhead patents are owned by the Marconi Company.

The apparatus has been improved by substituting for the spark gap a wheel with a number of studs on the rim, see Fig. 274, which pass very near two fixed studs attached to the ends of the secondary circuit when the wheel rotates. The action is much more regular than with the ordinary fixed gap.

The Telefunken system is very similar to that of Marconi, but the spark is quenched by spreading it over a number of small gaps of about one-hundredth of an inch each. This converts

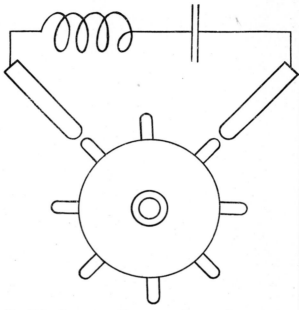

Fig. 274. DIAGRAM TO ILLUSTRATE ROTARY SPARK GAP.

the succession of impulses radiating from the aerial into a continuous wave having a period corresponding to the natural period of the aerial. In the Goldschmidt system, again, an alternating-current dynamo having a very high frequency is connected directly to the aerial circuit, which is tuned to correspond with it. A third system invented by Poulsen employs a special form of arc lamp to produce the vibrations. The poles are copper and carbon. They are placed in an atmosphere of methylated spirit or hydro-carbon vapour and subjected to the action of a powerful magnet. Under these conditions a circuit having its ends connected to the poles of the lamp has an oscillatory current corresponding to its natural period induced in it. In these three types of apparatus continuous waves are produced. Continuous waves are important because they are essential for wireless telephony.

We may conclude the remarks on the method of producing trains of electric waves with some account of their period and length. The number of sparks per second may vary from 25 to 1,000, but in large stations it is not usually less than 200. Each spark gives out a train of waves which may contain 50 or more ripples. This is just as though in the second experiment with water on page 374 a tap would produce 50 successive waves, and when these had been produced another tap was given. Now with 200 sparks per second and 50 waves in a train there would be 10,000 waves per second. Similarly, with 1,000 sparks per second there would be 50,000 separate waves in the same time. These ripples would flow outwardly at the speed of light— 186,000 miles per second, so that the waves would vary from say 30,000 feet to 6,000 feet in length. Such figures are difficult to comprehend. To split time into periods of $\frac{1}{50,000}$ of a second and to base practical calculations on it is a monument to man's faith in arithmetic. To imagine waves with crests even 6,000 feet apart is to conceive of something which in its immensity puts the ocean to shame. A slight movement of the hand will convert the energy of the dynamo into silent, unseen undulations, which can be detected by an operator over 2,000 miles away almost before the sender realizes what he has done. Such distances and times are outside the range of comprehension by the ordinary senses with which man is provided, but they are measured with certainty and accuracy by the instruments which he has invented for his use ; and his grosser knowledge enables him to manipulate with confidence the marvellous mechanism which is now being harnessed to his service.

THE DETECTION OF ELECTRIC WAVES

Sending signals across space is not of much value unless means can be devised of detecting and interpreting them at their destination, so we now proceed to examine the instruments at the receiving station. Marconi's first detector was a device due to Professor Branly, who discovered that metal filings in a small heap rested so lightly on one another that they offered considerable resistance to the passage of electricity, and a weak current was unable to flow through them. When, however, an electric wave fell upon them they became conducting, but shaking them up rendered them non-conducting again. On the supposition that the wave caused the particles to cohere and make continuous metallic contact, the device was called a coherer. In its original form it consisted of a heap of filings between the ends of two metal rods enclosed in a glass tube. The rods and filings formed part of a circuit containing also a battery and electric bell as in Fig. 275. When the wave fell on the coherer the current passed and the bell rang. Unless means were taken to prevent it, the bell continued to ring so long as the filings

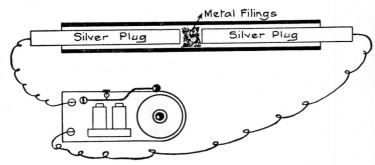

Fig. 275.   BRANLY COHERER AND BELL—THE LATTER IS DRAWN TO A SMALLER SCALE.

allowed the current to pass, so that even if the waves stopped the observer would not know it. To obviate this the hammer of the bell was so arranged as to strike the tube containing the filings, which were thus de-cohered and prepared to receive another signal. At the same time the gong of the bell was removed and the signals received through a telephone, Morse tapper, or any of the usual telegraphic receiving instruments. The use of the coherer, then, was to put into operation a local battery and telephonic or telegraphic receiving set, and it did this in accordance with the movements of a Morse sounder or tapper at the sending station.

The coherer was replaced in 1901 by Marconi's magnetic detector, an ingenious contrivance whereby the extremely feeble currents produced in an aerial at the receiving station are used to operate a telephone. It consists of a soft-iron band passed round two pulleys which keep it moving, as in Fig. 276. Over the band are the poles of two horseshoe magnets, and the band passes through a tube upon which is wound a fine coil of wire connected one end to the aerial and the other to earth. Over this coil is wound another, the ends of which are connected with the telephone receivers which the operator wears over his ears. As each train of waves reaches the aerial, it produces a click in the telephone, and with a high frequency of sparking at the sending station there will be a large number of clicks per second.[1] This causes a high-pitched musical note, and as the

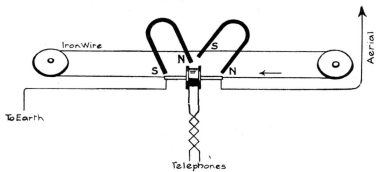

Fig. 276. DIAGRAM OF MAGNETIC DETECTOR.

radiation at the sending station is being broken up into long and short periods the telephone will reproduce these exactly in the operator's ear.

The magnetic detector is a really beautiful instrument, and in its simplicity and reliability it had much to do with the success of the Marconi installations in the early years of the new century. Since 1906, however, a host of new devices have appeared. Most of them have an advantage over the magnetic detector in that they store up the effect of several successive trains of waves, and then give it out as a more vigorous discharge to the telephone. The most modern types consist of a crystal or crystalline substance placed in the circuit through which small

[1] As the soft-iron wire moves under the magnet poles its magnetic polarity is changed. This change occurs very much more suddenly when electric waves are falling upon the wire. The slow change of magnetic polarity produces no audible effect in the telephone, whereas the sudden change produces a click.

currents induced by that in the receiving aerial will pass.   Silicon
or carborundum is used for the purpose, and in some cases
two substances, such as a copper point resting on iron pyrites.
A diagram of a typical arrangement is shown in Fig. 277.   The
aerial on the left hand is connected to earth through a coil which
is wound on the same axis as another coil in the telephone circuit.
The telephone circuit contains a battery and the detector,
and there are two condensers, one on each side of the detector.
The circuits are tuned so as to be in syntony with the aerial.
When a train of waves of the right frequency strikes the aerial,
currents are induced in the first portion of the circuit and charge
up the first condenser.   This tends to overflow through the
detector, which allows the second condenser to become charged,

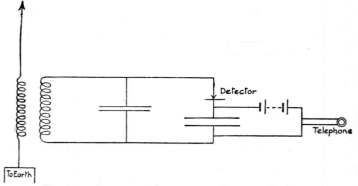

Fig. 277.   RECEIVING AERIAL WITH COUPLED CIRCUITS.

but does not permit the charge to pass out again.   It acts as
a trap, permitting the current to flow one way but not another.
The second condenser therefore discharges steadily through the
telephone, but at such a rate that the detector is enabled to
pass to it several charges before the first one has escaped.
Detectors of this kind are called integrating detectors, because
they sum up or add together a number of small impulses.
     The reader will have noticed that in the descriptions of the
sending and receiving apparatus, some emphasis has been laid
on the importance of the circuits at each end being tuned to the
aerial.   Now this problem of tuning is of such importance
and interest that it will be well to give a page to its examination.
Consider a pendulum composed of a weight suspended by a cord.

If the weight is pulled to one side and then released it swings backwards and forwards for some time. So long as it is not pulled very far to one side each to-and-fro motion takes the same time for completion. Even as the swings die away until the movement is hardly noticeable, the time taken for a to-and-fro movement remains the same. In this respect the swing of a pendulum and the oscillation in an electric circuit are the same. The natural period of the pendulum depends on the length of the string, and the period of an electric circuit upon its form and dimensions.

If the weight or bob of a pendulum at rest is given a series of very light taps at intervals corresponding to the time of swing, it is soon set vibrating, and if the taps are continued its movement becomes very large. On the other hand, if the taps are not timed to take place at the proper intervals, any motion first set up is soon stopped. The effect of a series of impulses properly timed to coincide with the period of any body capable of vibrating is very striking. Professor J. A. Fleming in his book on *Waves in Water, Air, and Ether* cites a case which illustrates this vividly. He was in a shipyard and noticed a long wooden mast lying horizontally on two supports near the ends. Making an estimate of the natural period of vibration of such a piece of timber, he pressed his finger on the middle of the mast and repeated this at what he conceived to be the right interval. In a short time the huge mast was vibrating so violently as to threaten to jump off its supports.

It is obvious from these facts that if the aerial has the same period of electrical vibration as one of the circuits in the sending or receiving apparatus, then a current oscillating in one of the circuits will set up powerful electrical oscillations in the aerial and *vice versa*. In every other department of engineering, resonance, as we have said before, has to be provided for at all costs on account of the enormous forces that are called into play. A regiment of soldiers marching over a suspension or other flexible form of bridge are thrown out of step lest their measured tread shall set up such vibrations in the structure as to destroy it. The alternating current at a central generating station may happen to coincide in period with the natural time of vibration of the circuit with disastrous results, and electrical engineers take great precautions against such an occurrence. But this phenomenon, so dangerous in nearly every other field, has been the salvation of long-distance wireless telegraphy. At present we are only concerned with part of the story. It is sufficient for us at this stage to be able to realize how at each station the whole of the apparatus for sending or receiving is adjusted to vibrate in sympathetic harmony, each part reinforcing and being reinforced by the other part, strengthening the feeble, softening down differences, and behaving in a manner which reflects the human genius that gave it birth.

Let us now carry the matter a step further, using this time tuning forks as an illustration. A tuning fork is made of steel or other highly elastic material, and the prongs are of such a size and stiffness that the same fork always produces the same note. The sound produced by a fork is not very loud, but it may be increased if the fork is fixed to a sounding-box. This is a wooden box, open at one end, and of such dimensions that the natural period of the column of air is the same as that of the fork. When the fork vibrates it sets the air in the box vibrating also, and the sound is louder. Suppose two forks have the same pitch, that is, they make the same number of vibrations per second and give the same note, and let one of them be set in motion. In a very short time it will be found that the other one is sounding, and if the first one be stopped by placing the finger upon it, the second one will be heard distinctly. Here the vibrations have travelled through the air, and, impinging on the second fork, have set it vibrating. If the second fork has a very different pitch it will not be affected by, nor will it affect, the other. But if the difference of pitch is small and the waves which the first fork emits are strong, there will be a " forced " vibration set up which, however, is not so powerful as the natural one.

Now two aerials behave in precisely the same way as the two tuning forks. If one is oscillating at a certain rate it is sending out waves of a corresponding length, because, as was shown on page 374, the wave-length is equal to 186,000 miles divided by the number of vibrations per second. These waves can be picked up most easily by an aerial having the same natural period of vibration and capable itself of radiating waves of the same length. Since the natural period of a circuit can easily be altered by altering the capacity, any circuit can be tuned so as to be in syntony with another circuit. It is clear that if every message was picked up by every receiving station there would be great confusion, and on that account each station

FIG. 278.—THE MASTS AT GLACE BAY WIRELESS STATION, NOVA SCOTIA.

To face page 384.

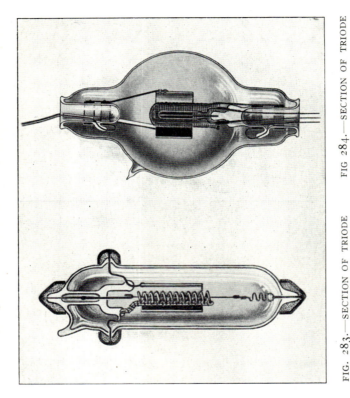

FIG. 283.—SECTION OF TRIODE VALVE-RECEIVER.

FIG 284.—SECTION OF TRIODE VALVE-TRANSMITTER.

To face page 385.

has a fixed wave-length for transmitting which is known to all the other stations. If a signal reaches a station which is not in proper tune with it, and the operator believes that he is being called, but cannot tell because the signal is not clear, he alters the capacity of his aerial until the sounds in the telephone reach their greatest distinctness.

THE GROWTH AND APPLICATIONS OF WIRELESS TELEGRAPHY

We shall now leave the theory to consider some of the applications and results, and return from time to time when further explanation is necessary. A short account has been given at the beginning of the chapter of the successive stages of progress. The first Marconi station was established at the Needles in the Isle of Wight in 1896, but it was only used for experimental purposes. The first paid Marconigram, however, was sent from there by Lord Kelvin to his friend Sir George Stokes. On Christmas Eve, 1898, a permanent apparatus was installed on the Goodwin Lightship, and twice in the succeeding year accidents were reported. Other land stations followed, and in 1900 the first move towards the conquest of the Atlantic was made by the erection of the more ambitious station at Poldhu in Cornwall. The aerials were supported by twenty masts each 210 feet high.

In the following year the Cape Cod station was commenced on the other side, but the masts both there and at Poldhu were blown down by heavy gales within a couple of months of one another, and were replaced by four wooden towers of the same height. By February, 1902, Marconi, on board the *Philadelphia*, was able to receive from Poldhu readable messages up to a distance of 1,550 miles, and Morse signals up to over 2,000 miles. Before the end of the year Transatlantic communication was an accepted fact, and arrangements had to be made to erect more powerful stations. In 1907 the new stations at Clifden Bay, on the west coast of Ireland, and Glace Bay, in Nova Scotia (Fig. 278), were opened for public service.

The problem of sending messages over long distances was facilitated by Marconi's invention of the *directional* aerial in 1905. He found that a horizontally bent aerial would radiate waves most strongly in a direction opposite to the free end. Thus at Clifden the aerial is earthed near the power-house, and stretches for three-quarters of a mile in the opposite direction to Glace Bay, the station on the other side. The receiving aerial is similar but longer. Separate sending and receiving aerials are necessary to enable messages to be sent and received at the same time.

The older stations, such as Poldhu, and, indeed, most of the newer ones, use alternating-current dynamos. At Clifden and Glace Bay, however, a direct current is used, and this is supplied by a battery of accumulators. The design of the apparatus for these large stations has presented a new series of problems to electrical engineers. The main circuit carries a current capable of performing work at the rate of 200 or 300 horse-power, and the pressure driving this electricity round the circuit is 20,000 volts. Yet this power has to be stopped and started three times in every second, and the apparatus must work continuously night and day for weeks without a breakdown.

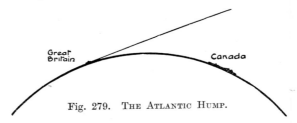

Fig. 279. THE ATLANTIC HUMP.

While communication across the Atlantic has been maintained regularly now for more than twenty years, it must not be imagined that there are no difficulties and that all the problems have been solved. In spite of a high degree of perfection in the instruments for producing electrical waves and in those used to detect them, the wave meets with many adventures on its way, and there is some uncertainty as to how it really gets there. One of the problems which, while it had been surmounted practically, evaded theoretical explanation, is the particular path pursued by the wave between the stations. When Marconi first made the attempt to put England and the American continent into communication, there were no scientific facts which pointed to success, but there were some which indicated the impossibility of surmounting the great aqueous hump of the Atlantic, 125 miles high, which lies between (Fig. 279). An electric wave is in effect a very long light wave travelling with the same velocity—186,000 miles a second—and possessing many other

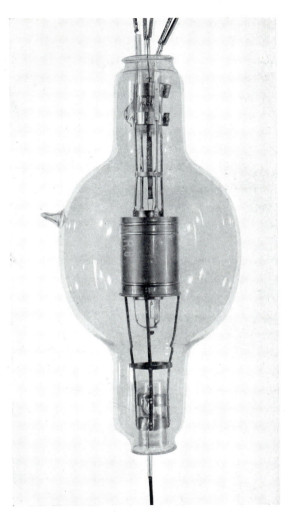

FIG. 285.—PHOTOGRAPH OF A TRIODE VALVE.

FIG. 289.—THE VALVES AT THE CARNARVON STATION.

To face page 387.

similar characteristics. Now light waves show a rooted objection to turning a corner. Save for a slight bending round the edges of objects, they pursue a straight path from origin to destination. If an electrical wave were endowed with equal rectitude, and were launched on its way to Canada from Poldhu, it would arrive there something like a thousand miles above the land. Signals hovering in the heavens above and having no tangible connection with the earth below would be rather useless ; from that height they could not even be collected by a kite. Fortunately, however, the waves come to earth themselves, and there is some evidence to show that they travel all the way through the air. Perhaps a more striking illustration of what the curvature of the earth involves is to be found in the fact that when receiving signals at Buenos Ayres from Clifden, a distance of 6,700 English miles, Marconi was detecting waves which had been deflected from their original direction by 97° !

Reference was made in a previous paragraph to the adventurous character of the waves' journey. Their progress is hampered in two ways, neither of which is quite understood. In the first place there is a falling off in their strength greater than any which was forecasted by electrical knowledge. Again, there is a marked difference between the distances at which signals can be heard by day and by night, and the variation is greatest at sunrise and sunset. Long and short waves do not behave in quite the same way as regards penetrating power, and they differ according to whether they are passing over land or water. Moreover, signals are more easily sent in a north and south, than in an east and west direction. Into a minute tabulation of these variations it is not necessary to enter. They are generally fairly constant, and in any case an operator has to send his message and risk getting it through.

There are, however, a number of other kinds of interference which arise from electrical disturbances in the earth's atmosphere. A flash of lightning is liable to give rise to a wave of enormous power which will set half the aerials on the earth vibrating in spite of the differences of pitch to which they are tuned. Thunderstorms are at their worst in the summer in temperate latitudes, but they occur to some extent all the year round, and those in the tropics are of extreme violence. As a consequence it is frequently almost impossible to decipher earthly messages owing to the imperious signals from the clouds.

Of the various methods adopted for choking off the "atmospherics" as the disturbances are called, one is to use receiving circuits which respond only to a narrow range of oscillations very different from those produced by a lightning flash. The employment of a high-pitched musical note in the telephone is also an advantage because its extreme regularity distinguishes it from stray waves.

THE THERMIONIC VALVE

The great value of the magnetic detector lay in the fact that it was a robust, reliable instrument not much affected by "strays", and in this respect superior to the coherer and crystal types. It is now largely superseded by a contrivance called a thermionic valve, invented by Professor J. A. Fleming, and improved by the inventor himself and other workers in the same field. Professor Fleming was led to his discovery by the fact that increasing deafness rendered it difficult for him to hear signals or messages on the telephone, and he was in search of some instrument which would act as a "relay", bringing into operation a local current which would enable the messages to be recorded, just as submarine telegraph messages are recorded by the mirror galvanometer and other devices.

In 1883 he had observed that a carbon lamp, in which the filament was weak in one place, so that its temperature at that point was higher, did not blacken evenly. There was a strip on the glass fairly free from deposit, and this was in such a position that it appeared as though the particles of carbon were shot off in straight lines, and had been intercepted by the other half of the horseshoe. About the same time Edison was making experiments with lamps of high efficiency. He fixed a metal plate between the limbs of the horseshoe filament, and discovered that when the filament was hot a current flowed between the plate and the positive limb, but not between the plate and the negative limb (Fig. 280). Investigating the Edison effect in 1889, Fleming found that a current would pass freely in one direction only between the plate and the limbs of the filament. The lamp would, therefore, act as a rectifier, and if an alternating current were applied between the plate and the film, the transient currents in one direction would be suppressed. A pulsating current would pass, but always in the same direction.

It had been shown by Becquerel in 1853 that air at 1,500° C. was a conductor, and by Guthrie in 1873 that a red-hot ball of iron lost a charge of negative electricity rapidly but not a charge of positive electricity. From 1880 onwards a great many investigations by Elster and Geitel combined with the later work of Professor (now Sir) J. J. Thomson had established the theory that at high temperatures matter ejected electrons or ions, about which more will be said in Chapter XX, and that they were primarily responsible for the conductivity of electricity

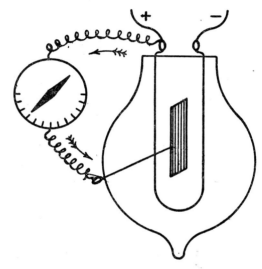

Fig. 280. THE EDISON LAMP.

through gases. Fleming was able, therefore, to explain the phenomena of the Edison lamp as an example of thermionic radiation.

It will be remembered that the crystal detector was in essence a rectifier, allowing current to pass only in one direction. Similarly, the thermionic valve will also act as a detector of electric waves. The ways in which it can be used are manifold, but the simplest, which was used by Fleming in 1904, is shown in Fig. 281. Here is shown the aerial, jigger, tuning condenser, valve, battery, and a resistance, the purpose of which is to regulate the temperature of the filament and hence the conductivity of the space between the filament and the plate. The explanation is as follows :—

When an electric wave strikes the aerial, there is a surge of electricity backwards and forwards in the coil S and round the circuit of which S forms a part. The upper plate of the condenser is, therefore, electrified alternately positively and negatively. When it is positive, there is at once a flow of

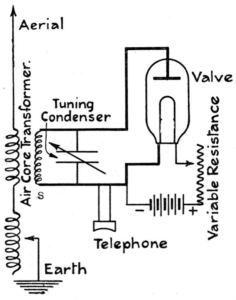

Fig. 281. DIAGRAM OF CONNECTIONS FOR THERMIONIC VALVE
IN WIRELESS TELEGRAPHY.

electrons or negative ions inside the valve from the filament to the plate which neutralize the positive electricity in the latter; and there is a current through the telephone. As the electricity induced by the wave surges back the top plate becomes negative, there is now no flow of negative ions, and no current passes through the telephone. At each semi-oscillation, there is a gush of electricity, and as the waves follow one another very quickly and at equal intervals the gushes form a musical note in the telephone, provided the gushes occur between the limits of,

FIG. 292A.—MODERN FORM OF AERIAL.

To face page 390.

FIG. 293.—PANEL OF 3-KILOWATT SET FOR WIRELESS TELEPHONY.

To face page 391.

say, 300 and 1,000 per second. It has been shown by Lord Rayleigh that the ear is most sensitive to interruptions in a telephone circuit when the note emitted corresponds to about 700 vibrations a second. As even the shortest depression of a Morse key or tapper will occupy many seven-hundredths of a second it will be clear that the dots and dashes of the Morse code will be easily distinguished in the telephone.

Though, for the sake of simplicity, the telephone has been shown as part of the main receiving circuit, it is more usual to couple it up than to insert it. Thus, if a coil of wire be inserted in the main circuit, and another coil, the ends of which are joined up

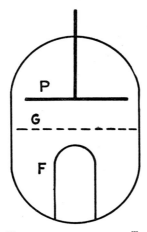

Fig. 282. DIAGRAM TO ILLUSTRATE TRIODE VALVE.

to the telephone, is wound over it, any changes in the main circuit will be reproduced in the telephone though there is no direct connection between them. It is this property of the transformer which renders wireless apparatus at once so elastic and so complicated ; and if the reader cares to examine a series of wiring arrangements of gradually increasing complexity he will find that this complexity is largely due to the separation and coupling of circuits.

Thermionic valves are made in a variety of forms, though they do not differ in principle. Generally, they contain a plate parallel with a filament or a metal cylinder surrounding a straight wire. An improvement was effected by Lee de Forest, who added a third electrode in the form of a grid—generally a coil of wire— between the plate and the filament (see Figs. 282, 283, 284, 285). The effect of this grid or third electrode, forming what is known as a triode valve, is extremely interesting and not difficult to follow. In the diagrammatic sketch of a triode valve, Fig. 282, F is the hot filament, G is the grid, and P is the plate. The passage of a current is simply a flow of electrons from the hot filament to the plate, and this flow is affected by an electric charge on the grid. If the grid is negative some or all of the

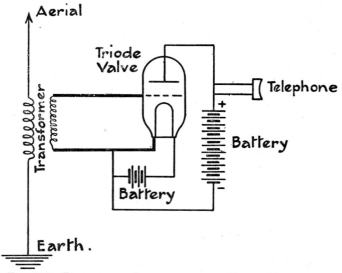

Fig. 286. DIAGRAM OF CONNECTIONS FOR TRIODE VALVE USED AS DETECTOR.

electrons travelling towards the plate will be repelled, and the current will be reduced or stopped altogether. But if the grid be positive the current will in general be strengthened.

Suppose now the triode valve is coupled up as shown in Fig. 286, in which the necessary condensers have, for the sake of simplicity, been omitted. The pulsations of electric waves in the circuit, coupled with the aerial, indicated by heavy lines, will cause rapid alternations in the grid potential, rapid alternations in the flow of electrons from hot filament to grid, and hence rapid alternations in the strength of the current in the

FIG. 294.—20-WATT SET FOR WIRELESS TELEPHONY.

To face page 392.

FIG. 295A.—MARCONI BEAM TRANSMISSIONS STATION AT DORCHESTER.

To face page 393.

telephone circuit.   If these alternations occur at the rate of, say, 500 a second, a musical note will be produced, and this musical note will be interrupted by the dots and dashes of the sending station.

So far the valve acts as a rectifier, just as the two electrode valve does.   But it has two more remarkable properties which the simpler instrument does not possess.   It will amplify the variations which are impressed upon it, and it will act as a generator of oscillations.   In the first case, it will, when the circuits are properly adjusted, amplify either the oscillations which reach it from the aerial or it will magnify the gushes of unidirectional current in the telephone circuit.   The first effect is called radio-amplification, and the second audio-amplification. Both effects are due to the fact that owing to the small capacity of the grid circuit a very small amount of energy suffices to change the potential on the grid and to produce a considerable alteration in the current which flows from the filament to the plate.   This energy comes from the battery between those two parts of the apparatus.   If radio-amplification is required, air-cored transformers are used for coupling, and if audio-amplification is required iron-cored transformers are necessary.

The amplification can be increased by arranging the valves in cascade.   The plate circuit of one valve is coupled to the grid circuit of a second, the plate circuit of a second to the grid circuit of a third, and so on.   In this manner the effects can be increased a thousand-fold, and oscillations far too faint to be detected in the older apparatus are now brought well within the range of hearing.

The use of the triode valve as an oscillator will be understood by reference to Fig. 287, from which condensers and resistances have been omitted for the sake of simplicity.   Note first the filament circuit, next the plate circuit, in heavy lines, and finally the grid circuit, and keep the two last clearly in mind.   Remember also that by means of resistances and condensers these two circuits are tuned so as to have the same period of vibration.   Then starting or increasing the current in the plate circuit will produce by induction between the coils L_1 and L_2 a momentary rush of electricity in the grid circuit *in the opposite direction*.   The grid will become negative and repel the negative electrons emitted by the hot filament.   Hence the current in the plate circuit will be reduced ; the interaction of the coils L_1 and L_2 will cause a rush of electricity making the grid positive ; the flow of electrons will be encouraged ; and the current in the plate circuit will be increased again.   But this increase will immediately result, through the coils L_1 and L_2, in a rush of electricity which renders the grid negative, and the current in the plate circuit again falls.   These changes occur over and over again at perfectly regular intervals—an oscillation in the grid circuit and a pulsating but unidirectional current in the plate circuit capable, if the pulsations are not too rapid, of operating a telephone.

Further, if the coil L_1 be coupled with an aerial of the same period, continuous waves will be radiated ; while, conversely,

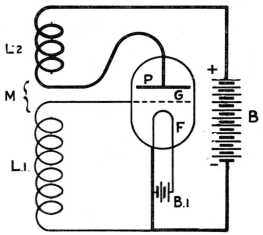

Fig. 287.   Diagram of Connections for Triode Valve used as Oscillator.   The Coils L_1 and L_2 are coupled.

if continuous waves of the same period are received by the aerial, pulsations will be set up in the plate circuit.   Hence a triode valve is an admirable receiver for continuous waves.

The high frequency alternating current dynamos used by Fessenden, and in the Telefunken and Goldschmidt systems, and the Poulsen Arc all produced continuous waves, but their progress was hampered by the absence of a good detector.   The disadvantage of the triode valve as an oscillator in wireless telegraphy is due to the relatively small sizes in which it has been constructed and the difficulties of constructing large ones.

During the war it was developed mainly for use on aeroplanes, and for wireless telephony over short distances. It is such a fragile piece of apparatus that the amount of energy which can be put into it is relatively small ; and though as we shall see in the next section large powers are not impossible, it is as a detector and amplifier that the instrument has hitherto found its greatest utility.

The most important improvement in valves, apart from capacity which will be considered later, is the " screened grid " or " shielded " valve. The ordinary triode type would not give a very high amplification over a wide range of wave-length without certain auxiliary devices. The electrode circuits had to be screened as much as possible from one another, and the capacity of the electrodes had to be compensated by the insertion of a condenser, which itself required adjustment for different wave-lengths. The number of valves which could usefully be employed in a receiving set was limited by these complications.

The Marconi type of shielded valve has four electrodes and is usually mounted in a horizontal position. The filament is surrounded by the ordinary control grid. The anode is a flat disc mounted parallel to the grid. Between the two is the shield, which consists of a flat grid of circular shape and of very fine mesh. The valve is double-ended. At one end are the external connections for the filament and control grid ; at the other end are the connections for the anode and shielding grid.

A two-stage high frequency amplifier with a shielded valve will give an average over-all magnification of about 750 over a range of wave-length in which the ratio of maximum to minimum is about 40 to 1. The shielded valve will operate with waves as short as 12 metres. The amplification on this length is small, but increases rapidly up to 50 for a wave-length of 600 metres. Thus a very high amplification may be reached in a single stage, and there is stability and ease of operation over a very wide range of wave-lengths. Compared with the triode valve, the screened grid valve represents an advance as great as the diode valve of Fleming on the magnetic detector.

The discovery of the generating property of the triode valve was made in 1913 by Lieut. Franklin and Capt. Round, of the Marconi Wireless Telegraphy Company, by E. H. Armstrong in the United States, and by Alexander Meissner, of the Wireless Telegraphy Company of Berlin, each working independently. It rapidly came into use in the heterodyne or " beat " method of transmission. If two musical notes which do not differ greatly in frequency are sounded together, a third sound is produced which has a frequency equal to their difference. Similarly if two electrical oscillations occur in the same circuit there is a reinforcement at intervals which depend upon the difference in the number of vibrations. Thus, let the receiving apparatus be tuned so that it is producing vibrations which are anything from 300 to 1,000 faster or slower per second than those which it is receiving from the aerial. Then reinforcements or beats will occur of a frequency which produces a musical note in the telephone, and this note can be cut up by the dots and dashes of the code. It is possible with this method so to arrange the apparatus that the strays, which were such a nuisance in the early days, are practically eliminated.

One vast improvement which has been made made is the degree of exhaustion of the bulbs. The older valves were, and still are, quite suitable for some purposes ; but for others a " harder " or more highly exhausted bulb is required. The conductivity is produced not merely by electrons from the hot filament, but also by others from the small quantity of gas which is present. In an individual case this may not matter, but it has been found that the only method of manufacturing them so that they shall be comparable in behaviour is to carry the exhaustion to the highest possible limit. This limit is so far beyond what could be achieved ten years ago that a short account of the method may be of interest.

Since the real difficulty is to get rid of the film of air or other gases which are persistently retained by all solids, and since this can only be accomplished by a high temperature, the bulb is heated during the operation in an electric furnace. The temperature—500° C. to 700° C.—required for the purpose is so high that the bulb might collapse if it were exposed to atmospheric pressure, so the furnace is exhausted by a pump. Most of the air is removed from the globe by a mechanical pump, and the remainder by a special form of pump, of which the Gaede is a good type, or by absorption in charcoal immersed in liquid air. The last traces of oxygen or water vapour are removed by a little magnesium attached to the anode. On heating this is vaporized, the oxygen or water vapour form magnesium oxide, which is a white, non-volatile solid, and the excess of magnesium is deposited as a silver coating on the interior of the bulb. The mirror-like appearance of the valve is, therefore, an accidental circumstance.

The Gaede pump, invented in 1913, is of extraordinary simplicity, having no valves or pistons, and consisting simply of a drum and "scraper". Some idea of the principle upon which it works will be obtained from Fig. 288. The drum is grooved and runs in a closed chamber with projections from the casing which nearly touch it. As the drum rotates the air is

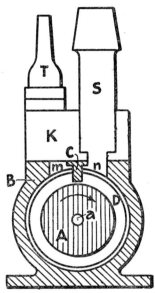

Fig. 288. DIAGRAM TO EXPLAIN ACTION OF GAEDE PUMP.

dragged round with it and a difference of pressure is formed on each side of the projection. This occurs in the first groove, and the compressed air passes into the next groove to be whirled round and still further compressed. Though the difference in a single groove may be small it is cumulative, and, combined with high temperature, a pressure as low as $\frac{1}{100,000,000,000}$ of a millimetre of mercury can be obtained. When this has been achieved, the conductivity of the valve depends almost entirely upon the production of electrons from the hot filament.

Owing to the difficulty of constructing valves capable of conveying a large quantity of energy the earlier continuous wave stations employed, and still employ, the Poulsen Arc as generator. Plant of this type is installed in the first station, at Leafield, of the projected Imperial Wireless Chain, and similar proposals are made for the stations at Nairobi in East Africa, Singapore, and Hong Kong. But elsewhere in England, Egypt, India, Australia, South Africa, and Canada, valve stations are proposed.[1] This decision has been determined largely by the successful trials in November, 1921, of the new Marconi Valve Station at Carnarvon (Fig. 289). This comprises 48 glass valves which deliver 100 kilowatts or 150 horse-power to the aerial, and has been used for a short time to transmit 150 kilowatts or 225 horse-power. Messages were sent to America and, during the most favourable hours of the day, to India and Australia. The glass valves cost about £15 each, or £720, and if used continuously would cost from £2,000 to £6,000 a year in renewals. The station is not quite so powerful as that specified for the Imperial Wireless Chain, 120 kilowatts, but it proved that thermionic valves could be used as transmitters for wireless signals over the longest distances that were likely to be required. Further, experiments by the British Admiralty at the Signal School, Portsmouth, showed that silica bulbs can be constructed which will convey a larger quantity of energy and enjoy a longer life. By 1923 a valve had been made which would transmit 1,000 kilowatts, or nearly 1,300 horse-power.

In America attention is being concentrated on the Alexanderson Alternator. " The only moving part of this machine is a steel disc mounted on a flexible steel shaft so that it can be rotated at high speed. Near its edge are cut slots or holes which are filled with non-magnetic material, and these slots as the edge of the disc revolves between the fixed field and the armature coils, change the total magnetic flow passing through the armature coils and so create in them an alternating electro-motive force." [2] This is far the most effective machine for the direct generation of continuous waves by mechanical means, and for the transmission of large quantities of energy it has obvious advantages over the fragile valve. The valve, on the other hand, is more elastic, and in the last few years the power which can be

[1] Report of Wireless Telegraph Commission, 1922. Cd. 1572.
[2] *The Times Engineering Supplement*, August, 1920.

put through them continuously has been greatly increased. The alternator, moreover, can only be used for transmission—not for reception.

PRACTICAL VALUE OF WIRELESS

In no direction was the practical value of wireless telegraphy so quickly recognized as in connection with shipping, and the history of disaster at sea, during the last twenty years is a striking tribute to the importance of the invention. Not a year has passed without wireless telegraphy bringing succour to some ship in distress, and though it has not in all cases availed to prevent loss of life, it has assuredly saved many people from certain death.

On the palatial steamers that plough the Atlantic the Marconi apparatus enables the travellers to keep in touch with their friends, to transact important business on either side of the water, and to secure a continuity of life which was formerly broken by a sea voyage. All the larger vessels now publish a daily paper on board, the news in which has been supplied by the same agencies that feed the newspapers on land. Information is flashed to meet or overtake the ship and caught up by her aerial as she pursues her way at 25 or 30 miles an hour.

In the case of cargo boats, the owners are able to communicate with the skippers at any point in their journey. They can advise the captain where to call for coal or cargo, while he on his part can get into communication with the authorities or his firm's agents at the port of call, and have every necessary or desirable preparation made for his arrival. Should an accident happen he can call for assistance, inform the owners, or relieve anxiety and suspense. At no time is he isolated from the world. The fortitude, courage, and daring of those " who go down to the sea in ships " has never been called into question, but it has, if any-thing, been emphasized by the receipt of messages from an operator at his post, to whom the bonds of duty were as bonds of steel, and who calmly operated the tapper until the water entered his cabin and brought him honourable release.

It is not enough that ship should be capable of communicating with ship, and all round the margins of the great oceans as well as inland are dotted groups of buildings with their tall masts which flash their messages across the waste of waters. These shore stations are more powerful than those on board ship. The latter vary from ½ to 10 horse-power and have a range of 50 to 500 miles. The former may employ several hundred horse-power and have a range across land and sea of 1,000 to 2,000 miles or more. As long ago as 1913 there were nearly 400 of them and more than a thousand ship installations. To-day there are many more. Each of them has definite " call-letters " and its range of wave-length is known. It is obvious that, with this vast increase in the number of stations and the relative lack of secrecy, great confusion was likely to arise. In this country the fact received early recognition, and the Government assumed the responsibility, through the Postmaster-General, of prohibiting unauthorized installations. An aerial can only be set up under licence, and private transmission is rarely allowed. The United States was much later in taking action, but after the confusion that arose from private messages in regard to the *Titanic* disaster, restrictions were imposed. International Conventions were held in 1903, 1906, and 1912, at which rules governing the working of stations have been drawn up. For commercial purposes standard wave-lengths of 300 and 600 metres were adopted, while wave-lengths of 600 and 1,600 metres were adopted for naval and military purposes. To-day the wave-length for long distance messages ranges from 10,000 to 20,000 metres, or 6 to 12 miles. The original call-letters in case of disaster at sea were C.Q.D., and this was from the first acknowledged to have preference over all others. At the International Convention of 1906 the signal was altered to S.O.S., the initial letters of " saving of souls ", though the identity is accidental, not intentional.

As a practical illustration of the improvements which have been effected during the last ten years, it may be remarked that the 1½ kilowatt (= 2·25 horse-power) set, which was the standard size for marine work, had a range of 400 miles. With the same expenditure of power and the use of valves, a ship crossing the Atlantic is to-day in touch with America right up to Liverpool bar.

Before leaving the applications to shipping, it will be interesting to examine one important invention or group of inventions which have enormously increased the value of wireless communication. It was originally impossible for a ship or a shore station to tell from what direction the signals were reaching her; and a number of attempts were made to devise an apparatus which would reveal not merely the signals but also the direction from which they came.

Marconi tried to rotate his horizontally bent aerial (see p. 385) about a vertical axis, but this was not possible on board ship. A more effective plan was proposed by S. G. Brown and improved by André Blondel. If an aerial consists of two vertical wires, half a wave-length apart, and connected by a horizontal wire as in Fig. 290 (consider one only) then a wave travelling in the direction of the horizontal wire will induce an upward current in one vertical wire and a downward current in the other. A current will flow, therefore, from one vertical to the other, along the horizontal wire. A wave reaching the aerials at right angles to their plane will tend to produce an upward or downward

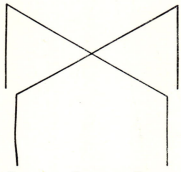

Fig. 290.   DIAGRAM TO ILLUSTRATE DIRECTIVE AERIALS.

current in both vertical wires. In this case no current will flow through the horizontal wire at all. Should the wave approach from any other direction, a current will flow in the horizontal which will vary in strength with the direction, being greater as the line of approach coincides with the plane of the aerial. In 1907 Bellini and Tosi patented the application of two of these aerials placed at right angles to each other, so that the waves would produce their maximum effect in one when they produced no effect in the other. An important part in the invention was the radiogoniometer, by means of which the effect in two aerials was compared. This consisted of two coils, one in circuit with each aerial and mounted at right angles to one another, as in Fig. 291. A third coil, called a search coil, connected with the

D d

receiving apparatus, was mounted inside the others, so that it could be rotated into a position parallel with either of them. The direction of the plane of this coil when the loudest signals were received indicated the direction but not the sense of the incoming waves. The great defect was that the apparatus could only be constructed to give reliable results with waves of short length. C. E. Prince improved the arrangement by using one *directive* and one *directional* aerial and by a device which enabled the relative strengths to be exactly compared. The form of

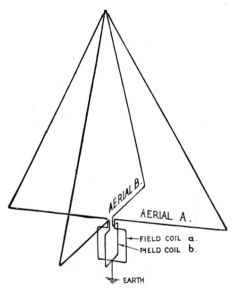

Fig. 291. MODERN FORM OF DIRECTIVE AERIAL.

aerial now used on land stations by Marconi's Wireless Telegraph Company is shown in Fig. 291. The search coil is rotated to the position of minimum effect because this can be detected with the greater accuracy. Further, the sense of the message can be recognized.

In 1922 the direction finder was the means of saving life at sea, in two precisely similar cases, when the ordinary means of wireless communication had failed. In each case a ship lost in a fog, unable to determine exactly her position, and sinking, sent out a wireless call for help, and gave the latitude and longitude in which she believed herself to be. Several ships hastened to the spot, but could not find her. As one of the vessels turned back to her original course the wireless operator noticed that the S.O.S. signal was becoming stronger. The use of the direction finder enabled the captain to steer in the direction of the call, the vessel in distress was found (in one case 90 miles away from the position given) and the men on board were saved.

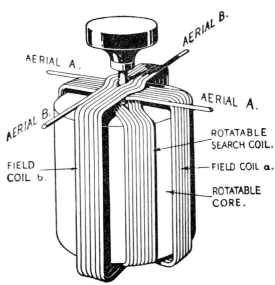

Fig. 292.   DIAGRAM TO EXPLAIN THE RADIO-GONIOMETER OR DIRECTION FINDER.

For communicating overland in civilized countries, and from land to land along the recognized routes of ocean travel, Wireless Telegraphy is in direct competition with the telegraph line and cable. But there are great stretches of country where it has a clear field. If a map of Africa be examined, it will be seen that Great Britain has immense possessions in the south which extend from the Cape to Rhodesia ; and that in Egypt and the Southern Soudan she exercises an influence from the northern coast nearly half-way down the continent. Similarly, Germany has possessions on the east and west coasts. Readers will be familiar with Cecil Rhodes' dream of a Cape to Cairo Railway cutting the Dark Continent in two halves, and will see from time to time in the newspapers an announcement of the opening of a new section. It is probably true that in most cases the telegraph precedes the railway, or, at all events, accompanies it. But it is unlikely that this will be so in Africa. The Cape telegraph line now goes far north into Rhodesia, and the Egyptian telegraphs stretch south-wards into the Soudan. Between the two is a vast area of country containing the great lakes, peopled by savage tribes, roamed by herds of elephants, and affording sustenance to the white ant. For the present, at any rate, this track is to be bridged North and South, East and West by Wireless Telegraphy, which, so long as it has a fairly civilized origin and destination, is unaffected by the savagery of the land over which its messages are sped. The trackless forests of the Amazon, where it would be well-nigh impossible to maintain a clearing, are being threaded by wireless messages worked on the Marconi, Poulsen, and Telefunken systems.

WIRELESS TELEPHONY

The problem involved in the transmission of speech without wires can be stated simply ; the methods by which this problem has been solved in the Twentieth Century are less easy to understand. Sound, as everyone knows, consists of vibrations which are propagated through the air by wave motion. The sounds of speech are of two kinds—vowels and consonants. For every vowel there is a definite kind of wave, but it is not a wave produced by a simple vibration. The pitch of the note is due primarily to the vibration set up when air is expelled from the lungs through the vocal cords, but it is profoundly modified by the mouth, and the fundamental sound is accompanied by harmonics. A simple explanation will make this clear. The note emitted by a plucked violin string depends upon its length. If the length is halved the number of vibrations per second (i.e. the pitch) is doubled. If only a third is taken, the pitch is three times as high. The string may vibrate as a whole and in parts at the same time. The note given by the whole string is called the fundamental ; those given by the parts are called harmonics. The vowel sounds contain both. The consonantal sounds are

FIG. 295B.—INTERIOR OF DORCHESTER BEAM TRANSMISSION STATION.

To face page 404.

FIG. 295C.—MARCONI-MATHIEN MULTIPLIER APPARATUS.

To face page 405.

never continuous. They modify the beginning or end of those which represent vowels.

The telephone transmitter consists of a thin metal disc or diaphragm with granules of carbon packed loosely behind it. A current of electricity passes through these, and when a person speaks into the mouthpiece the disc vibrates, the resistance offered by the granules alters, and variations in the strength of the current are produced which flow along the line to the receiver. The receiver consists of an electro-magnet round which the line current flows. One end of the magnet lies just behind a metal disc, and variations in the line current cause the disc to vibrate, producing sounds more or less similar to those which gave rise to the current variations.

It was explained in the last section how strains of waves of high frequency could be made to produce a musical note in a telephone, and how this note could be broken up by the dots and dashes of the Morse Code. But the object to be achieved in Wireless Telephony is something far more delicate. The waves must not, even momentarily, be destroyed. They must be caused to wax and wane with the waves of sound corresponding to human speech. Trains of wave which themselves vary in strength will not serve the purpose. They must be *continuous*, and produced by apparatus so sensitive that variations in the current in a telephone circuit can be impressed upon them ; and they must be received by an apparatus so sensitive that it immediately responds to these variations. An apparatus suitable for these purposes is found in a three-electrode valve, used as an oscillator at the transmission end, and a series of similar valves used (*a*) to amplify and (*b*) to rectify the waves at the receiving end.

Wireless telephony has reached its highest development in the small power sets used on aeroplanes during the war. The transmitters weighed less than 10 lb., and were supplied with current at 600 volts from electric generators weighing only 18 lb. The generators were driven by a small windmill fixed outside the aeroplane, at a speed of 3,000 to 4,000 revolutions per minute. The aerial consisted generally of from 100 to 300 feet of wire, which could be trailed behind or wound up on a drum when not in use. A weight at the end of the wire kept it straight. For directing the fire of artillery the range was limited to about 10 miles, so as not to interfere with operations on another part of the front, but for long distance reconnaissances a range of

200 miles was possible. A microphone which was not affected by the roar of the engine had to be designed, and many other difficulties had to be overcome before success was attained.

The earliest notable example of long distance wireless telephony is that carried out by the Western Electric Company in 1914.[1] They set up, at the United States Naval Station at Arlington, a wonderful piece of apparatus consisting of the following parts : a small triode valve used as an oscillator was coupled through a transformer to the grid circuit of a tube of rather larger size, which acted as a magnifier of the oscillations. This grid circuit was coupled up with a microphone (telephone transmitter), so that the plate circuit produced magnified oscillations which could be modified by the human voice. The plate circuit was next coupled to the grid circuits of a number of amplifying valves, and these were coupled to the grids of over 500 triode valves. By this means speech was transmitted regularly to the Eiffel Tower, 2,300 miles, and occasionally to Honolulu, 5,000 miles away. With the far more perfect apparatus now available a greater distance could be covered with a less complicated arrangement and a smaller expenditure of power.

Thus the Marconi Wireless Telegraph Company manufacture a 20 kilowatt (27 horse-power) set with which speech has been transmitted 1,500 miles. A 3 kilowatt valve panel for both telegraphy and telephony is shown in Fig. 293. The switchboard is in the middle. On the left are five oscillation valves and on the right three modulating valves. A still smaller set absorbing only 20 watts is illustrated in Fig. 294. This has a range for telephoning of 15 miles, with a 30 ft. aerial, though 45 miles is possible for telegraphy. It is provided with a small electric generator to be worked by hand, though it can be used equally well with a small transformer attached to a town supply. The valve filaments are heated by a 6-volt accumulator. The whole apparatus is supplied either in a portable or a stationary form, and the former can be carried either by hand or on a pack saddle.

One of the chief disadvantages of wireless telephony is the fact that you cannot " cut in " when a man is speaking to you. Before " listening " can become " speaking " a switch must be thrown over. This disadvantage has been overcome in the connection partly by land line and partly by wireless which the

---

[1] W. H. Eccles in *Nature*, 24th June, 1920.

never continuous. They modify the beginning or end of those which represent vowels.

The telephone transmitter consists of a thin metal disc or diaphragm with granules of carbon packed loosely behind it. A current of electricity passes through these, and when a person speaks into the mouthpiece the disc vibrates, the resistance offered by the granules alters, and variations in the strength of the current are produced which flow along the line to the receiver. The receiver consists of an electro-magnet round which the line current flows. One end of the magnet lies just behind a metal disc, and variations in the line current cause the disc to vibrate, producing sounds more or less similar to those which gave rise to the current variations.

It was explained in the last section how strains of waves of high frequency could be made to produce a musical note in a telephone, and how this note could be broken up by the dots and dashes of the Morse Code. But the object to be achieved in Wireless Telephony is something far more delicate. The waves must not, even momentarily, be destroyed. They must be caused to wax and wane with the waves of sound corresponding to human speech. Trains of wave which themselves vary in strength will not serve the purpose. They must be *continuous*, and produced by apparatus so sensitive that variations in the current in a telephone circuit can be impressed upon them ; and they must be received by an apparatus so sensitive that it immediately responds to these variations. An apparatus suitable for these purposes is found in a three-electrode valve, used as an oscillator at the transmission end, and a series of similar valves used (*a*) to amplify and (*b*) to rectify the waves at the receiving end.

Wireless telephony has reached its highest development in the small power sets used on aeroplanes during the war. The transmitters weighed less than 10 lb., and were supplied with current at 600 volts from electric generators weighing only 18 lb. The generators were driven by a small windmill fixed outside the aeroplane, at a speed of 3,000 to 4,000 revolutions per minute. The aerial consisted generally of from 100 to 300 feet of wire, which could be trailed behind or wound up on a drum when not in use. A weight at the end of the wire kept it straight. For directing the fire of artillery the range was limited to about 10 miles, so as not to interfere with operations on another part of the front, but for long distance reconnaissances a range of

200 miles was possible. A microphone which was not affected by the roar of the engine had to be designed, and many other difficulties had to be overcome before success was attained.

The earliest notable example of long distance wireless telephony is that carried out by the Western Electric Company in 1914.[1] They set up, at the United States Naval Station at Arlington, a wonderful piece of apparatus consisting of the following parts : a small triode valve used as an oscillator was coupled through a transformer to the grid circuit of a tube of rather larger size, which acted as a magnifier of the oscillations. This grid circuit was coupled up with a microphone (telephone transmitter), so that the plate circuit produced magnified oscillations which could be modified by the human voice. The plate circuit was next coupled to the grid circuits of a number of amplifying valves, and these were coupled to the grids of over 500 triode valves. By this means speech was transmitted regularly to the Eiffel Tower, 2,300 miles, and occasionally to Honolulu, 5,000 miles away. With the far more perfect apparatus now available a greater distance could be covered with a less complicated arrangement and a smaller expenditure of power.

Thus the Marconi Wireless Telegraph Company manufacture a 20 kilowatt (27 horse-power) set with which speech has been transmitted 1,500 miles. A 3 kilowatt valve panel for both telegraphy and telephony is shown in Fig. 293. The switch-board is in the middle. On the left are five oscillation valves and on the right three modulating valves. A still smaller set absorbing only 20 watts is illustrated in Fig. 294. This has a range for telephoning of 15 miles, with a 30 ft. aerial, though 45 miles is possible for telegraphy. It is provided with a small electric generator to be worked by hand, though it can be used equally well with a small transformer attached to a town supply. The valve filaments are heated by a 6-volt accumulator. The whole apparatus is supplied either in a portable or a stationary form, and the former can be carried either by hand or on a pack saddle.

One of the chief disadvantages of wireless telephony is the fact that you cannot " cut in " when a man is speaking to you. Before " listening " can become " speaking " a switch must be thrown over. This disadvantage has been overcome in the connection partly by land line and partly by wireless which the

[1] W. H. Eccles in *Nature*, 24th June, 1920.

Marconi Company have recently established between London and Amsterdam. The waves which carry the signals to Holland are of a slightly different length from those which carry the signals back, and the transmitting and receiving stations (Fig. 295) at Southwold and Zandvoort are separated. The experiment, conducted with the co-operation of the British and Dutch Post Office authorities, has been very successful, and other similar links have been established elsewhere.

BROADCASTING

The development of wireless telephony enormously increased the utility of radio-communication for publicity, entertainment, and education. While few people could interpret the dots and dashes of the Morse code, millions could purchase a receiving set and listen to words and music poured into the ether from transmitting stations all over the world. The public broadcasting of entertainments began in Canada in 1920. In 1922 the British Broadcasting Company was formed, and since 1923 regular programmes have been broadcast from an increasing number of stations. For those living within a few miles of the transmitting station the receiver was extraordinarily simple and cheap. An aerial, two coils, mounted on one axis diametrically so that the angle between them could be altered, a crystal-detector and a pair of headphones was all that was required. Such an instrument was called a variometer, and with it many thousands of people were able to enjoy the music or talks provided and to hear a resumé of the day's events.

But the instrument makers, and especially the valve makers, were quick to see the possibilities. The programme of a distant or foreign station required amplification in order to render it audible. Condensers, transformers, high-tension batteries and accumulators, and valves were improved in design and construction and, owing to the demand, produced more cheaply. And though each receiving set would supply several headphones, these were rather expensive, and, to many people, uncomfortable. So the loud speaker was perfected. This was, in the first instance, merely a telephone receiver, or microphone, with a horn attached. Its main disadvantage that the diaphragm and the horn had a definite vibration period of its own, and the instruments resounded more strongly to certain notes. It would carry us too far to describe how these disadvantages have been overcome. The reader will be aware, however, that these instruments can be made to give a sufficient volume of sound to fill a small room without unpleasantness, and also to enable 20,000 people to hear a speech in the open air. The microphone used in small loud-speakers requires very little power to work it, but the volume of sound is correspondingly small. The horn acts as an amplifier. In some instruments a large volume of air is set in motion by a pleated paper disc, which is operated by the diaphragm, and this type is replacing the horn type gradually.

BEAM WIRELESS

The most recent development in wireless transmission is beam wireless. It will be remembered that the ordinary aerial emits waves in all directions. Marconi devised a directional aerial (see p. 385), but this is not so effective as the beam system. By this system we mean a directed, non-divergent train of waves, and it involves a special form of aerial and feeders, together with short waves. Like most other development, it began in a tentative way some years before its possibilities were fully appreciated. Short waves were actually being used in 1906–7, and short-wave valve generators were used in 1919. They were generally found to be less reliable than long waves. Serious experiments were begun in 1922. Franklin used two parallel wires, one of which acted as a reflector. When this was tried with short waves they were found to be more regular and reliable than was supposed, and to require less energy for a given distance than long waves.

Beam transmission depends primarily upon the fact that when two or more aerial wires are set in line the electro-magnetic radiation is almost entirely at right angles to the plane of the wires. A series of such wires, in one plane, will radiate from both faces. If the plane is east to west, the radiation will be northwards and southwards. If the plane is north to south, the radiation will be eastwards and westwards. The longer this plane or " sheet " or wires and the more wires there are per unit of length in the sheet the more complete is the radiation at right angles to it. The length of the plane or sheet horizontally is always several wave-lengths.

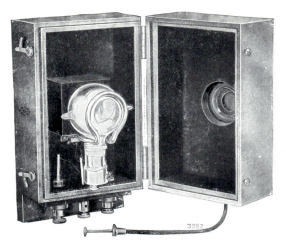

FIG. 295E.—A PHOTO-ELECTRIC CELL.

FIG. 295H.—A NEON LAMP.

To face page 408.

FIG. 297.—THE SAME SUBJECT PHOTOGRAPHED ON AN ORDINARY
PLATE (UPPER) AND A PANCHROMATIC PLATE (LOWER).

To face page 409.

Now as the plane or sheet will radiate in two directions, so a reflector is required. This is supplied by a similar sheet of wires, but more closely spaced, and suspended one-quarter or three-quarters of a wave-length behind the transmitting aerials. Such an aerial will radiate a beam which only diverges about 10°, and the beam is received by an aerial of precisely similar construction. A service has been established between England and Canada, England and Australia, England and South Africa, England and India, Australia and Canada, Portugal and her African colonies, and between Brazil and Europe. Fig. 295A shows the transmitter house, masts, and aerials of the Marconi Beam transmitting station at Dorchester, which is the largest station of this type in the world, and Fig. 295B shows the interior of the transmitter hall at this station.

The establishment of beam transmission has been followed by an ingenious invention which enables simultaneous transmission of speech (or music) and telegraphy between the same stations. This is the Marconi-Mathien Multiplex apparatus, which picks out the telephone from the telegraph signals at the receiving station. The effectiveness of this apparatus was demonstrated in 1927 at the Bridgewater Beam Station, when some of the experts danced to the strains of a band playing in Montreal while telegraphic messages were also passing between the two stations. Fig. 295c shows the apparatus which was employed.

PHOTOGRAPHY AND TELEVISION

The transmission of pictures by telegraphy is not exactly new, but it is only recently that the principal daily papers have adopted it as a more or less regular practice. The first attempt was Bain's chemical printing instrument of 1847. At the transmission end were metal letters. Metal brushes, consisting of a number of springs, with a line wire to reach spring, were passed over the letters. While a spring rested on any portion of a letter a electric current flowed through it and its corresponding line wire. At the receiving end similar brushes passed over a sheet of paper impregnated with a substance that decomposed and showed colour when a current passed through it. In that way the letters were reproduced.

Five years later Bakewell developed an instrument for transmitting writing or drawings. At the transmission end was a cylinder covered with a sheet of tinfoil, and the writing or drawing was made on the tinfoil in shellac ink, which is a non-conductor. As the cylinder rotated a metal needle pressed lightly upon it, and the cylinder rose, so that the point traced out a spiral. At the receiving end was a similar cylinder, covered with a sheet of chemically prepared paper. This cylinder rotated at the same rate as the other, with a metal needle moved spirally over its surface, just as the sending needle moved. An exact copy of the original writing or drawing was reproduced on the paper.

Space will not permit of a review of the methods which were adopted by successive inventors, and we must pass on to the principles involved in the modern process. The transmission of a photograph involves the transmission not of an outline, but of light and shade. For this mechanical devices are inadequate, and optical means must be employed. Consider first how a photograph can be broken up for analysis. If it is re-photographed through a plate ruled with fine lines, these lines will be thickened where the photograph is darkest, and will appear thinner where they cross the lighter areas. The photograph can, therefore, be reproduced by drawing a number of parallel lines of varying thickness. Or if a photograph is re-photographed through a screen with a number of parallel lines at right-angles to one another it will appear as a number of dots, thick and heavy over the dark portions, and faint or widely dispersed over the lighter areas. A photograph can, therefore, be regarded as consisting of dots varying in density and distribution. If we can reproduce a number of parallel lines or dots corresponding to those into which the original has been analysed the problem will be solved.

The first successful inventors made use of the peculiar property of selenium, discovered in 1873. In the dark this substance offers a high resistance to the passage of electricity, but when light falls upon it the resistance decreases. If light from different parts of a photograph, obtained by rotation and rise as in Bakewell's instrument, is directed upon selenium, then its electrical resistance will vary according to the light and shade of the photograph. If a current is flowing through the selenium the strength of the current will vary in a corresponding way. If this varying current passes along a line wire to a distant station it may be used to vary the light of a lamp. And if this light be caused to traverse a piece of photographic paper in the same way as the original photograph is being analysed at the receiving end, a copy of the photograph will result. This is the principle employed in Korn's apparatus, which was in practical use for some years.

But selenium has the disadvantage that it does not respond very quickly to the light changes, and a more sensitive analyser was required. This has been found in what is called a photo-electric cell. When Hertz was making his experiments in 1887 with electric waves, he noticed that a spark passed more readily across a gap when it was illuminated. Subsequently, it was discovered that certain bodies, such as sodium, potassium, rubidium, etc., emit electrons (see Chapter XX) rather freely when exposed to light, and these electrons, which are particles of negative electricity, constitute, when in motion, an electric current. A photo-electric cell consists of a vacuum tube coated on the inside with one of these light-sensitive metals. The current which passes through it varies with the strength of the light which falls upon the metal (sodium, potassium, or rubidium), and as the photograph is " explored " the light and shade signals are delivered to the line. At the other end this varying current operates a string galvanometer. This consists of a fine wire suspended between the poles of a powerful magnet. The fine wire carries the weak but varying current, and with each variation it moves in the strong magnetic field. In its steady position it closes an opening through which light would pass. With each movement it allows light to fall on a piece of sensitized paper, wound on a cylinder, and moving exactly in the same way as the original photograph moves at the transmitting end.

The photograph may be analysed by dots, a spot of intermittent light being flashed up and down or across the picture, and a mirror galvanometer may be used to communicate the light and shade to the sensitized paper ; but the principle is the same. It takes from 5 to 20 minutes to telegraph a picture in this way, and the photograph has then to be developed and fixed.

Television presents a more difficult set of problems. It does not involve photography. An image of a distant person or object is to be projected upon a screen. The person or object has to be analysed by a line or spot of light in the same way that the photograph is analysed in photo-telegraphy. That is not a difficult

Fig. 295D. A Photograph Consisting of Lines of Varying Thicknesses.

task. But in producing the image the last line or spot must be recorded before the first has faded from view. In Chapter XIX it is explained that as an image falling upon the retina—the sensitive screen at the back of the eye—persists for only one-tenth of a second, the pictures on the cinematograph screen must follow one another at a rate of not less than ten a second in order that a continuous impression may be formed. Similarly in television, the whole of the picture must be traced not in lines or in spots in less than that interval if the reproduction is to be seen as a whole. In other words, if the minimum time in which a photograph can be telegraphed is five minutes, television

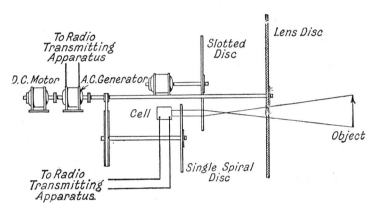

Fig. 295F. DIAGRAM OF BAIRD'S TRANSMISSION ARRANGEMENT.

requires that the process shall be speeded up 3,000 times. The primary difficulty is a mechanical one. A secondary difficulty arises in the quantity of light required to illuminate a large area. The images produced so far could be contained on a postcard.

Let us examine the general arrangement. Firstly the object must be brightly illuminated. Secondly there must be a photo-electric cell. Thirdly the light from each small area successively of the object must fall upon this cell. Alexanderson of the Radio Corporation of America accomplishes this by a set of plane mirrors arranged, with slightly different angles of tilt, round the edge of a rapidly rotating wheel or drum. Baird, whose work has excited the most interest in this country, uses three discs, one with a series of radial slots round the edge, one with a spiral slot, and the third with a series of lenses arranged in a spiral. The radial slots render the light from the picture intermittent, the spiral slot determines the portion of the object from which the light is reflected, and the lenses focus this light upon the photo-electric cell, see Fig. 295E. The varying current from the transmitter operates a Moore lamp (see the Moore tube, pp. 115–17), in which the vacuum tube is a spiral inside a globe, see Fig. 295H, at the receiving end. The gas in the tube is neon, which gives a brilliant orange-red glow. The light from this lamp

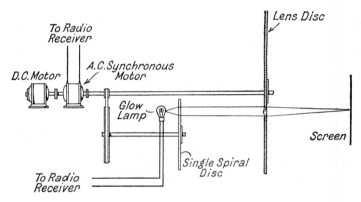

Fig. 295G. DIAGRAM OF BAIRD'S RECEPTION ARRANGEMENT.

passes through a system of discs and lenses, see Fig. 295G, similar to those at the transmitting end, and falls upon a screen. Up to the time of writing the image has not exceeded postcard size, whereas what inventors are aiming at is a life-size picture of the object, person, or scene.

For wireless television this apparatus needs only to be associated with a wireless transmitting set at one end and a wireless receiving set at the other. The success which has been so far achieved is a very wonderful development of the physical investigations of the last thirty or forty years, and it represents the work of many men in many places at many times.

Sufficient will have been said to indicate the wonderful progress which has been made since the birth of the invention in 1896, and to explain to some extent the simpler principles upon which the improvements have been based. The reader will be able to see, in imagination, the powerful waves spreading outwards from the transmitting aerial, and, in some as yet inexplicable way, curving round the surface of the earth. He will see them arriving at their destination so exhausted by their journey of 2,000 or 3,000 miles across space that they are barely able to reveal their presence. But he will know that, however feeble they may be, so long as they exist at all they retain undiminished a property of which time alone can rob them. They still vibrate at the same rate as they did when the transmitting aerial flung them out across continent and ocean to carry their message of joy or sorrow, life or death, between remote places on the earth. If he realizes this he will see them converting the world into a whispering gallery, caught up here and there by a receiving aerial tuned to thrill to their music, reinforced and strengthened in sympathetic circuits, and reconverted into human speech or the universal language of the Morse Code in the telephone at the observer's ear.

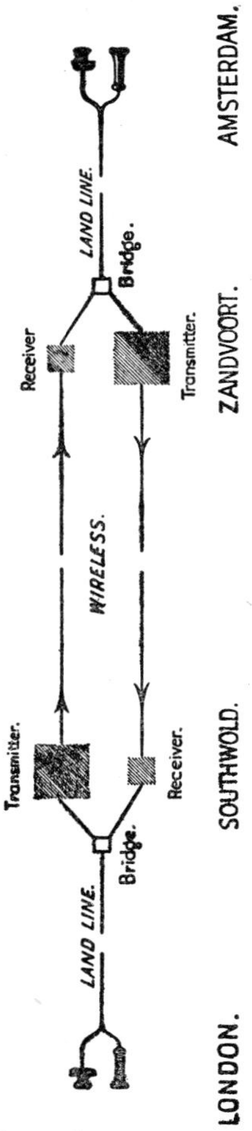

Fig. 295. DIAGRAM OF THE LONDON-AMSTERDAM CIRCUIT.

CHAPTER XIX

SOME APPLICATIONS OF PHOTOGRAPHY

PROBABLY no group of discoveries and inventions is more familiar through its methods and results than those which enable pictures of the external world to be reproduced faithfully and in any quantity desired. The work of the professional photographer, the picture post-card, the illustrated magazine, are found in every home, and the record of well-loved features, of happy hours, and the contemplation of beauty of form, of light and shade, are available to rich and poor alike. Spare half-hours spent in the picture palace open the door to the secrets of nature, and annihilate distance by reproduction of scenes from every quarter of the globe. Finally, the enormous growth of photography as a hobby has made hundreds of thousands, young and old, acquainted with the methods of taking, developing toning, and fixing the impressions which rays of light make upon the sensitive plate.

For the last reason, as well as from considerations of space, no attempt will be made in this chapter to given instructions for taking photographs ; but such space as can be spared will be devoted to a description of some of those newer achievements of the science which have, as yet, hardly come within the scope of amateur effort. A brief review of the photographic process for the benefit of the uninitiated will be followed by an explanation of photography in colour, and some applications of the photography of motion.

THE PHOTOGRAPHIC PROCESS

When light passing through a lens falls upon a suitably placed screen, a picture of objects in front of the lens is formed. The same effect can be obtained by passing the light through a pinhole in an opaque screen, instead of through a lens. The screen upon which the picture falls is of glass, collodion, or paper, and is covered with a thin film of gelatine containing, in extremely fine particles, certain salts of silver. A liquid containing another liquid in such fine particles that a milky appearance is produced is called an *emulsion*, and the emulsion for photographic plates is prepared by mixing two solutions, containing :—

(*a*) Gelatine, ammonium bromide, and potassium iodide ;

(*b*) Silver nitrate and ammonia.

A fine precipitate of silver bromide and silver iodide is formed, and when the liquid is poured on a sheet of glass or other material and allowed to dry the particles of silver compound are distributed evenly over the plate.

If the two solutions are mixed in the cold the resulting plate is slow in taking the picture, but still quite fast enough for ordinary snapshot photography. Keeping the first liquid at 120° F. while the second is added produces a plate very much more rapid in action, while if the mixture is kept at 130° F. for an hour there is a further marked increase in the speed. The time required for the light to impress the plate is so small as to be hardly conceivable. In some of the experiments to be described later the exposure is not much more than $\frac{1}{10,000,000}$ part of a second !

The effect of the light is to decompose the silver bromide and iodide at those points upon which it falls. The lighter parts of the object photographed reflect the most light, and where the image of these falls the greatest amount of decomposition occurs. At first the picture is not visible ; it has to be "developed" by immersion in a bath containing one of the numerous substances sold for the purpose. It is then fixed by immersion in another bath so that light has no further action upon it. The picture, however, is a negative—the light portions of the original are dark in the picture, and vice versa. To obtain a positive, a piece of sensitized paper is placed behind the negative and exposed to light, and the impression is fixed either with or without "toning". The latter process consists in soaking in a bath containing a gold or platinum salt, which converts the silver print into one of gold or platinum.

A photograph obtained on a plate prepared in the way described represents only approximately the lights and shades of the original, because the activity of the rays varies with the colour. The plate is affected most readily by blue or violet, and a red object cannot be photographed against a black background. The plate would be affected to a very little greater extent by the red coat of a soldier than by the light coming from a black curtain behind him.

In order to understand not only how this difficulty is avoided but also how others which are dealt with later are overcome, it is necessary to consider the nature of colour. Probably all readers are aware that if a ray of light falls upon a prism, or wedge-shaped piece of glass, it is bent from its original direction, and spread out into a band of colour. Red, orange, yellow, green, blue, indigo, and violet always appear in this order, the last named suffering the greatest deflection (Fig. 296). If the band is passed through a similar prism with its wedge in the opposite direction the colours re-combine to form white light. Or, if each colour is received upon a small mirror so mounted that it can be twisted to reflect the light which falls upon it to the same spot, white light is again obtained.

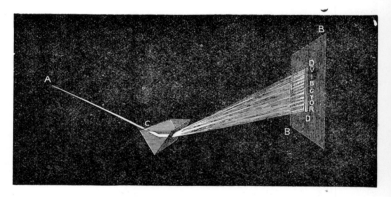

Fig. 296.   DECOMPOSITION OF WHITE LIGHT.

All the properties of light are explained by supposing it to consist of waves or ripples in a medium which exists throughout all space and in all material things—a medium which can neither be measured, nor weighed, nor detected by any of the senses through which a knowledge of the external world is acquired. A wave of definite wave-length—that is, with a definite distance from crest to crest—produces a narrow line of colour ; and a group of waves whose lengths are nearly equal produces a band of colour corresponding to one of those in the spectrum. The smallest waves that produce light are those corresponding to violet, and are no longer than $\frac{3}{10.000}$ of a millimetre or

$\frac{3}{254,000}$ of an inch. The red waves are about $\frac{3}{4,000}$ of a millimeter or $\frac{3}{100,000}$ of an inch in length.

But though these are the only waves which affect the eye, there are larger and smaller waves at either end of the visible spectrum. The former have relatively small photographic activity, but they *can* affect a photographic plate, and by interposing a trough containing potassium bichromate and a plate of cobalt glass between the lens and the sensitive plate, Professor R. W. Wood has succeeded in taking photographs of objects by the infra-red light which they reflect. Similarly, a quartz lens coated with a very thin layer of silver is opaque to ordinary light, but allows ultra-violet waves to pass, and permits of a photograph being taken by their aid alone.[1]

The band of colour which can be detected by the eye corresponds, in fact, to a short range of waves which belong to a whole series ; and bears much the same relation to the whole of the radiation from a luminous body that an octave does to the whole gamut of a piano. At one end of the series are the short, rapid ultra-violet waves whose length has just been given, which produce no visible effect, but which are exceedingly active in promoting chemical change. From these the series passes through waves of gradually increasing length until in the infra-red they give rise to all the phenomena of heat. And beyond these are the still longer waves which are used in Wireless Telegraphy.

Now so far as the correct representation of light and shade in an ordinary photograph is concerned, the greater activity of the blue and violet tints throws the picture out of balance, and the problem has been to produce a plate equally sensitive throughout the spectrum. This has been achieved by using a dye, either in the sensitive emulsion or in a screen which is placed between the lens and the plate, which filters the light, and delivers each colour only in such quantity that equal photographic effects are produced in equal times. Such are orthochromatic, isochromatic, and panchromatic plates, which are now obtainable from dealers in photographic materials. For the ordinary purposes of photography the invention of these plates constitutes the most important advance since the intro-

---

[1] There are admirable examples of both effects in Garrett's *Advance of Photography* (Kegan Paul).

duction of the dry plate. Fig. 297 shows the result of photographing the same subject on an ordinary plate (upper), and on a panchromatic plate (lower). It will be observed that not only do some of the brightly coloured calceolarias appear very dark on the former, but the geranium is hardly visible against the background, and the stripes on the petals of the cinerarias are completely lost.

PHOTOGRAPHY OF COLOUR

From the very beginnings of the art of Photography attempts have been made to secure pictures as faithful in their representation of colour as of form and light and shade, and these attempts have been crowned only with a limited amount of success. Of the half-dozen methods which have been devised, that of Professor Gabriel Lippmann stands alone in scientific accuracy.

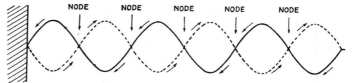

Fig. 298. REFLECTED WAVES.

In 1891 he showed that if a sensitive plate formed one side of a trough with the gelatine surface inwards, and the trough contained mercury or quicksilver, a photograph of the spectrum and of coloured objects could be obtained. In order to understand how this is effected, it is necessary to consider how the tiny light waves act when they fall upon a reflecting surface.

If a rope is attached at one end to a wall, and the other end is held in the hand, a quick up-and-down movement will send a pulse or ripple along the rope, and when this ripple reaches the other end it will be reflected. If the pulses are repeated at proper intervals the direct waves will coincide exactly with the reflected waves as in Fig. 298. At equal intervals portions of the rope will be still, and between these there will be portions in violent movement. Just in the same way the waves of light will form within the film layers of rest and of violent movement alternatively, and the latter will be active in causing decomposition of the silver salt. There will thus be formed alternate layers of decomposed and undecomposed silver compound, and

BLACK IMPRESSION OF YELLOW PLATE.

3.—SHOWING PROGRESSIVE RED OVER YELLOW.

1.—FIRST WORKING, YELLOW.

4.—THIRD WORKING, BLUE.

2.—SECOND WORKING, RED.

5.—FINISHED PRINT

FIG. 299.——HOW A THREE-COLOUR PRINT IS MADE.

To face page 420.

the distance apart of the layers will depend upon the wave-length of the light which formed them. For red they will be farther apart, for blue they will be closer together, and for green they will be at intermediate distances. If after fixing, white light falls upon the plates it is analysed by the successive layers in each part of a picture, and only those waves whose lengths coincide with the distance between the layers can escape from the film. All others are suppressed.

The evidence upon which the explanation is based is as interesting as the achievement itself. If the film be warmed by breathing upon it, it expands, and the distances between successive layers are increased. The waves composing white light are now sorted out differently ; those corresponding to the original colours are suppressed, and the colours in the picture change. Mr. E. Senior and others have cut thin sections of the film and examined them under the microscope. But though evidence of layers was obtained in this way the power of the microscope was insufficient properly to separate them. A more effective proof was obtained by Professor S. R. Cajal, of Madrid, who caused the gelatine sections to swell by placing them in water, and then photographed them under the microscope.

A well-known writer has said that you can fool some of the people all the time, you can fool all the people some time, but you can't fool all the people all the time ; and this well describes the advantages and disadvantages of Lippmann's method of colour photography. The spectrum and some objects can always be photographed, but for many purposes the method is, unfortunately, unreliable.

All other processes are based on the Young-Helmholtz theory of colour vision, according to which the human eye is sensitive to only three fundamental colours—red, green and blue. Every tint that can be recognized is composed of one of these or of a mixture of two or all three of them ; and all three in certain proportions produce white light. It is therefore necessary to photograph only the red, green, and blue portions of a coloured object in order to secure a picture which represents the original colours so far as they can be detected by the unaided eye. Unfortunately, the only methods which have been devised involve the use of dyes and coloured glasses, and the difficulty of securing always the same tint renders it impossible to obtain more than a close approximation.

In 1892 Frederick Ives, of Philadelphia, adopted the plan of taking three photographs through red, green, and blue glass screens respectively, and in 1893 he patented two pieces of apparatus for viewing the pictures so formed. In one of them, the pictures, placed side by side in a lantern with a triple front, were projected on a screen, and by means of a lever, were caused to fall on the same disc. This superposition of the red, green, and blue portions of the photograph gave a beautiful picture quite near enough to the actual tints to satisfy any but the most captious critic. The only defect in the particular instrument used by the writer some thirty years ago was an objectionable fringe of colour to the white portions ; but this was only notice-able at close quarters, and was probably due to the fact that the three discs did not exactly register on the screen.

Each of the screens used in taking the photographs transmits a broad band of colour, so that the variety of colour in the object shall all be utilized as far as possible on the sensitive plate. But for throwing the picture on the screen advantage is taken of lack of sensitiveness of the eye, and each screen transmits only a narrow band.

Another process was invented by Mr. Sanger Shepherd about the same time. Three photographs were taken in the same way, through screens of appropriate colour, and then stained with dyes. The three plates were then bound together in the form of a lantern slide, which could either be used in the lantern or viewed by being held up to the light.

While Ives and Sanger Shepherd were experimenting with methods involving three photographs, Professor Joly, of Dublin, was engaged upon a plan which required only one. His screen was covered with a very large number—350 to the inch—of red, green, and blue lines ruled in dyes on a glass plate. Each line had to be in contact with the one on either side of it and there had to be no overlapping. The photograph was taken and viewed through the same screen. The lines were so narrow that they could only be detected by close inspection. At a little distance they merged into one another and individual colours were lost.

Suppose a red button was being photographed. The light from the button falling on the sensitive plate, would only reach it through the red lines. If the image of the button on the plate was an inch in diameter it would be crossed by nearly 120 red,

FIG. 300.—GENERAL VIEW OF CINEMATOGRAPH.

To face page 422.

FIG. 301.—MECHANISM OF CINEMATOGRAPH PROJECTOR.

To face page 423.

120 green, and 120 blue lines, so that the photograph would really be in red lines about $\frac{1}{60}$ of an inch apart. On viewing the fixed plate through the screen, the photograph itself cuts off the green and blue, and allows only the red light to pass. The process was given up in 1898 because of the difficulty of securing a sufficient number of lines to the inch.

Within the last few years a screen of this kind with 600 lines to the inch has been constructed by Mr. T. H. Powrie and Miss Florence Warner, of Chicago, and it is known commercially as the Florence plate. The method is extremely ingenious. Lines about $\frac{1}{300}$ of an inch wide are ruled in black ink on a glass plate, with spaces $\frac{6}{100}$ of an inch in width. A plate covered with a film of gelatine containing bichromate of potash is exposed under this screen, and where the light falls through the spaces the gelatine is rendered insoluble in warm water. The plate is then washed, fixed, and dipped in green dye, which is absorbed by the fine gelatine line which remains. Another film of bichromated gelatine is run over the plate, and a second exposure made with the black line on the screen covering the green line. This leaves a narrow line of the new gelatine exposed. The plate is treated in the same way as before, but with a red dye. There are now green, red, and colourless lines on the plate. A fresh film of gelatine is run on, a further exposure made with the black lines covering the green and red lines. The third line is now stained blue, and a Joly screen is produced with lines only about half as wide.

The process which is most widely used at the present time is that patented by Lumiere et Cie., and is known as the autochrome process. Three quantities of starch are stained with red, green, and blue respectively, and then intimately mixed so that the colour of the mass is neutral. But if a few of the minute grains of which the starch is composed were examined under the microscope, they would be found to be transparent globes of red, green, or blue according to the original batch from which each had come. The dry grains are dusted over the plate in a single layer and pressed, or else the spaces are filled in with a fine black powder. The layer is secured by a waterproof varnish, and the sensitive emulsion is poured over the top, thus forming plate and screen in one. The smallest detail in a photograph which is visible to the naked eye will be covered by a multitude of grains of all colours, and whatever the colour of

424 DISCOVERIES AND INVENTIONS

the original may be, sufficient light passes through the appropriate grains to affect the plate.

Another very interesting method is that of the Paget Prize Plate Company, to whom the writer is indebted for information. The screen is in this case separate from the sensitive plate, and is covered with a number of minute squares of red, green, and blue. It is prepared by coating a clean glass plate with a special collodion, which is then stained with a red dye. Portions of the plate are then coated with a " resist ", after which it is placed in a bath and the uncoated portion bleached. It is then placed in a green dye, which replaces the red which has been dyed out. A further series of " resist " squares is printed on the plate, and the uncovered green is bleached. Finally, the plate is re-dyed with blue. The result is a finished screen with all its colours in one plane, without any overlap, and no white or black. Very effective copies for viewing directly or by the lantern can be made, and all kinds of coloured objects can be faithfully and brilliantly reproduced.

A most important application is the production of the beautiful coloured illustrations which appear in modern books and magazines. The process is based on that of Ives. Three photographs are taken of an object or scene, and a block is made from each. When these blocks are stained with ink of the requisite colour and impressed in succession on the paper, the object or scene is reproduced in colours strikingly near to the original (see Fig. 299). The trouble of taking three separate photographs is sometimes avoided by using in the first instance a Lumiere plate. The three blocks are then made from the same photograph by interposing appropriate screens.

THE PHOTOGRAPHY OF MOTION

Not many people are aware that the first step towards the photography of a succession of movements were taken as long ago as 1872. In that year Mr. Muybridge, a Californian, obtained twenty-four successive photographs of a trotting horse. His plan was to arrange twenty-four cameras in a line opposite a white screen. Stretched between each camera and the screen was a thread, and as the horse passed it tightened and broke the thread, and in so doing operated the shutter of the corresponding camera.

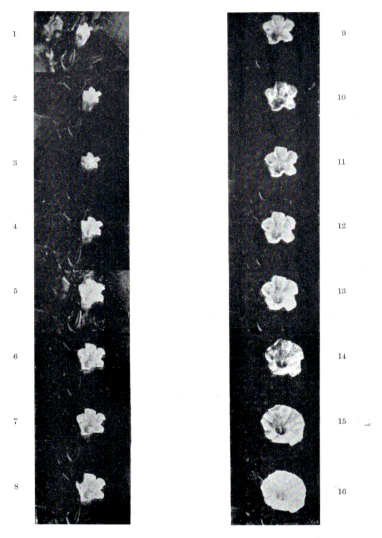

FIG. 302.—OPENING OF A FLOWER (CONVOLVULUS).

To face page 424.

FIG. 303.—THE KINEMACOLOUR PROJECTOR.

TO face page 425.

In 1882 Dr. Marey, of Paris, constructed the beautiful apparatus known as Marey's pistol.[1] It was, indeed, very like a revolver, but the drum which in the fire-arm carries the cartridges, in this case carried a circular glass plate coated with sensitive emulsion and wholly enclosed. The only direction from which light could reach it was down the barrel. When this pistol, charged with its sensitive plate, was pointed at any object, and the trigger pulled, the plate rotated about its centre in a succession of jerks, and as it paused for a moment after each step a photographic impression of the object was made near the rim.

No real advance in the photography and reproduction of motion was possible until improvements in the manufacture of celluloid provided a long thin strip of sensitized material upon which a succession of many pictures could be obtained. The stimulus which led to this was the need for a film to replace glass plates in a magazine camera, thus reducing the weight and permitting a larger number of snapshots to be taken. And when success was attained there was one man at any rate—Thomas Alva Edison—who was ready to take advantage of it. At the World's Fair at Chicago in 1893 machines were exhibited which worked upon the penny-in-the-slot principle. A nickel ($= 2\frac{1}{2}$d.) was dropped into a machine, and with eyes glued to a small opening the observer saw for about half a minute a complete set of movements illuminated by a small electric lamp.

The principle of this and all later machines is that an image thrown upon the retina—the wonderful screen at the back of the eye—persists for about a tenth of a second after the stimulus which produced it has passed away. A picture can be formed on a photographic plate far more rapidly than this, and the number of pictures that can be taken in a second is only limited by the speed at which a shutter can be made to flash the light upon successive portions of the film as it is wound rapidly from one roller on to another. For all ordinary purposes it is sufficient to take sixteen photographs a second and submit them to the observer at the same rate.

It does not seem to have occurred to Edison to project the pictures on a screen, and the subsequent development of moving pictures as we know them to-day is mainly due to Mr. R. W. Paul, the scientific instrument maker, of London. According

[1] M. Janssen, the astronomer, had used a similar instrument to record the transit of Venus, in 1874.

to Mr. F. A. Talbot,[1] Edison did not patent his invention in England, and Mr. Paul's attention was drawn to the matter by a man who asked him to make films for him. The possibility of projecting them by means of a lantern soon appeared, and one night in 1895 the attention of the police was called to loud cries proceeding from a building in Hatton Garden. On entering they found that what they had suspected to be a grim tragedy was a joyful demonstration which attended the first successful attempt to show moving pictures on the screen. The show was repeated for their benefit, and they were the first persons other than Mr. Paul and his assistants to become familiar with the new invention.

The terms cinematograph, bioscope, vitagraph, merely indicate different mechanical devices for obtaining the movement of the film. This is $1\frac{1}{2}$ inches wide, and is pierced with holes along both edges. The teeth of wheels something like chain wheels and called sprockets, fit into these holes and control the movement. At first this was continuous and a rotating shutter in front of the lens allowed each picture to fall upon the screen for a short time, but the best effect is obtained by intermittent motion by which each picture is allowed to come to rest before it is disclosed by the shutter. The general arrangement of a kinetoscope is shown in Fig. 300, and the mechanism in greater detail in Fig. 301.

The manufacture of films has become an enormous industry, and Messrs. Hepworth, of London, Lumiere, Pathé Freres, Gaumont, and other firms employ thousands of operators. The subjects come in from resident operators in all part of the world. They are developed and fixed in special machines which pass them through the necessary baths and dry them. They are then copied and dispatched to the picture houses.

During the last few years the demand for the picture play has enabled each film company to maintain in regular employment a company of actors and actresses. Huge studios in which an appropriate setting can be arranged have been built, and all the paraphernalia of the stage is recorded by the film. But the performance lacks one of the principal features. The human voice which, after all, does so much to make or mar the drama, was absent, and the action proceeded to the accompaniment of the orchestra, which harmonized more or less with the emotions

[1] *Moving Pictures* (Heinemann).

FIG. 304.—PORTION OF CINEMATOGRAPH FILM OF TRYPANOSOMES
OF SLEEPING SICKNESS AMONG BLOOD CORPUSCLES, × 400.

To face page 426.

FIG. 306.—M. LUCIEN BULL'S APPARATUS FOR OBTAINING A CINEMATOGRAPHIC RECORD OF INSECTS IN FLIGHT.

To face page 427.

depicted on the screen. This has now been supplied in the " talking " pictures with which everyone is familiar.

Not the least interesting records are those which have been obtained of the habits of animals, and the growth of plants. To secure the former the haunts of beast and bird have been invaded, and the camera has penetrated the dark recesses of the tropical forest where formerly a gun would have been regarded as the only weapon that could safely be used. In registering very slow motions such as the transformation from caterpillar to chrysalis, and chrysalis to butterfly, the growth of a plant, or the unfolding of a flower (Fig. 302), photographs are taken at long intervals and then thrown on the screen in rapid succession. Many of the trick pictures in which, for example, a knife cuts up a loaf of bread and a sandwich is made without visible hands, are the result of a large number of separate photographs in which the setting is changed between each, the film being covered meanwhile by the shutter.

It was hardly to be expected that inventors would be satisfied with pictures in black and white, and some of the earlier films were coloured by hand. But when longer films came into vogue this was too expensive, and instead of painting in each picture by hand, stencils were adopted, and though the same amount of delicacy was not possible, there was colour. But even this process soon became expensive with a film 1,000 feet long containing more than 12,000 pictures.

As early as 1899 a method was devised by Greene, whereby the photographs were taken through red, green, and violet screens and flashed on the screen successively through screens arranged in the shutter. But while sixteen a second is sufficient for black and white, a three-colour process of this kind requires forty-eight pictures a second, and there were mechanical difficulties in securing this. The film must be panchromatic, and can only be developed in darkness.

The difficulties of a three-colour process led Albert Smith to propose two colours only—red and green. The method was patented in 1906, introduced commercially in 1907, and improved in 1911. This is the famous Kinemacolour process. Pictures are taken alternately through red and green screens, and projected through a rotating disc having two opaque sectors, one transparent red and one transparent green sector (see Fig. 265). Blue is not entirely absent owing to the green containing a little, but indigo and violet are not reproduced, and the reds and greens are emphasized.

Green's process was later revived under the name Bicolor, and the Ives three-colour process was being applied to moving pictures by the use of three films. It is only a dozen years since the first colour films were exhibited, and those whose memory enables them to look back over the thirty-five years since the first picture show will realize the progress that has been made and comprehend something of the promise of the future.

If one wishes to know something of the fidelity and speed of the modern photographic plate the greatest achievements will be found in the laboratories of scientific workers, who use the camera to record observations that the eye cannot distinguish nor the mind, without difficulty, conceive. The tiny bacteria, those low forms of vegetable life, some not more than $\frac{1}{25,000}$ of an inch in diameter, which exercise a powerful influence in health and disease, are photographed with ease. A minute drop of the liquid or slice of the jelly in which they are cultivated is placed on a glass slide under a high-power microscope, and the image, hundreds of times larger than the object, is thrown upon a sensitive plate. When this is developed the investigator has a record which he can examine at leisure and use for comparison without undergoing the strain that microscopic observation involves.

The special services which the microscope and the camera render to the steel maker and the engineer have been detailed in Chapter VIII. With their aid the minute internal structure of metals is revealed and permanently recorded. In association with the chemist, the microscopist and the photographer have built up during the last fifteen or twenty years a body of knowledge that exercises an influence upon the most delicate instrument of precision, and the most gigantic structure conceived and erected by the engineer. The tiny waves of light falling on the polished or etched surface of a piece of steel reveal those variations of level which are due to the varying hardness or chemical composition of the constituents. And the examination of samples of proved strength and reliability affords a standard by which untried materials can be judged.

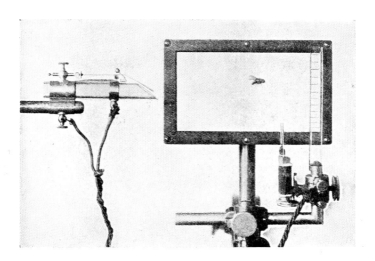

FIG. 307.—APPEARANCE OF A BEE IN FLIGHT.

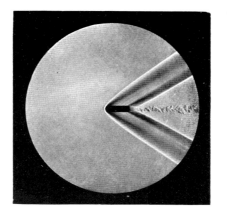

FIG. 310.—WAVES AND EDDIES IN AIR
FORMED BY A BULLET.

To face page 428

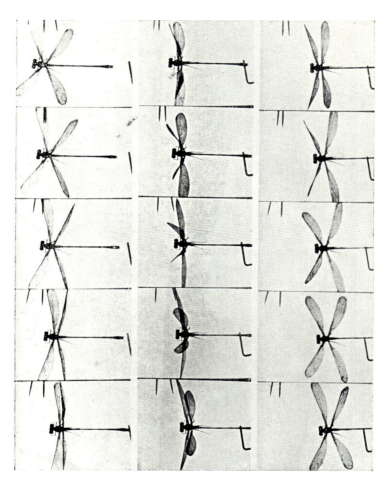

FIG. 309.—PORTION OF CINEMATOGRAPH FILM OF DRAGON FLY.

To face page 429.

Some of the most remarkable results in the photography of bodies in motion have been obtained at the Marey Institute in Paris, which was established to continue the methods of inquiry —mainly, in physiology and medicine—to which Dr. E. G. Marey had devoted his life. From the numerous investigations which have been carried on at this institute, two are selected for notice —one in which the objects studied are extremely minute, and the other in which the movements are extremely rapid.

In few subjects has such remarkable progress been made in recent years as in the study of diseases—particularly those which are due to living organisms. While many diseases are caused by the tiny members of the vegetable world called bacteria, others have been found to be due to equally minute forms of animal life called *trypanosomes*.[1] A particular organism found in the blood of patients when suffering in a particular way, and at no other time, is assumed to be the primary cause, and a cure can only be found by a study of the organism itself.

Blood is a colourless fluid containing myriads of microscopic particles called corpuscles—red and white—so small that in $\frac{1}{20.000}$ of a cubic inch there are nearly 5,000,000 of the former and 6,000 of the latter. To the red corpuscles the blood owes its colour, and they serve to carry the oxygen round the body and to remove the waste products that are formed in the tissues. The function of the white corpuscles remained for many years a mystery, until it was found that they waged war upon the germs of disease. Neither the red corpuscles nor the *leucocytes*, as the white corpuscles are called, are living creatures, and the leucocytes act as though they suffocated or poisoned such of their enemies as became entangled within their substance.

Such facts as these have been established by patient and laborious work with the microscope—work which has often had to be conducted in those unhealthy districts in the tropics where disease is rampant, and its causes present in overwhelming array. The application of photography was not so simple as it appears at first sight, because the germs are extremely sensitive to light and heat ; the concentration of radiant energy upon the drop of liquid or jelly in which they grew in the beam from the lamp was often sufficient to kill them in a few seconds, leaving nothing but their dead bodies for examination. The heat could

[1] The trypanosomes are only a sub-group of the *protozoa*, of which several other sub-groups exercise a similar effect.

be cut off by interposing a trough of water, but the transparency of the objects rendered them difficult to observe, dead or alive, and often it was necessary to kill them and stain the remains so that they could be more easily examined.

The use of photography is of great value in obtaining a record which can be examined and compared with others at leisure; but it only represents a momentary glance, as it were, and can only be supplementary to continuous and persistent observation. For these small objects are in constant movement; they are increasing or decreasing, creating great changes in the liquid or jelly in which they are immersed, and entering into conflict with leucocytes if present in blood.

The investigator requires exact information on these matters, and more particularly he desires to study the behaviour of these organisms in the presence of various substances, amongst which he hopes to find a cure. And to these ends he has called in the services of the cinematograph. After many experiments Dr. G. Comandon, working in conjunction with Messrs. Pathé Freres, succeeded in obtaining records which enabled the processes to be examined over and over again, on a scale thousands of times greater than the actual size. The portion of a film reproduced in Fig. 304 can be projected on the screen so that it is magnified sixty times; and as the pictures on the film itself are already 400 times larger than the actual size of the objects, the total magnification is some 24 thousands.

In these investigations the difficulties which had to be overcome arose almost entirely from the minute character of the objects whose movements it was desired to record, for the slightest vibration would throw them out of focus. But in the experiments on the flight of insects, by M. Lucien Bull, now to be described, the objects were large enough to require little or no magnification, but their movement was so rapid that not even a revolving shutter would permit of a sufficiently short exposure. Instead of one-sixteenth of a second, the pictures had to be taken at intervals of a few thousandths, and for this purpose a series of electric sparks had to be employed.

The general arrangement of the apparatus is shown diagrammatically in Fig. 305 and its actual appearance in Fig. 306. Referring to Fig. 305, the sparks are produced by an induction coil A, and occur between two poles of magnesium at E. This metal produces a light very rich in ultra-violet rays, and therefore

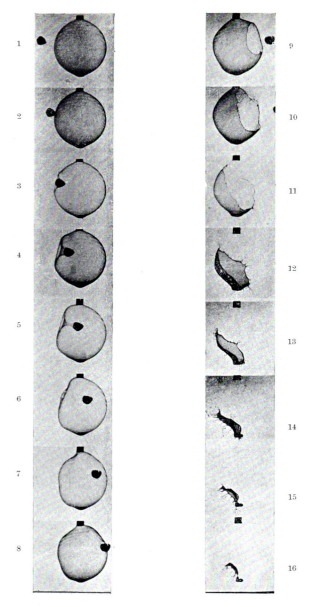

FIG. 311.—BREAKING OF A SOAP BUBBLE BY A BULLET.

To face page 430.

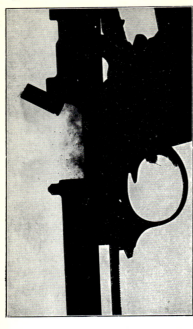

FIG. 312.—INSTANTANEOUS PHOTOGRAPHS SHOWING EJECTION OF SPENT CARTRIDGE FROM AN AUTOMATIC PISTOL.

Continued on next page.

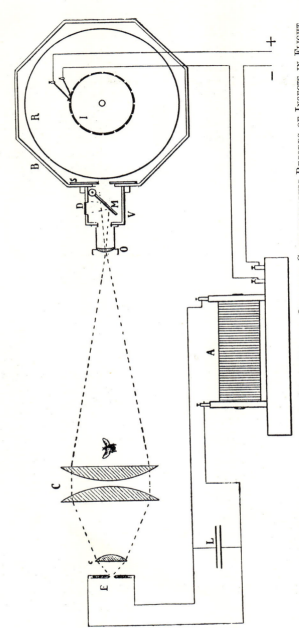

Fig. 305. Diagrammatic Arrangement of Apparatus for Obtaining Cinematographic Records of Insects in Flight.

enables an effect to be obtained on a photographic plate or film with a very short exposure. The film is fixed to the rim of a drum R which is rotated at high speed by an electro-motor which can be seen in Fig. 306. The drum is enclosed in an octagonal box upon one face of which are fixed the lenses. D is a small window the light from which is reflected towards the insect by the mirror M. For as the work has to be done practically in the dark, there must be some means of controlling the direction of flight, and advantage is taken of the fact that all insects fly towards the light.

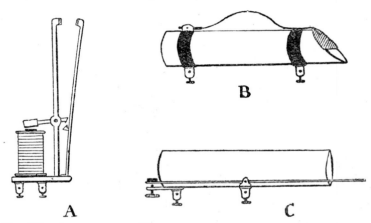

Fig. 308. APPARATUS FOR HOLDING AND RELEASING DIFFERENT SPECIES OF FLYING INSECTS.

The " make-and-break " of the coil is not accomplished by a vibrating spring but by a rotating interrupter I, fixed on the shaft of the drum. This ensures a definite number of sparks, and therefore a definite number of impressions, per revolution. Ordinarily the insect appears in the picture as a silhouette, as shown in Fig. 307, and parts of the wings which it may be desired to observe are not easily seen owing to the flatness of the picture. But M. Bull avoided this by arranging a stereo-scopic front combined with an ingenious shutter device which enabled him to take pictures showing proper perspective. The double front for this purpose is shown in Fig. 306.

The spark-gap is one millimetre long, and the number of sparks

FIG. 312.—INSTANTANEOUS PHOTOGRAPHS SHOWING EJECTION OF SPENT CARTRIDGE FROM AN AUTO-MATIC PISTOL.

To face page 432.

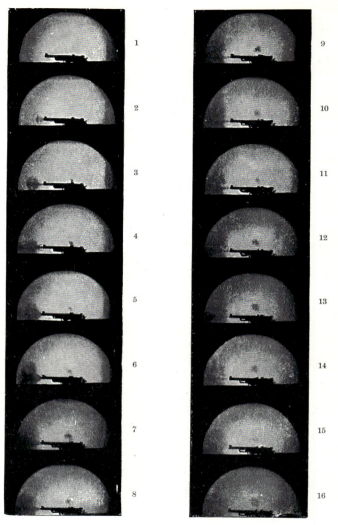

FIG. 314.—A MANNLICHER PISTOL BEING FIRED AND
EJECTION OF THE SPENT CARTRIDGE.

Continued on next page

To face page 433.

2,000 per second. The diameter of the drum was 34·5 centimetres, or about a foot, and with a film 1·08 metres, or a little over 3 feet, long it is possible to obtain fifty-four successive pictures of the usual cinematograph size. The speed of the film was 45 feet a second, so that the total time during which the movements of the insect could be recorded was one-fifteenth of a second. In view of the rapid movement of the wings this is amply sufficient to enable a detailed analysis to be made.

Dragon flies and house flies were held in a pair of tongs shown at A in Fig. 308. The limbs tend to fly apart, but are prevented by a small catch. On closing the circuit, which includes the electromagnet and the shutter (S in Fig. 305), the insect is liberated and at once commences its flight. As all insects become sluggish during confinement they have to be used in a fresh condition, or they do not start at once. But while this method is satisfactory for the insects mentioned, a different plan has to be adopted for hymenoptera, such as bees and wasps, which hesitate for a moment before taking flight. The one finally adopted for these is shown at B, Fig. 308, and consists of a glass tube about 2½ inches long and wide enough to allow the insect to crawl through it easily. The front is closed by a light flap of mica attached to the end of a metal arm which closes the circuit between the two metal bands. The circuit is first broken by a switch, then the insect is introduced. As it lifts the flap in crawling out, the switch is put on, and as the insect flies, the flap falls, completes the circuit, and operates the shutter.

For insects belonging to the coleoptera, such as beetles which hesitate for a still longer time, a third form of release had to be provided. This (see C, Fig. 308) consisted of a lever pivoted at the centre of a similar tube, along which the insect crawls. The hand switch is first broken, then the insect is introduced. During the first half of its journey it presses the back end of the lever upon the contact, and when it passes the centre the lever tips up. The hand switch is now put on, but the shutter cannot act until the insect rises from the front end of the lever and allows the back end (which is the heavier) to fall. In this way the insect unconsciously " pulls the trigger " which enables the picture to be taken.

Fig. 309 shows a series of fifteen pictures of a dragon-fly. The small points appearing on the margin of each picture are the prongs of a small tuning-fork the pitch of which is known

F f

and which therefore serves as a measure of the times between successive impressions.

Perhaps the most remarkable examples of accuracy and delicacy of the photographic record, however, occur in connection with investigations of the flight of projectiles and the adjustment of fire-arms. About forty years ago Professor C. Vernon Boys employed an electric spark to obtain photographs of flying bullets. They were fired from a pistol, and caused two wires to come into contact, whereby a circuit was closed, and a spark occurring at another gap, a silhouette was obtained on a photographic plate. In this way the various effects produced by a bullet striking and penetrating a glass plate were recorded. At the first contact the surface of the glass was powdered ; then a rounded disc was forced out, and only after the bullet had emerged on the other side did the plate shiver and crack.

Incidentally it was observed that certain black lines appeared in the photograph which corresponded to no visible part of the apparatus. These turned out to be waves in air similar to those which a ship makes in moving through water. They form a V with its apex a little in front of the nose of the bullet, and if an obstacle such as a piece of wood or glass is placed so that one of the limbs impinges upon it, the wave is reflected exactly in the way that theory predicts.

It is interesting to note in this connection that about six or seven years later, Professor R. W. Wood, of the Johns Hopkins University, Baltimore, extended these experiments to the instantaneous photography of waves of sound. Two spark-gaps were arranged, so that while the crack of one caused the aerial disturbance, the other cast a shadow of the wave on a photographic plate. By increasing gradually the interval between the two sparks he was able to trace the wave in ever-expanding circles, just like the ripples in the surface of water when a stone is thrown into a pond. He followed the wave through a small hole in a screen and showed that the wave front beyond formed the surface of a sphere, and similar results were obtained when two or three perforated screens were placed in line. And finally he showed the reflection of the wave at plane, spherical, and parabolic surfaces.

Fig. 310 is from a photograph by Mach of the air-wave produced by a bullet in full flight. The eddies in the wake are very noticeable, and may be compared with the photograph of water

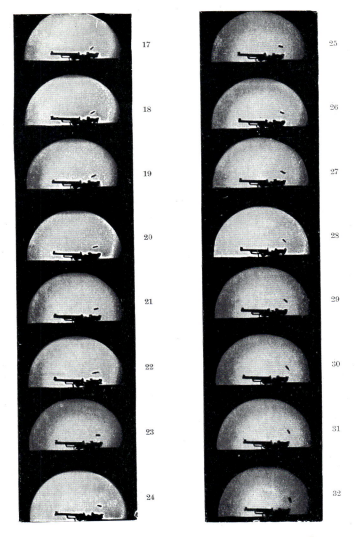

FIG. 314.—A MANNLICHER PISTOL BEING FIRED AND
EJECTION OF THE SPENT CARTRIDGE.

To face page 434.

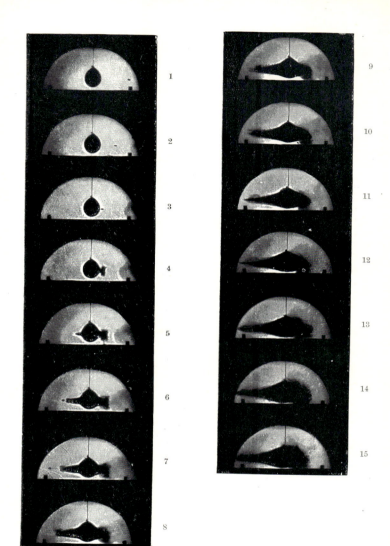

FIG. 315.—PASSAGE OF A BULLET THROUGH A SUSPENDED
BAG OF WATER.

To face page 435.

flowing past an obstacle in Figs. 218–22. It will be observed that the apex of the waves causes the bullet to exert an effect before it actually touches an object. In some experiments the spark is produced by the bullet breaking a thin strip of copper stretched across its path, and it has been shown that the strip is broken by the air-wave. The bullet and strip do not come into contact. When the method of making the bullet or projectile produce its own spark or series of sparks had been devised, the next step was to record successive movements on a film, and this was accomplished (Fig. 311) by Dr. J. Athanasiu in 1903. But the most remarkable experiments in this direction have been made by Professor C. Cranz of the Berlin Academy of Military Technology, who has employed a variety of methods.

So far as slower motions are concerned, a series of sparks can be obtained at regularly recurring intervals by the rotation of a wheel with a number of metal strips on the rim, which come into contact with a fixed metal brush and discharge a battery of Leyden jars. But for motion of the greatest rapidity, the only method is to make use of an apparatus like that employed in wireless telegraphy, which gives a series of sparks at intervals corresponding with the natural period of electrical vibration of the circuit.

As a beautiful example of a connected series of photographs of a piece of mechanism moving far too rapidly to be followed by the eye, we may consider the illustrations in Fig. 312 which represent the ejection of the spent cartridge from a self-acting pistol. The photographs belong to a set of twenty-five, but the seven reproduced show very clearly the path of the empty cartridge. It will also be noticed that specks of unburnt powder are thrown out, and in the last two pictures the new cartridge can be seen rising up in line with the barrel into which it will be thrust by the return of the breech-block. By such photographs the gun-maker can ascertain whether the mechanism is acting properly, for if the new cartridge is not quite horizontal when the breech-block returns, it may jam, and prevent the gun working at a critical moment.

But Professor Cranz's greatest achievement was that of obtaining a cinematographic record of the whole process of firing a gun, including the flight of the bullet through the air and through various obstacles. The apparatus, which was designed in 1909, is shown in diagrammatic form in Fig. 313. An alternating-current generator supplies electricity to a transformer, which is connected to the terminals of a spark-gap placed in front of a large concave mirror. Between the generator and transformer is a pendulum-break, shortly to be described, and between the transformer and the spark-gap is a condenser which gives capacity to the secondary coil of the transformer and determines its period of vibration. Opposite to the mirror is a steel roller, carrying a film, which can be rotated at a regular and known speed. It is 50 centimetres (nearly 20 inches) in circumference, and 28 centimetres (about 11 inches) long, and the speed of rotation can be so great as to give the film a velocity

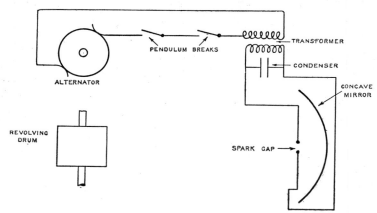

Fig. 313. PROF. CRANZ'S APPARATUS FOR CINEMATOGRAPHY OF FLYING BULLETS.

of 140 metres per second. This in British units is 420 feet per second, or nearly 300 miles an hour. Usually the velocity is 90 metres or nearly 280 feet a second, or about 180 miles an hour. Even this velocity is three times the speed of an express train and one-fifth of the speed of a bullet as it leaves the muzzle of a rifle.

The pendulum-break which has been mentioned consists of a metal pendulum which is held up or released by an electromagnet. Below it are three curved rods of metal, each of which carries a contact piece. When the experiment is ready the pendulum is released by a hand-switch, and operates the contacts in succession. The first contact fires the gun, the second

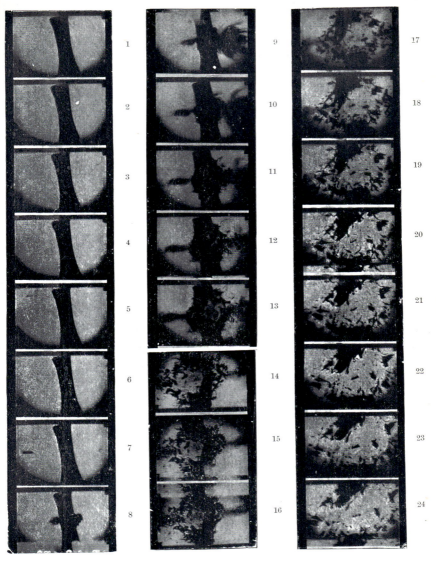

FIG. 316.—PASSAGE OF A RIFLE BULLET THROUGH A SUSPENDED BONE.

To face page 436.

FIG. 321.—THE COOLIDGE
X-RAY TUBE.

To face page 437.

starts the sparks, and the third stops them. The sparks follow one another at the rate of 5,000 a second, and each spark produces a picture, so that 500 pictures can be taken in a tenth of a second, with the alternator making 2,500 alternations per second. Machines have, however, been built which give 50,000 alternations and therefore 100,000 sparks per second.

Some of the results obtained with the apparatus described are illustrated in Figs. 314–16. In Fig. 314 is shown a series of photographs of the firing of an older form of Mannlicher automatic pistol. The first action is to drive out air from the muzzle and a small volcano is observable at the breech. The bullet appears for the first time on the fourth picture, and by the sixth it has passed out of the field of view. Later photographs of the series show the spent cartridge being ejected. The method is of value in ascertaining whether the breech opens before or after the bullet has left the muzzle ; for the latter circumstance would be attended by danger to the user from the backward rush of hot gases.

Fig. 315 shows a series of pictures taken while a shot is passing through a suspended rubber bag of water, and exhibits a curious result. As soon as the bullet has entered, the water is expelled through the opening in the opposite direction to that of the shot. A similar effect is observed when a shot is fired into earth, an effect which in both cases is quite unexpected. When the bullet emerges through the other side of the bag the water follows in its wake, and later pictures convey some idea of the violence of the after effect.

A bullet fired into a suspended bone from a pistol with a small charge drills a hole clean through without breaking it. The effect of a shot from an infantry rifle with a full charge is a complete shatter of the bone *after* the bullet has passed through. The earlier pictures in Fig. 316 show the powdering up of the bone and the projection of the particles backwards, as the bullet enters. When the bullet has passed out of the field of view the bone begins to splinter up, the shock evidently requiring an appreciable time to act upon the particles of which the bone is composed.

Space will not permit of a description of many other interesting results which have been obtained by Professor Cranz, Lieutenant Becker, and others, in the study of the motion of projectiles. Apart from all these applications, photography is employed to record continously all those changes about which man desires information, in cases where personal observation would be neither so continuous nor so infallible. The varying height of the barometer, the temperature, the duration of sunshine, the changes in temperature and volume within the cylinder of an internal-combustion engine, tremors in the earth's crust, and the rhythmic beat of the human heart, are capable of being registered in the almost immeasurably thin film on a roll of sensitized paper.

But enough has perhaps been said to show that Photography has not only undergone a very considerable development in its methods, but has materially assisted discovery and invention in fields widely separated from its own. Its processes are adaptable to the most subtle interpretation of an æsthetic scene, and to the accurate record of scientific phenomena. It is a hobby, a fine art, and a method of precision ; and in all three departments of activity it is unrivalled.

CHAPTER XX

RADIUM, ELECTRICITY AND MATTER

WHEN the element radium was discovered about twenty years ago, it excited widespread interest and brought fame to Madame Curie its discoverer. Radium bromide is a white salt, and the few ounces which have been prepared are worth a king's ransom. It is, therefore, fortunate that a very small quantity will serve anyone's purpose, and having paid so much for it no one wastes it. As a matter of fact, it wastes itself, and in that peculiarity lies its value. The insignificant quantity which has been produced is gradually disappearing, but it has already enabled discoveries to be made which have changed the aspect of physical science.

There are certain periods in the history of science when a discovery or group of discoveries alters the whole trend of thought. Startling as the new acts and phenomena may be, they are overshadowed by the important and far-reaching character of the ideas they suggest and by the influence they exert in modifying views which have come to be regarded almost as irrevocable as the laws of the Medes and Persians. Old mental pictures, fruitful in indicating the direction of further experiment and reasoning, are wiped out, and for a time scientific men are busy painting with tentative and hesitating strokes the new picture of the physical universe.

Of such a nature are the discoveries which have risen out of a study of the properties of radium. For this element, occurring very widely, though in minute quantities, in the earth's crust, is of relatively small importance considered alone. But attempts to explain its properties have shed a new light on the elusive phenomena of electricity and shaken the foundations upon which our most elementary notions of chemistry were laid. Moreover, like most of the profound problems of physical science, the real solution was not attained by a single series of experiments, nor by the work of a single individual. The obstacles have been overcome and the knowledge revealed by a whole army of workers, whose patient investigation and steady, persistent endeavour constitute the silent unseen force which is manifested, but hardly measured, by its results.

439

In order to give some idea of the nature and meaning of radioactivity it will be necessary to pursue several lines of enquiry, and afterwards to correlate them. With so much explanation—offered lest the reader should be wearied by apparent irrelevance—let us consider

THE DISCHARGE OF ELECTRICITY THROUGH GASES

Air and other gases are, at ordinary temperatures, and when dry, non-conductors of electricity. This does not mean that electricity cannot be induced to pass through them at all, but that an enormous electromotive force is required for the purpose. On this fact is based the possibility of transmitting electrical power over long distances by means of bare wires, for, though some leakage does take place, it is mostly through the solid supports or the thin film of moisture which covers them in wet weather. In order that a spark may pass between two balls one inch apart in air, an electromotive force of something like 100,000 volts is required. If points instead of balls are used a discharge takes place more readily, with a hissing sound.

A highly rarefied gas conducts more easily. If it is contained in a tube which can be gradually exhausted, the electrodes by which the alternating current from an induction coil enters may be placed several inches apart. At first there is no discharge, but as exhaustion proceeds a broad band of light appears between the electrodes, which, as the pump is worked, widens until it fills the tube. The colour depends upon the nature of the gas. At one stage there is a flickering appearance owing to the concentration of the light in thin layers which fill the tube from end to end. If in this condition the tube is sealed off, it forms one of the well-known vacuum tubes sold by electrical dealers, which give such beautiful effects when connected up with an induction coil or influence machine.

As the vacuum becomes higher a dark space forms round one of the electrodes—called the cathode—and this space increases as the quantity of gas in the tube becomes less, until it fills the whole tube, the walls of which glow with a faint greenish light. Finally, when the exhaustion is pushed to the fullest extent the electricity refuses to pass, showing that the gaseous matter originally in the tube was necessary to convey electricity through it.

The broad band of light first formed is produced when the pressure falls to about 10 millimetres of mercury ; the striæ or flickering layers are most brilliant at 3 millimetres ; while the dark space fills the tube at about 0·03 millimetre.   There is then present less than $\frac{1}{25,000}$ of the amount of air required to fill the tube at atmospheric pressure.

From the middle of last century these effects excited considerable interest, and many beautiful experiments were devised. Hittorf placed a small mica cross in the tube in front of the cathode, and found that the end of the tube covered by the cross did not glow.   This indicated that something was projected from the cathode which travelled in straight lines.   Ten years

later Sir William Crookes carried out a remarkable series of investigations.   Instead of Hittorf's cross he placed a small wheel with vanes, mounted on an axle in the middle of the tube.   The fact that this wheel rotated when the dark space reached it showed that actual particles

Fig. 317.
CROOKES' RAILWAY TUBE.

of matter were projected across the space between the cathode and the walls of the tube.   The fixed wheel was replaced by one having its axle resting on two glass rails running the length of the tube (Fig. 317), and the wheel rolled from end to end. By reversing the direction of the current through the tube the motion of the wheel was reversed.

Crookes was led to the view that matter in a fine state of division, such as the attenuated gas in the tube, possessed special properties, and he gave to it the name "radiant matter". He showed that if a magnet was held near the tube the stream of particles could be bent out of its original direction so that the wheel did not then turn.   If the stream was concentrated upon a small piece of platinum by means of a concave cathode, the platinum was raised to a red heat by bombardment, while various substances which possess the property of fluorescence (see page 443) glowed in similar circumstances with their characteristic colours.

Of these properties the most important is the deflection by a magnet (Figs. 318 and 319) ; for this deflection is just what would occur if the stream were composed of tiny particles carrying charges of negative electricity. Such a stream would be equivalent to a current of electricity, and the interaction between the magnetic field of this current and that of the magnet would cause the more movable one to twist round so that the lines of force of the two coincided.

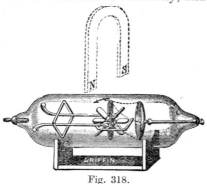

Fig. 318.
CROOKES' MILL-WHEEL TUBE, SHOWING ALSO DEFLECTION BY A MAGNET.

A further property was discovered in 1894 when P. Lenard constructed a tube with a thin aluminium window at the end opposite to the cathode, and found that the rays would penetrate it. Outside the tube they caused a cloud to form in moist air, and, by rendering the air a conductor, discharged an electroscope.

It may be well here to devote a few words to the gold-leaf electroscope which, while one of the simplest and commonest pieces of electrical apparatus, has proved in relation to radioactivity to be one of the most delicate instruments of research.

Fig. 319.   DEFLECTION OF CATHODE RAYS BY A MAGNET.

It consists ordinarily of a pair of strips of gold leaf attached to the bottom of a metal rod. This rod is fixed by means of paraffin wax, ebonite, or other nonconductor in the neck of a flask, or in the top of a box with glass sides. A good type for ordinary purposes is shown in Fig. 320. A more suitable form for measurement is one in which the rod terminates in a metal plate, and a single strip of gold leaf is attached at the top edge so that it hangs as a hinged flap.

When the instrument is electrified the leaf is repelled from the plate, and falls as the charge leaks away. A graduated scale enables the rate of loss of charge to be measured by the rate at which the leaf falls, and this gives a measure of the conductivity of the air. In this way Lenard showed that the rays are absorbed by various substances at rates which are proportional to their densities. This explains why, of the commoner metals, the light metal aluminium is the most suitable for a window, and a heavy metal like lead as a screen. It justifies, moreover, the use of aluminium for anode and cathode ; a heavy metal like platinum is, it will be remembered, rendered incandescent by the bombardment of the rays.

Further investigations, mainly by Sir. J. J. Thomson and his pupils, have proved that these " cathode rays " produce the same

Fig. 320. CHATTOCK'S GOLD-LEAF ELECTROSCOPE.

effects whatever the gas in the tube. By an ingenious arrangement into which we cannot enter here Sir J. J. Thomson measured the mass of the flying particles which compose the stream, and showed that in all cases they were about $\frac{1}{1.700}$ of the weight of a hydrogen atom. More recent measurements by Professor Millikan, of Chicago, conducted with extraordinary precautions to secure accuracy, give the fraction as $\frac{1}{1.840}$. They are projected from the cathode with a velocity which may reach one-third the speed of light, or 60,000 miles a second, and as the results are always the same whatever the gas in the tube the conclusion is inevitable that the radiant particles are common to all matter. In 1897 Sir. J. J. Thomson advanced the view that these *electrons*, as they are called, are actually present in the atom—that a group of electrons, in fact, composes the atom—and that they are torn off during the passage of the electric current. Additional evidence was forthcoming through a series of investigations then proceeding in France, but before describing these it will be convenient to consider another kind of radiation produced when the flying electrons strike an object in their path.

X-RAYS

In 1896, Professor W. C. Röntgen was using one of the tubes which have been described when he found that some photographic plates which were in a drawer in the bench were fogged. From this he was led to the discovery of the X-rays which are produced by any solid body when it is bombarded by a stream of radiant matter or electrons. Usually a disc of metal is mounted centrally in a bulb, and the electrons are directed upon it by passing an electric discharge through the bulb. The rays affect a photographic plate, cause many substances to *fluoresce* or shine with a light of characteristic colour, and discharge an electroscope. They are absorbed to a greater extent by dense substances than by light ones, passing readily through paper, flesh, etc., but with less ease through bone and metals. This difference of penetrative power rendered them of immense value in surgery. When they were passed through the hand, arm or thinner parts of the body a shadow of the bones and of any denser substance such as a ring or an embedded needle, was cast upon a fluorescent screen or a photographic plate beyond.

The source of electricity was at first an induction coil with the ordinary " make-and-break " in the primary circuit or an influence machine. But much research has been carried out on the best strength of current, voltage, and frequency, and a great deal of ingenuity has been employed in devising means of interrupting the primary current in the coil. It was found, too, that that amount of residual gas in the bulb affected the penetrating power of the rays, and the result of many investigations has been not only greater uniformity but an increased power of the X-ray tube. The most effective form is that invented by Mr. Coolidge, and is shown in Fig. 321. It possesses three special features : First a high exhaustion rendered possible by the more effective air-pumps of the present day (see p. 397).

FIG. 322.—X-RAY PHOTOGRAPHS OF BUTT
WELDS IN MILD STEEL.

To face page 444.

FIG. 323.——HEXAGONAL PRISMS OF CALCITE TERMINATED BY
RHOMBOHEDRA.

FIG. 323A.——SCALENOHEDRAL CRYSTALS OF CALCITE :
"DOG-TOOTH SPAR."

To face page 445.

Secondly, a heavy tungsten anode, which will withstand severe electrical bombardment without fusing. Thirdly, a coil of tungsten wire as cathode which can be heated by a separate current and emits the electrons necessary to render the space conducting. The wire is surrounded with a molybdenum tube or hood to focus the cathode stream on the anode.

With modern apparatus the whole body can be photographed, and important discoveries with practical results have been made. Thus the alimentary canal, or tract through which food passes and which extends the whole length of the body with many twists and turns in the intestines, has been rendered visible. This is accomplished by giving the patient a meal of bismuth carbonate —say 2 ozs. of the carbonate in 10 ozs. of porridge, which is so opaque to X-rays that a shadow of the tract as the food passes through it may be thrown upon the plate. It is found that local disease of the canal is accompanied by general trouble—you cannot have a clean stomach and appendicitis at the same time. Another valuable application is to a joint such as the knee, for it is possible to discover at an early stage the existence of tubercular disease. And its original use for detecting the presence of foreign bodies such as a needle or a bullet was extended to hundreds of thousands of cases during the war.

Perhaps an even more striking example of the increase in penetrative power is furnished by its use in engineering and metallurgy. Flaws in forgings and castings and imperfect welds in joints can readily be detected provided the metal is not more than say two inches in thickness. Fig. 322 is an X-ray photograph of a butt weld in mild steel and shows blow-holes due to ineffective workmanship.

For a long time the nature of the rays was in doubt, but as more and more of their effects became known it was evident that they were electromagnetic or light waves, with a wavelength very much smaller than the waves of ultra-violet light— in fact, about one-thousandth of the length. But they differ from ordinary light waves in not being continuous. They consist of short trains of waves, or pulses, produced by the impact of the electrons upon the anode. And just as waves of light enable us to investigate the structure of bodies through which they pass— just as they enable us to picture the mechanism of refraction, dispersion, and polarization—so also the more minute wavelets in the X-rays provide us with an instrument of surpassing delicacy for the discovery of secrets which the grosser light rays fail to reveal. To understand how this is accomplished we must first explain.

THE PHENOMENA OF DIFFRACTION

Though the statement that light travels in straight lines is true so far as propagation in a continuous medium of uniform density is concerned, it *does* bend round the edge of an opaque object, and the larger the waves the greater is the bending. The red rays are bent more than the yellow, the yellow more than the green, the green more than the violet. This explains why an opaque object held between the eye and a distant source of light appears to be bordered by coloured fringe. Each of these is a small spectrum, but the colours are so close together that they cannot be clearly distinguished. The same effect is observed when light coming from a narrow slit is examined.

Now suppose that light of one wave-length only issues from a slit and then by means of a lens passes as a parallel beam through a number of slits formed by ruling, say, 6,000 to 14,000 lines with a diamond on a glass plate. If the light from this grating, as it is called, is focussed on a screen, bending will cause the rays to travel different distances from the slits to reach any one line on the screen. When two waves which have travelled distances differing by half a wave-length (or any odd number of half wave-lengths) fall on the same line, the crest of one will coincide with the crest of the other, they will be mutually destroyed, and that line on the screen will be dark. But if the waves differ by a whole wave-length (or an even number of half wave-lengths) crest will coincide with crest, trough with trough, they will reinforce one another and there will be light on the screen. We shall have, therefore, alternate bands of light and darkness, and as each slit has two edges these will be arranged symmetrically on each side of a centre line formed by the light which passes directly through the slit.

If the experiment be repeated with white light, each of the bright lines will be broadened into a band of colour, and these bands are spectra. Because they are produced by diffraction at the edges of slits they are called diffraction spectra, and they are said to be of the first, second, third, etc., order according to their position with reference to the centre line. They become fainter as they become more remote.

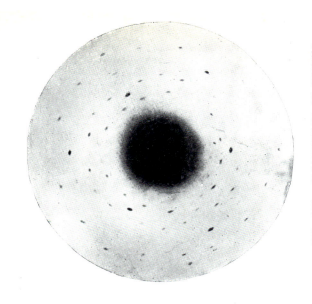

FIG. 325.—X-RADIOGRAM OF CRYSTAL OF QUARTZ, PERPENDICULAR TO AXIS.

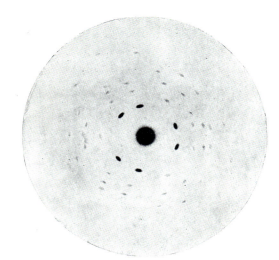

FIG. 324.—X-RADIOGRAM OF CRYSTAL OF ZINC BLENDE, PERPENDICULAR TO OCTAHEDRON FACE.

To face page 446.

FIG. 332.—MODEL SHOWING ARRANGEMENT OF THE ATOMS IN A CRYSTAL OF ALUMINIUM OXIDE (RUBY).

FIG. 331.—MODEL SHOWING ARRANGEMENT OF THE ATOMS IN A CRYSTAL OF ZINC SULPHATE.

FIG. 330.—MODEL SHOWING ARRANGEMENT OF THE ATOMS IN A CRYSTAL OF POTASSIUM OR SODIUM CHLORIDE.

To face page 447.

It will be clear that a diffraction grating can be used instead of a prism in a spectroscope, and that when mounted in this way it will be possible to measure the amount of bending which has taken place for any point in any of the spectra which are not too faint to be seen. And since the amount of bending which a ray undergoes depends entirely upon the wave-length and thickness, the wave-length may be determined. By this method, the lengths of the waves which form the visible spectrum have been calculated (see p. 418). For rays of long wave-length, beyond the red end of the spectrum, the eye is replaced by the Bolometer, an instrument of great delicacy invented by Professor S. P. Langley for absorbing and measuring heat waves. It consists of two blackened strips of platinum foil, $\frac{1}{10}$ mm. $\times \frac{1}{1000}$ mm., placed in the arms of a Wheatstone bridge. When radiation falls on one of these its resistance alters and the galvanometer is deflected. For rays beyond the violet end the eye is replaced by the sensitive film of a photographic plate because these rays exert a powerful photographic action.

Instead of a ruled glass plate, a reflecting grating, made by ruling fine lines upon a metal mirror, is often employed. The action is precisely similar. The waves which reach the eye or a line on a screen have travelled different distances from the strips and produced alternate bands of light and darkness.

The phenomena have so far been studied as ocurring in a regular way, but there are many cases in Nature in which diffraction occurs by light waves striking small particles and becoming " scattered ", which is only another way of saying that they are bent in various directions. Thus, when a beam of light passes through a smoke or dust-laden atmosphere, or through a liquid in which small particles are suspended, the track of the beam can easily be seen by looking at it at right angles. The amount of scattering depends upon the wave-length and the size of the particles. Short waves are more freely scattered than long ones, and the blue of the sky, which is always deepest at right angles to the sun, is due to the scattering of short waves by the small particles in the atmosphere. The particles which will effect this are too small to be seen by the naked eye, so the fact that scattering occurs furnishes information as to a structure which is invisible. In Chapter XI it was explained how this method has been used to detect the presence of particles beyond the range even of microscopic vision, and the actual existence of which has only been proved in recent years by the ultra-microscope.

Nearly every substance known can, under certain conditions, be obtained in a crystalline state (Fig. 323) and crystals never fail to excite admiration and wonder by their colour, brilliancy or regularity of form. Such regularity is an outward and visible sign of an internal order and arrangement which has hitherto been concealed from us. Light passes through or is reflected from the plane faces, but just as the red rays pass through the dusty atmosphere while only the blue rays are scattered, so the whole of the waves of the visible spectrum are too coarse to be affected by the atoms and molecules of which a crystal is composed. And we might have remained ignorant of these facts for all time if it had not occurred to Von Laue, in 1912, that as X-rays were, in effect, light waves of much shorter wave-length, they might be affected by atoms in the same way that dust particles scatter the shorter light waves from the sun. And surely enough when Friedrich and Knipping in 1913 passed a narrow beam of X-rays through a thin plate of crystal on to a photographic plate they secured a pattern which depended upon the form of the crystal they employed.

Figs. 324 and 325 are photographs produced by the passage of the beam through zinc blende and quartz. The large spot in the centre is due to the portion of the beam which passed directly through the crystal, but the smaller spots all round it are produced by rays which were deflected by the atoms then encountered. The regular arrangement of these smaller spots indicates a corresponding regularity of the atomic positions, but it is not possible to say from an inspection of the flat pattern exactly what this arrangement is. Nevertheless this regularity suggests analogy with diffraction by means of a grating rather than with the scattering of irregularly arranged particles in the atmosphere, and it occurred to Professor W. H. (now Sir William) Bragg to apply similar methods of investigation. So he devised the X-ray spectrometer for the purpose. In the hands of Sir William Bragg, his son Professor W. L. Bragg, and other workers, this instrument has made wonderful progress.

The X-ray spectrometer is an instrument in which an X-ray tube takes the place of a source of light, and a crystal takes the place of a prism or grating. The rays are directed upon a cleavage plane of a crystal at a certain angle, and are reflected from it. The reflected ray is received in an ionising chamber. When it enters this chamber the air is ionised, and the fact is detected by means of an electroscope. The arrangement is shown diagrammatically in Fig. 326.

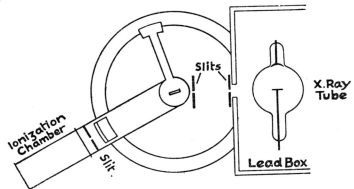

Fig. 326.   X-Ray Spectrometer.

Now that the facts that crystal is bounded by plane faces suggests that the atoms or molecules are arranged in layers parallel to each cleavage plane, so that when the waves which constitute the X-rays fall upon each layer they will be partially

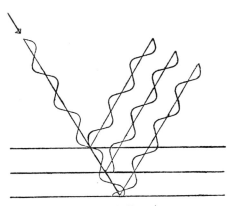

Fig. 327.   Diagram to Explain Interference in a Crystal.

reflected. The waves reflected from each layer (see Fig. 327), will reinforce or destroy one another according to the even or odd number of half wave-lengths they have travelled. This number will vary according to the angle at which the incident ray strikes the face, the distance apart of the two layers of molecules, and the wave-lengths. If the wave-length is known then since the angle of incidence can be measured it is easy to calculate the distance between the atomic layers. Further, if the process be repeated for the other two faces it is possible to construct a model showing the arrangement of the atoms in space.

By such methods the lattices shown in Figs. 328 and 329 have been devised. Another method of building up crystal models is to use small spheres in contact as in Figs. 330, 331, and 332. In these examples it is not intended to convey the idea that the

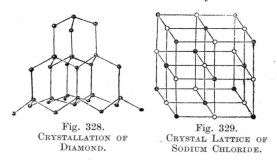

Fig. 328.
CRYSTALLATION OF
DIAMOND.

Fig. 329.
CRYSTAL LATTICE OF
SODIUM CHLORIDE.

atoms are simple solid masses completely filling the spaces allotted to them. Each sphere represents merely a space which is not occupied wholly or partially by any other atom. Symmetrical as crystals are when viewed externally by ordinary light waves, the existence of law and order is much more strikingly manifested when they are analysed by the almost infinitesimal wavelets from an X-ray tube, and a beauty which is intelligible becomes more entrancing than it was before.

RADIOACTIVITY

While bodies generally become luminous only when heated, there are a number of substances which can be induced to emit light without any rise of temperature. This light is usually of a characteristic colour. Thus, when a strong beam of sunlight

FIG. 335.—ENLARGE-
MENT OF A PORTION
OF FIG. 334.

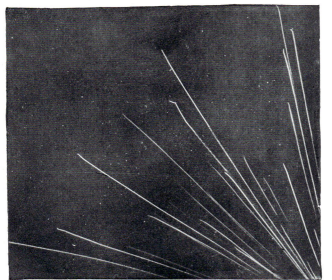

FIG. 334.—PHOTOGRAPH SHOWING TRACK OF AN
X-PARTICLE IN AIR.

To face page 450.

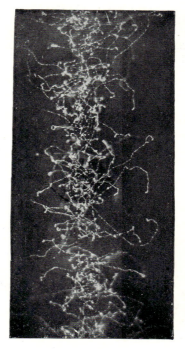

FIG. 336.—PHOTOGRAPH SHOWING
TRACK OF A BEAM OF X-RAYS
IN AIR.

To face page 451.

is passed through sulphate of quinine in water a beautiful blue glow suffuses the liquid. Similarly, fluorescein gives a brilliant green glow ; uranium glass, a canary yellow substance, appears green in strong light ; and so on. This phenomenon is known as *fluorescence* and the glow ceases as soon as the light is cut off.

A number of other substances possess a similar property when excited by exposure to light, but retain it after the light has been removed. Thus Balmain's luminous paint, which is composed of calcium sulphide, has long been a source of juvenile amusement, because any object painted with it and exposed to strong light will continue to shine for some time in the dark. In order to distinguish this from the property described in the last paragraph, the term *phosphorescence* is used. The distinction between them appears, however, to be one of degree only.

The glow on the walls of a vacuum tube when the dark space fills it is a case of fluorescence, and is apparently caused by the bombardment of the glass by the negative electrons. The Röntgen rays, again, are capable of producing brilliant effects, and the screen upon which the shadows are cast is usually coated with barium platinocyanide, a yellow salt which glows with a greenish light under the rays.

Shortly after the discovery of the X-rays by Professor Röntgen, Professor E. Bequerel repeated and extended some experiments made by Nièpce de St. Victor thirty years before, and demonstrated that the salts of uranium, which are capable of phosphorescence, will effect a photographic plate in the dark. He also showed that uranium caused the discharge of a gold-leaf electroscope. Further investigation revealed quite a number of substances possessing this property, and they were said to be *radioactive*, or to possess the property of *radioactivity*.

During the next four years, Professor and Madame Curie examined a large number of minerals containing uranium, and found that their radioactivity varied considerably. They came to the conclusion that the cause was a substance or substances occurring in the minerals in minute but varying quantity. Finally by a long and tedious process, they extracted radium from Austrian pitchblende and found that it possessed a radioactivity over 1,000,000 times greater than the uranium salts which had previously been used.

One ton of pitchblende contains about 0·37 gramme, or less

than $\frac{1}{70}$ of an ounce of radium, and only half of this can be obtained owing to losses in the process of extraction. In appearance and properties radium salts are very much like those of barium, and compounds of the two elements have to be separated by means of slight differences in solubility. When a solution of radium and barium bromides is cooled, the radium bromide separates out first, and this process has to be repeated over and over again until tests with the electroscope show no increase in activity. It is on account of the tediousness of this process that radium salts are worth many times their weight in gold.

The temperature of radium compounds is always about 1·5 C. higher than that of their surroundings. They decompose water, yielding oxygen and hydrogen, and this fact, together with other properties, indicates the liberation of an enormous amount of energy from an apparently inexhaustible store. Exact measurement shows that radium is continually producing sufficient heat. to raise its own weight of water from the freezing-point to the boiling point every hour.

In addition to exercising photographic action and rendering the air conductive, the rays from radium cause phosphorescence in a large number of substances ; they discolour paper and glass and cause many chemical changes to take place which ordinarily require special conditions. When allowed to fall on any part of the body for a considerable time they cause painful sores which are deep seated and difficult to heal. On this account radium salts are being used in an attempt to destroy cancer and other growths which can usually be removed only by the surgeon's knife. The emission of the rays is independent of the temperature or of any influence which man can bring to bear. It has started when he first finds it, and goes on uninterruptedly, in spite of him. It presents new phenomena to his mind, new problems for his reason to grapple with, and gives a striking stimulus to his imagination.

The total quantity of radium which has so far been obtained does not amount to more than a few ounces. Yet so powerful is its radiation that the electroscope is able to reveal the presence of a quantity smaller than the most accurate balance can measure, or the spectroscope, hitherto the most delicate of all instruments, detect. Most hospitals possess some of it, and for twenty years a host of workers have been investigating its ways. Into these let us enquire a little more closely.

FIG. 338.—OUTSIDE VIEW OF THE LATEST
FORM OF THE SHIMIZU'S APPARATUS
FOR PRODUCING CLOUDS OF IONS.

To face page 452.

THE CAUSE OF RADIOACTIVITY

It has been found that the radiation is of three kinds which are known as α (alpha), β (beta), and γ (gamma) rays. The α-rays are deflected slightly by powerful magnetic forces and have but slight penetrative power. A few layers of paper or an inch or two of air will cut them off entirely. The experimental evidence points to the view that they are atoms charged with positive electricity and shot off from radium with a velocity of nearly 20,000 miles a second.

The β-rays are strongly deflected in the opposite direction to the α-rays by much weaker magnetic forces. They carry a charge of negative electricity and have a velocity which, in some cases, approaches that of light—185,000 miles a second. In penetrative power and in practically every other respect except their greater velocity they are similar to the cathode rays in a Crookes' tube, and they are, therefore, believed to be electrons with a negative charge.

The α-rays are not deflected by a magnet, and they penetrate many bodies which are opaque to ordinary light. They affect a photographic plate, excite fluorescence, and behave exactly in the same way as Röntgen rays ; and like them their precise nature was, for some time, a matter of speculation.

The track of both α- and β-particles in air can actually be traced by a very beautiful method devised by Mr. C. T. R. Wilson in 1909. Many years ago Mr. John Aitken found that if air or any other gas was saturated with water vapour, then cooling the gas did not cause separation of this moisture and the formation of cloud unless fine particles of dust were present. It was subsequently shown that as the diameter of a drop of water at any given temperature decreased, the tendency to evaporate increased. The rate of evaporation is clearly proportional to the area of the surface. Now the volume of a sphere of radius r is given by the formula

$$V = \tfrac{4}{3} \pi r^3,$$

and the area of surface by the formula

$$A = 4 \pi r^2.$$

The volume of a sphere decreases, therefore, as the cube, and the surface as the square, of the radius. If the radius decreases from three to two the volume will decrease from twenty-seven to eight, while the area will decrease from nine to four, so that there ratio of area to volume will increase from nine and twenty-sevenths to four-eights, or from one-third to one-half. A small globule of water gradually decreasing in size will, when its radius reaches a certain lower limit, flash off into vapour. Conversely, the formation of drops from vapour is difficult unless there is some solid object, of more than this limiting curvature, upon which the drop can form.

Aitken showed that the fogs of towns were due largely to the smoke around the small particles of which the water could be deposited. The water-drops in fog or mist fall slowly to the ground. Their own weight pulls them down and the friction

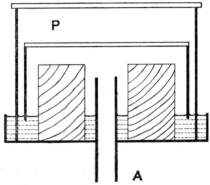

Fig. 333.   DIAGRAM OF WILSON'S EXPANSION CHAMBER.

of the air on their surfaces offers resistance to their motion. The relation between volume and surface discussed in the preceding paragraph shows that the more minute the particles the greater will be the retarding influence of friction and the more slowly will they subside. Sir George Stokes showed how to calculate the rate of subsidence from the size of the drops and the viscosity of the air, so that if the rate could be measured the size of the drops could be calculated.

Now a charged atom or electron is capable of acting as a nucleus upon which moisture can condense. If air is charged with water vapour and, quite free from any radioactive or electrical influence, is caused to expand suddenly no mist is formed. But if it be exposed to a radioactive substance a cloud immediately forms on expansion, and by measuring the rate at which the upper surface of the cloud falls and knowing the amount of water precipitated from the extent of the expansion, the size and number of drops can be calculated.

The chamber in which the expansion was produced is shown diagrammatically in Fig. 333. When the space underneath the piston P was connected through the tube A with a larger vessel containing air at lower pressure, the piston fell, the air in the chamber expanded, and a cloud was formed. The method enabled both α- and β-particles to be counted and a separation to be made. For if the volumes of the air before and after expansion was in the ratio 1 to 1·25, the α-particles alone were instrumental in producing cloud, while if the ratio was 1 to 1·31, the β-particles also became involved.

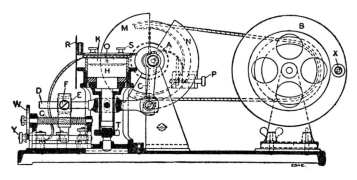

Fig. 337. SECTION OF SHIMIZU'S APPARATUS FOR PRODUCING CLOUDS ON IONS.

Three years later, Mr. Wilson succeeded in rendering visible the track of a single α-particle. As the particle pursues its way through the gas it leaves a path of ions—either from itself or other atoms—and on expansion this is marked out by a streak of cloud which can be recorded on a photographic plate. Figs. 334 and 335, show the effect which is produced. Fig. 336 shows also the ions which are formed by a beam of X-rays traversing the gas. It has been found that a single α-particle produced from 2,000 to 6,000 ions per millimetre and a β-particle from twenty to thirty. The reason for the greater effect of the α-particles in rendering a gas through which they pass conductive is, therefore, explained.

The apparatus has been modified recently by Mr. Takeo Shimizu, who finds that sudden expansion is not necessary to produce the cloud. In Fig. 337, K is the expansion chamber and the piston H is given a rapid oscillatory motion by the crank H, operating through the rod D. The piston and the glass top of the chamber are covered with gelatine which will absorb a large quantity of water and keep the air saturated. From 50 to 200 expansions can be produced per minute. An outside view of the apparatus is given in Fig. 338.

It is necessary to distinguish between the properties of particles shot off from a radioactive substance with a velocity of 20,000 and 60,000 miles or more per second, and of those particles when their progress has been stopped by collision with the molecules of a gas under ordinary conditions of temperature and pressure. Their path is then zig-zag, and though their velocity at any given instant is far greater than any we are able to obtain by mechanical means, the moving charges vary so rapidly in direction that the electrical and mechanical effects observed under low pressure are not possible.

THE NATURE OF RADIOACTIVITY

If a radium salt is dissolved in water and then evaporated to dryness, or if the dry salt is merely heated for a few hours, the activity is decreased. The emission of β- and γ-rays is stopped altogether for a time, and the quantity of α-rays is reduced to one-fourth. Yet apart from its radioactive properties the radium compound is in no way changed. It has the same appearance, and, so far as the balance is a test, the same weight as it had before. In course of time it recovers its activity, and this invariably occurs no matter how many times the operations may be repeated.

But some material substance escapes during the process. A gas—originally called *radium emanation*—can be collected, and this on examination is found to possess exactly the same amount of activity that the radium has lost. The quantity obtained is excessively minute—so small as to be outside the range of the most accurate balance, and barely measurable in the most accurate apparatus for determining volume. From a gramme of pure radium the emanation would amount to only 0·6 of a cubic millimetre—which is about the size of a pin's head. Yet, says Mr. Soddy in his fascinating book on *Matter and Energy*, " the rays from far less than a thousandth part of this quantity will cause zinc sulphide to floresce so brilliantly as to be plainly visible in an absolutely dark hall to a thousand people." Moreover, he adds, if one-thousandth of this quantity " were mixed uniformly with the air of a very large hall, say of 100,000 cubic feet—or 3 tons by weight—of air, no delicate instrument such as is customarily employed in the measurement of radioactivity could be worked in the hall, and the amount in a single cubic inch of the air could still be detected by a sensitive gold-leaf electroscope."

The heat evolved from the emanation is in proportion to its activity ; the amount obtainable from 1 gramme evolves heat three-fourths as fast as 1 gramme of the element itself, while the latter evolves heat at only one-fourth of the rate before the separation. This means that 0·6 of a cubic millimetre produces heat at such a rate that it will raise 1 gramme of water from the freezing to the boiling-point in an hour. If a cubic inch of the gas could be obtained it would be equivalent in heating power to a powerful arc lamp ! Has ever a more astounding statement than this been made in the whole history of science ? A cubic inch of gas producing spontaneously so much energy that it would fuse any vessel in which it happened, for a time, to be confined !

But the marvels of these new substances are not yet exhausted. The radioactivity of the emanation decays at the same rate as that at which the radioactivity of the original radium is recovered. After four days only half of the original activity remains, and within a month the emanation has disappeared from the tube and has been replaced by helium. This is a gas which for many years had been known to exist only in the sun, where it was recognized by a line in the spectrum corresponding with that of no terrestrial element. When Sir William Ramsay and Lord Rayleigh discovered argon in the air, the former began an exhaustive search for this or other gases in the minerals of the earth's crust, and among the gases evolved on heating certain rare minerals he found helium.

The activity of the emanation as measured by the electroscope effect is due largely to the α-rays, which are now known to be electrons with a positive charge, and many times larger than the negatively charged electrons which constitute the β-rays. The fact that the rate of recovery of radium is the same as the rate of decay of the emanation, suggests that the latter is formed by disintegration of the former. When the radium breaks down into emanation, a-particles are produced, and in the inter-atomic commotion that ensues the negative electrons are flung off, and the tremor which spreads outwards through space produces the effect of the γ-rays. The relatively greater activity of the emanation is explained by the statement that one-fourth of the a-rays are produced by the change of radium into the emanation, and three-fourths by the change of the emanation into helium, while the whole of the β- and γ-rays are formed in the second process. Both these changes are occurring in the original radium, and it is only when the accumulated emanation has been expelled that the separate influences can be distinguished.

Helium is not the sole product of the emanation. A minute trace of a radioactive solid is left behind which is called polonium and a whole series of substances are formed successively with an evolution of helium at each stage. Some of these exist for but a brief period, the successive changes occurring so rapidly that they can with difficulty be followed. But there is reason to believe that a final stage is reached at which the product is stable and radioactivity ceases.

The most interesting rays, therefore, are those composed of a-particles. Their mass is more than a thousand times greater than the mass of the β-particles, and their loss must be most important in the destruction of the radium atom. By a marvellously ingenious process, Rutherford has succeeded in counting the number of particles emitted in a given time, and he can detect the loss of a single member of the stream. By measuring the rate at which they are radiated he has been able to calculate the length of time which would be required for radium to become extinct, and has been able to estimate it at about 25,000 years. From careful estimates which have been made it would appear that the whole of the radium in the earth must have disappeared long ago unless there was some regular source of supply. And that could only be the case if some substance with a much slower rate of change was producing radium as a result of its decomposition.

Now it has been shown that helium is a decomposition product of radium, and before radioactivity had been discovered Sir William Ramsay had found helium always present in minerals from which radioactive substances were subsequently obtained. It seemed probable, therefore, that radium itself was an intermediate product, formed by the disintegration at a slower rate of some other constituent of the minerals in which it is found. Moreover, as uranium and thorium—elements with the heaviest atoms known to chemists—are invarably present in these minerals, and possess a feeble radioactivity which suggests a slow rate of change, it seems feasible to imagine that there is a continuous process by which these elements are breaking down into others with lighter atoms of more or less stability. And this would go on until a stable substance was formed possessing no radioactive properties whatever.

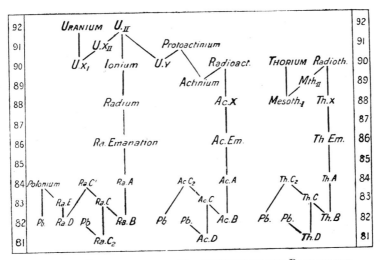

Fig. 339.   DIAGRAM SHOWING RELATIONSHIP OF THE RADIOACTIVE ELEMENTS.

It is important to note in this connection that the ratio of radium to uranium in minerals is very nearly constant—about 1 to 3,000,000—and this is what would be expected if uranium were the parent of radium. Moreover, if the average life of radium is 25,000 years, the average life of uranium would be 3,000,000 times as long or 7,500,000,000 years. The rate of change is so slow that though experiments have been going on for years the change from uranium to radium has not yet been

detected in the laboratory.   If a uranium salt—the nitrate in
this case—is shaken up with water and ether, the lower aqueous
layer contains uranium and gives α- and β-rays, and the upper
ethereal layer gives α-rays only.   The substance which uranium
appeared to be producing is known as uranium X.   The uranium
X lost its radioactivity at the same rate that uranium proper
regained it—a result precisely similar to that of radium and its
emanation, but the period in this case was from 6 to 12 months.
The process could be repeated over and over again.

Experiments of this kind led to the conclusion that uranium
broke down into uranium X and helium, that uranium X broke
down into a radioactive solid called ionium, ionium was continuously and spontaneously passing into radium, that radium
passed through its emanation (the inert gas niton) into polonium,
and that finally polonium passed into lead.   Further experiment
has revealed many stages and the complicated character of the
changes so far as they are at present known is shown in Fig. 339,
which is taken from an article by C. G. Darwin in *Nature* of
16th September, 1920.   In the table a movement of two points
downwards on the vertical scale corresponds to the loss of an
α-particle from the atom, while a movement of one point upwards
corresponds to the loss of a β-particle.   The final product is lead,
and as this element is not radioactive the transformations do not
proceed any further.

ELECTRICITY AND MATTER

We are now in a position to consider some of the views which
are now exercising such a profound influence on scientific thought.
For centuries man has been accustomed to regard matter as
being made up of distinct particles, and since the time of
Lavoisier, the chemist, had become more and more convinced that
the relative quantities which pass into or out of chemical combination represented truly the smallest particles of whose
existence it was possible for the mind to conceive.   And so far as
the balance was able to testify, the chemist was justified in his
attitude.

An instrument of far greater accuracy was the spectroscope,
for it enabled elements to be detected when they were distributed
throughout such an enormously greater quantity of other
materials that unless their presence had been suspected the

ordinary methods of chemical analysis would have passed them by.

Radioactivity is a property of matter which until the last twenty years or so was unknown—a property, moreover, which is susceptible of the most delicate measurement. The step from the balance to the spectroscope—great as it was—is smaller than the step from the spectroscope to the gold-leaf electrometer. And by its aid it has been found that a few relatively rare elements are undergoing a process of disintegration into bodies of lower atomic weight.

Now though these processes are confined to a few substances, Sir J. J. Thomson and his pupils have, by experiment on the discharge of electricity through all gases, produced certain effects which are precisely similar to those which occur spontaneously during radioactivity. The β- and γ-rays have precisely the same properties respectively as the cathode stream and the Röntgen rays. So there is strong reason to believe that any view of the constitution of the radioactive substances is true of those which are not radioactive. At the same time the enormous amount of energy liberated during radioactive disintegration gives an idea of the magnitude of the forces which must exist in the interior of an atom.

But the most striking feature is the fact that when the ultimate particles of matter are revealing their properties, it is not their masses which we are able to measure, but their electrical charges. For without such charges they would apparently possess only attributes which defy measurement—at any rate, on so small a scale as is performed by the electroscope. The study of radioactivity, therefore, combined with that of electrical discharges in high vacua, leads to an electrical theory of matter in which the properties of each individual substance are determined merely by the number and arrangement of ultimate particles which are common to all the stuff of which the world is composed.

Let us now go into the question a little more closely and endeavour to picture the complicated unseen processes which the results of scientific experiment and reasoning have enabled the pioneers in the new field of knowledge to describe in such detail. It is an elementary fact that the chemist, relying on his balance, had found that when two or more elements combine with or replace one another, they do so in definite proportions. Thus 1 gramme of hydrogen combines with or replaces 8 grammes of oxygen, 31·5 grammes of copper, 28 grammes of zinc, 39 grammes of potassium, 16 grammes of sulphur, and so on. This indicates that they consist not of a continuous material, but of a material consisting of tiny grains, all of the same weight, which act individually in chemical change. And this granular property of matter is forced on us by a host of common experiences. For if a minute quantity of a powerfully scented substance is liberated in a room there is soon no corner in which it cannot be detected by the sense of smell. The quantity necessary for this purpose may be weighable, because a balance has now been constructed which will weigh $\frac{1}{10,000}$ of a milligramme or $\frac{1}{2,800,000}$ of an ounce ; but, even so, the amount in a cubic foot of space would be far too small to affect an instrument even of this delicacy.

Again, if all the air is pumped out of a vessel as far as that can be done with the most powerful and effective pump, and a quantity of gas occupying the smallest measurable volume at ordinary temperature and pressure is admitted, it will expand throughout the whole space, so that it can be detected by the spectroscope. It is scarcely conceivable that the small quantity has a jelly-like structure, and has filled the vessel *continuously*. Rather is it easier to imagine that the gas is composed of small grains or atoms, which spread out until there is an equal number of them in every cubic inch. Such atoms would be the particles which are concerned in chemical change.

In order to distinguish between substances that can and substances that cannot be split up, the chemist calls the former compounds and the latter elements. At various times in the history of chemistry substances formerly thought to be elements have been found to consist of two substances, and in this way new elements have been added to the list until the number stands at eighty or thereabouts. But in every case the decomposition has been effected by ordinary chemical or physical methods. True, some compounds are unstable under ordinary conditions and decompose spontaneously, but an element always has the same properties when pure, and neither decomposes spontaneously, nor can it be decomposed by any means available.

Having stated that the atom of an element was the smallest indivisible particle of matter the chemist went no further. In his experiments it behaved as a whole ; so far as he could judge it was solid throughout, and he had no reason to believe that it consisted of smaller particles bound together by forces so great that his methods were of no avail to render them asunder. Moreover, the spectroscope showed that these same elements existed in the sun and stars, in which the range of temperature was far wider than on the earth, and in that respect the spectroscope supported the evidence of the balance.

But the spectroscope is more delicate than the balance. It distinguishes elements, not by the relative weights of their atoms, but by the waves set up in the all-pervading ether of space by atomic or intra-atomic vibrations ; and it is difficult to account for some of the phenomena revealed by the spectroscope if the atom is regarded as a single solid particle. These phenomena suggest a complex structure, and a complex structure is necessary to explain the facts of radioactivity and the conduction of electricity in gases. The problem then arises : Since α- and β-particles appear to be present in all atoms, how many are there of each in any particular one and how are they arranged ?

Now since the β-particle has a mass $\frac{1}{1.840}$ of the mass of an atom of hydrogen there cannot be more than 1,840 β-particles in an atom of that substance. Further in other elements there cannot be a larger number of these particles than 1,840 times the atomic weight. The number of atoms in any given volume of a substance is known, and the number of particles can be calculated by noting the scattering of X-rays or β-particles. But a more accurate result is obtained by using a beam of α-particles. These are helium atoms with an atomic weight of 4 and, therefore, 70,000 times the mass of β-particle. When they are projected through a plate they are either bent out of their path or reduced in velocity, and by measuring the bending or the reduction in velocity it has been shown that the number of electrons in any atom is about half the atomic weight.

In these experiments it was noticed that a few of the α-particles were deflected widely and others were thrown right back. They had evidently come under the influence of heavier bodies and in some cases had actually collided with them. This led Rutherford to suggest that the whole mass of the atom resided in a nucleus of not more than $\frac{1}{1.000,000,000,000,000}$ centimetre diameter, carrying a positive charge equal to half the atomic weight, and surrounded by a cloud of electrons which just neutralized the positive charge. This view has been confirmed by experiment, and the electrons are supposed to rotate round the nucleus like the planets round the sun.

Sir Ernest Rutherford has carried the transmutation of elements still further by bombarding them with X-rays. In this way he has knocked hydrogen out of nitrogen, and is proceeding to break up the complex atoms of other elements in the same way. Thus in 1921 he and Chadwick succeeded in obtaining hydrogen by the X-ray bombardment of sodium, boron, fluorine, aluminium, and phosphorus. There is now reason to believe that the old hypothesis of Prout, that all elements were built up out of hydrogen, may be correct, only in Prout's day the existence of electrons was not known. Since the helium nucleus is able to

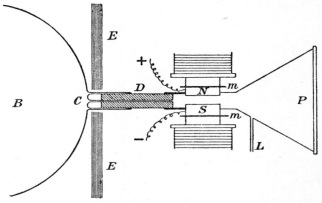

Fig. 340. Sir J. J. Thomson's Apparatus for Studying the Properties of α-Particles.

break up the atoms of other elements it is evidently a very stable arrangement, and that explains why, in the table of p. 459, the changes should take place by loss of helium rather than of the lighter element.

But if the atoms of other elements have nuclei composed of hydrogen atoms, they should all be whole numbers. In many cases this is not so, and the atomic weight elements which depart from this standard have been determined over and over again in order to discover, if possible, a source of error in the older experiments. But it has been all to no purpose. The atomic weight of chlorine remained at 35·45 and in the case of some elements the result of greater refinement and accuracy was to remove that atomic weight still farther from a multiple of unity. Experiments on radioactivity and the discharge of electricity in gases showed, however, that the properties of an element might be modified by the gain or loss of a β-particle which could not be detected by the most delicate balance, and Sir J. J. Thomson showed how α-particles differing in weight could be detected. In order to obtain α-particles he took advantage of the fact that when an electron is shot forward from the cathode of a Crookes' tube there is a sort of recoil, and an α-particle is shot backwards. If the cathode is perforated a stream of α-particles is produced behind it and moving in a direction opposite to the β-stream. Further, as these α-particles carry positive electrical charges, they can be deflected by the action of electric and magnetic forces just in the same way as the ordinary cathode stream.

The apparatus is shown in Fig. 340. The cathode is of aluminium with a fine copper tube from $\frac{1}{10}$ to $\frac{1}{100}$ of a millimetre in diameter running through it. Through this tube the α-particles pass into the vessel on the right, being exposed when necessary to the action of electric and magnetic forces on the way. At the end of this vessel is a photographic plate. In the absence of electric and magnetic forces the particles strike the plate at a point exactly opposite the end of the fine tube. Under the influence of electric and magnetic forces they are bent and form a parabolic curve on the plate. The amount of bending depends on the mass, and as the α-particles from neon give two curves, he expressed the view that these were two kinds of neon atoms differing in atomic weight.

Many years ago it was suggested by Professor Schützenberger and by Professor Crookes that the atomic weights which had been determined might really be mean values, and that the atoms of any particular element might not be all of the same weight. The first indication that this might be true came from a study of the radioactive elements. Sir Ernest Rutherford and his colleagues showed that the final product, lead, could be reached in several ways—see, for example, Figs. 341 and 342, which are taken from an article by Dr. Stephen Miall in *The Chemical Age*. It will be clear that lead produced from uranium has an atomic weight lower than that produced from thorium. Atoms like this, possessing different atomic weights but precisely similar in chemical properties, were called *isotopes* by Professor Soddy. By the analysis of lead compounds from different radioactive minerals, Soddy, Richards, Hönigschmid and others have proved that these isotopes exist.

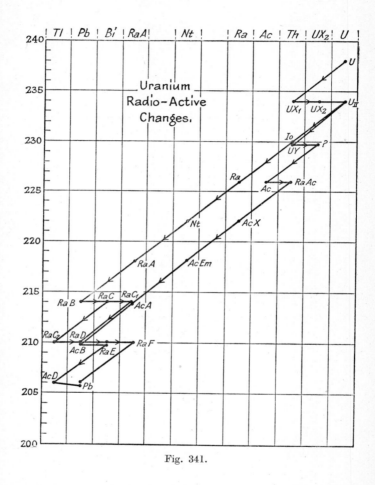

Fig. 341.

Dr. H. Aston, working in Sir J. J. Thomson's laboratory at Cambridge, has very much improved the original apparatus and can now separate positive or α-particles with an accuracy of one in a thousand. In this way he has shown that if the atomic weight of oxygen is 16 that of hydrogen is 1·008, and that the atoms of helium, carbon, nitrogen, oxygen, fluorine, phosphorus,

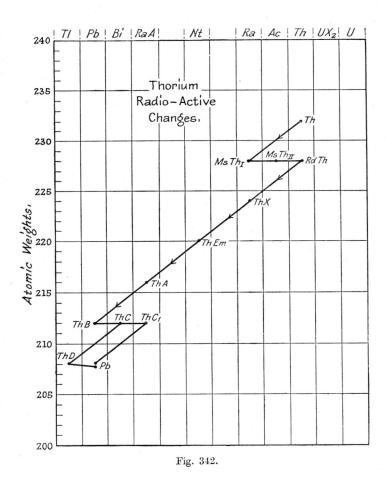

Fig. 342.

sulphur, arsenic and iodine are in each case all alike. On the other hand boron has two isotopes weighing 10 and 11, neon two of 20 and 22, silicon two of 28 and 29, chlorine two of 35 and 37, argon two of 36 and 40, bromine two of 79 and 81, krypton six, zenon five, and mercury six. With the exception of hydrogen all are whole numbers. Since theory shows that in close packing the effective mass may be decreased, and as hydrogen has been expelled from a number of the elements, it seems probable that all atoms are built up from hydrogen nuclei, though for reasons which have been given the grouping of the hydrogen atoms is often that of the helium nucleus, and it ought to be added that we do not know yet what the hydrogen nucleus is.

Since isotopes do not differ from one another chemically they cannot be separated by any process of chemical analysis. An attempt was made by Aston to separate the constituents of neon by repeated absorption by charcoal in a vessel immersed in liquid air. This failed. He then tried to separate them by diffusion through pipeclay. The rate at which a gas diffuses through material of this kind varies inversely as the square root of its density, so that the heavier constituent should diffuse more slowly than the lighter one. The operation is slow and tedious and though a difference of density of 0·7 per cent was obtained, it was not wholly convincing. More recently Brönsted and Hevesy, of Copenhagen, have tried to separate the isotopes of mercury by distilling it under very low pressure and have obtained two positions of densities, 0·999980 and 1·000031, taking the density of mercury as unity. Other workers are endeavouring to separate the two hydrochloric acids by diffusion.

The result of all these investigations is the view that the electron is Nature's unit of negative electricity and that the charge on the nucleus of a hydrogen atom is the corresponding unit of positive electricity for which the name " proton " has been proposed. The relation between this and the electron in an atom has been well expressed by Aston in a lecture at the Royal Institution on February 11th, 1921.[1] " The chemical properties of an element depend solely on its atomic number N, and this is the value of the positive charge on the nucleus in terms of e, the charge on an electron. Since the hydrogen appears to consist of one proton and one electron in any electrically neutral atom of atomic number N there is a nucleus consisting of K + N protons and K electrons, and around this nucleus are N electrons. As the weight of the electrons is almost immeasurably small, the atomic weight is K + N. The protons and electrons are, there-

[1] *Nature*, May 12th, 1921.

fore, the standard bricks out of which all matter, living or dead, is built up."

The electrical theory of matter sheds a new light also upon the mechanism of conductivity, for a current of electricity is merely the flow of electrically-charged particles or ions. If a current of electricity is passed through a gas, a movement of the ions takes place—positive in one direction, negative in the other. But except under very low pressures there are many collisions, freedom of motion is restricted, and recombinations occur. As the electromotive force is increased the resistance increases, until the electromotive force rises to such an extent that a spark discharge takes place. This in itself produces ions, and explains the well-known fact that once an arc has formed the poles may be drawn apart to a distance across which the electricity would not previously flow.

In liquids and metals, again, the passage of the current can be increased indefinitely, because the positive and negative ions have fewer opportunites of recombining. The only limit is set by the heating effect. The result of concentrating the cathode stream upon a piece of platinum has already been described, and a similar bombardment taking place in a metal wire gives rise to those molecular vibrations which reveal themselves in an increase of temperature. In liquid conductors some chemical action usually takes place, and the ultimate violence of this may limit the quantity of electricity which can be conveyed in a given time.

While radioactive substances have the most powerful ionizing effect, hot bodies, X-rays and the minute waves of ultra-violet light, and probably chemical action have the same result. The influence of ultra-violet light waves has been used by Professor W. H. Eccles to explain why the waves used in wireless telegraphy curve round the surface of the earth instead of spreading outwards into space. The upper layers of the atmosphere are relatively dry and invariably positively electrified. It has been suggested that the ultra-violet rays in sunlight, which are largely absorbed on their way to the earth's surface, separate the positive and negative ions, and that the latter are removed by acting as nuclei for the formation of water drops which appear as clouds and ultimately reach the earth as rain. The surface of the earth becomes negative, and the electric wave flows the path between that and the upper positive layer. This result could not have been foretold when Marconi actually transmitted wireless messages across the Atlantic.

Wonder at the experimental results is increased rather than diminished by a consideration of the magnitudes involved. The atom is about $\frac{1}{10,000,000}$ of a millimetre in diameter and a millimetre, we may remind the reader, is $\frac{10}{254}$ or roughly $\frac{1}{55}$ of an inch. The weight of an atom of hydrogen is $\frac{1}{1,000,000,000,000,000,000,000,000,000}$ of a gramme, and a gramme is $\frac{1}{28}$ of an ounce. The electron weighs about $\frac{1}{1,840}$ of the weight of an atom of hydrogen. But what the electron lacks in size and weight is made up in speed, for these infinitesimal particles travel with velocities that range from 10,000 to 60,000 miles a second ; or if emitted from radium with nearly the velocity of light, 185,000 miles a second. Sir Oliver Lodge, in his book on *Electrons,* points out that a body weighing one milligramme, travelling with the velocity of light, would possess energy amounting to 15,000,000 foot-tons, and Sir William Crookes remarked that a gramme moving with the same speed would be capable of lifting the whole British Navy to the top of Ben Nevis. So we may consider ourselves fortunate that the bodies are so small that they can be handled in glass vessels, and used in the neighbourhood of a strip of gold leaf, which represents about the most delicate product of human hands.

After all, these magnitudes are not more remote from the ordinary dimensions of familiar objects than are the magnitudes in astronomy. The shadowy electron with its electrical charge is hardly smaller than the sun or the solar system are larger. But it is perhaps the more wonderful that the same mind can grasp the one as easily as the other, that the mechanism of the human mind can range, arithmetically at least, from the infinitely great to the infinitely small. Enough, however, for the present. Looking back over the preceding pages we see how, in the past twenty or thirty years, man has made gigantic strides in the utilization of those natural resources amidst which he lived for so long in semi-blindness and dim comprehension. We have noted some of the new materials he has discovered, the new uses to which materials long familiar have been applied, the machines that have been invented, the great ships which have been built for crossing the sea and navigating the ocean of air. We have glanced at some of the instruments he has used in subduing to his imperial will the intractable elements of earth, air, fire and water. And, finally, we have been permitted to peer beneath the surface, to catch a glimpse of the delicate and intricate mechanism of the Universe, and to realize in the dash of atom and electron something of the stupendous energy which, until the Twentieth Century, had been beyond the range of human understanding.

INDEX

473